LARGE-SCALE MINES AND LOCAL-LEVEL POLITICS

Between New Caledonia and Papua New Guinea

LARGE-SCALE MINES AND LOCAL-LEVEL POLITICS

Between New Caledonia
and Papua New Guinea

EDITED BY COLIN FILER AND
PIERRE-YVES LE MEUR

Australian
National
University

PRESS

ASIA-PACIFIC ENVIRONMENT MONOGRAPH 12

ANU PRESS

Published by ANU Press
The Australian National University
Acton ACT 2601, Australia
Email: anupress@anu.edu.au
This title is also available online at press.anu.edu.au

National Library of Australia Cataloguing-in-Publication entry

Title:	Large-scale mines and local-level politics : between New Caledonia and Papua New Guinea / edited by Colin Filer, Pierre-Yves Le Meur.
ISBN:	9781760461492 (paperback) 9781760461508 (ebook)
Series:	Asia-pacific environment monographs ; no.12.
Subjects:	Mines and mineral resources--New Caledonia.
	Mines and mineral resources--Papua New Guinea.
	Local government--New Caledonia.
	Local government--Papua New Guinea.

Other Creators/Contributors:
    Filer, Colin, editor.
    Le Meur, Pierre-Yves, 1962- editor.

Cover design and layout by ANU Press. Cover photograph by John Burton: Peakame Taro airing a grievance at the Apalaka resettlement village, Porgera mine, 9 June 2006.

Contents

Tables

Figures

Boxes

Acknowledgments

Initial drafts of most of the chapters in this book were first presented as papers at a conference on 'Mining and Mining Policy in the Pacific: History, Challenges and Perspectives', which was hosted by the Pacific Community and the French National Research Institute for Sustainable Development (Institut de Recherche pour le Développement) in Nouméa from 21 to 25 November 2011. Although this book is not to be read as the proceedings of that conference, since many other papers were presented, the editors would like to acknowledge the contributions made to its organisation and funding, without which this book would never have eventuated. Additional organisational support was provided by the University of New Caledonia (Université de la Nouvelle-Calédonie), the New Caledonian Institute of Agronomy (Institut Agronomique Néo-Calédonien), the French Agricultural Research Centre for International Development (Centre de Coopération Internationale pour la Recherche Agronomique pour le Développement), the National Centre for Technological Research on Nickel and Its Environment (CNRT Nickel) and CORAIL (Coordination pour l'Océanie des Recherches sur les Arts, les Idées et les Littératures). Additional financial support was provided by the Pacific Fund (Fonds Pacifique) of the French Ministry of Foreign Affairs, the Congress of New Caledonia, the authorities of South Province and North Province, and the mining companies Société Le Nickel, Koniambo Nickel SAS and Vale SA.

with the financial support of :

The case studies presented in the chapters on New Caledonia (Chapters 2–6) are derived from a research project on 'The Politics of Nickel Between Local and Corporate Governance: Mining and Industrial Trajectories in New Caledonia', which was funded by CNRT Nickel over the period from 2010 to 2014.

Contributors

Nicholas A. Bainton is an Associate Professor in the Centre for Social Responsibility in Mining at the University of Queensland.

Glenn Banks is a Professor and Head of the School of People, Environment and Planning at Massey University.

John Burton is a Professor in the Department of PNG Studies and Director of the Centre for Social Research at Divine Word University in Madang, Papua New Guinea.

Christine Demmer is a Researcher at the Centre National de la Recherche Scientifique, Centre Norbert Elias, Marseille.

Colin Filer is an Associate Professor in the Crawford School of Public Policy at The Australian National University.

Sonia Grochain is a Researcher at the Institut Agronomique Néo-Calédonien, Pouembout, New Caledonia.

Susan R. Hemer is a Senior Lecturer in the Department of Anthropology and Development Studies at the University of Adelaide.

Phillipa Jenkins has recently graduated with a PhD from the Crawford School of Public Policy at The Australian National University.

David Kombako is a Lecturer in the Division of Anthropology, Sociology and Archaeology at the University of Papua New Guinea.

Matthias Kowasch is an Assistant Professor in the Institute of Geography and Spatial Research at the University of Graz.

Dora Kuir-Ayius is a Lecturer in the Division of Social Work at the University of Papua New Guinea.

Pierre-Yves Le Meur is a Senior Researcher in the Research Unit GRED (Gouvernance Risque Environnement Développement) at the Institut de Recherche pour le Développement, Montpellier.

Claire Levacher is a Post-Doctoral Researcher at the Institut Agronomique Néo-Calédonien, Pouembout, New Caledonia.

Joyce Onguglo is a former Project Manager with ANU Enterprise in Canberra, now a private consultant specialising in project management and social impact assessment.

David Poithily is a former Researcher at the Institut Agronomique Néo-Calédonien, Pouembout, New Caledonia.

Anthony J. Regan is a Fellow in the State, Society and Governance in Melanesia Program at The Australian National University.

Bill F. Sagir is a Senior Lecturer in the Division of Social Science at the University of Goroka, Papua New Guinea.

Jean-Michel Sourisseau is a Senior Researcher at the Centre de Coopération Internationale en Recherche Agronomique pour le Développement, Montpellier.

1. Large-Scale Mines and Local-Level Politics

COLIN FILER AND PIERRE-YVES LE MEUR[1]

Two Political Economies

Papua New Guinea (PNG) and New Caledonia (NC) have two key things in common. First, they both belong to the geopolitical region known as Melanesia, which was originally defined (by Europeans) in terms of the racial, linguistic and cultural characteristics of its indigenous population. Second, they have comparable levels of economic dependence on the extraction and export of mineral resources. In the past decade, the extractive industry sector has accounted for roughly 80 per cent of the value of PNG's exports and more than 90 per cent of the value of NC's exports. In most other respects, these two places are profoundly different.

PNG achieved its independence from Australia in 1975; NC is still technically part of France. Under the terms of the Nouméa Agreement of 1998, NC ceased to be an 'overseas territory' and became a 'special collectivity', and a referendum on full independence must be held by the end of 2018. Despite the current difference in political status, we shall refer to both of them as 'countries'.

---

1   We thank Claire Levacher and Jean-Michel Sourisseau for their comments on a draft of this introductory chapter, but we take full responsibility for any mistakes remaining in this final version.

PNG is much bigger than NC. In 2014, NC had a resident population of 268,767. The estimated population of PNG in that year was 30 times larger. The population of NC is roughly the same as the population of the Autonomous Region of Bougainville, whose government could hold its own referendum on full independence from PNG at the same time as the NC referendum.

PNG is also much poorer than NC. PNG's gross national income per capita is less than US$2,500; the equivalent figure for NC is 15 times that amount. The distribution of cash incomes is highly unequal in both countries, but even poor people in NC would count as reasonably wealthy people in PNG. If cash incomes are left out of the equation, and human development is measured only by the standard health and education indicators in the United Nations Human Development Index, currently only 15 countries are worse off than PNG out of the 187 independent countries included in the rankings (UNDP and GoPNG 2014: 3).

The vast majority of people in PNG are descendants of the population that existed when the country was first subjected to European colonial administration in the nineteenth century, although the very small proportion of European or Asian descent wields a disproportionate influence over the country's business activities. NC is more like Tasmania, in the sense that it was first established as a penal colony and then as a settler colony, but a much larger proportion of the indigenous population survived this process of colonisation. Indigenous (Kanak) people now account for less than half of the resident population of NC, but still outnumber those of metropolitan French descent. The balance of the population in NC consists mainly of people who originate from Southeast Asia or other French Pacific territories.

Between 75 and 80 per cent of the population of PNG still live in traditional rural village communities, while the rest are distributed between large and small urban centres, peri-urban settlements and 'rural non-villages' such as oil palm estates and resettlement schemes. More than 60 per cent of the NC population live in urban areas, mostly in and around the capital of Nouméa. Indigenous people constitute the majority of NC's rural population.

Given these similarities and differences, we can now pose the questions addressed in this volume as follows: is the relationship between large-scale mines and local-level politics in PNG and NC quite similar, either because of the way that the mining industry nowadays deals with local communities, or else because of the way that Melanesian communities deal with the mining industry? Or is the relationship quite different because of all the other differences between the two countries? To refine these questions, we need to briefly consider the way in which the mining industry is organised and regulated in these two countries, and the way in which the division of each country between political and administrative units, or their differing degrees of political independence, is related to the definition of a mine-affected area or a set of mine-affected communities. We may then proceed to develop a single conceptual framework through which to compare the relationship between large-scale mines and local-level politics in different political and institutional settings.

Large-Scale Mining Operations

A large-scale mining operation is here defined, for the sake of argument, as one that produces mineral commodities with an average value of more than US$100 million a year for a period of at least ten years. Although the title of this volume features the concept of a 'large-scale mine', we do not think of a 'mine' as a place where mineral resources are extracted from the ground, but rather as a group of excavation and processing activities that are integrated into a single whole by means of a network of corporate relationships between the producers. In PNG, each large-scale mining operation occupies and constitutes a distinctive territorial enclave within the country's borders, with one place of excavation, one processing plant and one route by which the product leaves the country. In NC, the situation is somewhat different because one of its three processing plants derives its raw material from a number of distinct excavation sites located in different parts of the country. Furthermore, some of this raw material is actually exported to a fourth processing plant located overseas that is predominantly owned by a company based in NC, so it is ownership and control of one or more refineries that constitutes the core of each large-scale mining operation.

In calculating the contribution of extractive industry to PNG's exports, we have included a petroleum project that has been exporting oil on this scale since 1992, and this has recently been supplemented by a liquid natural gas project whose contribution to the value of PNG's exports will be much greater. However, for the purpose of the present volume, we have excluded oil and gas projects from our definition of a 'mining operation': first because the mining and petroleum industries are subject to different forms of regulation in PNG; and second because NC does not yet have an operational petroleum industry with which to make a comparison.

PNG has hosted seven large-scale mining projects in the period since independence, five of which are still in operation.[2] The Panguna gold and copper mine operated from 1972 to 1989, when it was forcibly closed by civil unrest on the island of Bougainville. The Misima gold mine (in Milne Bay Province) operated from 1986 to 2004. The Ok Tedi gold and copper mine (in Western Province) began production in 1984; the Porgera gold mine (in Enga Province) in 1992; the Lihir gold mine (in New Ireland Province) in 1997; the Hidden Valley gold mine (in Morobe Province) in 2009; and the Ramu nickel and cobalt mine (in Madang Province) in 2013. The Ok Tedi, Porgera and Hidden Valley mines are all scheduled to close within the next decade. A large-scale seabed mining project has been approved for development but is not yet operational. Four other prospective large-scale mining projects (in Milne Bay, Morobe, Madang and West Sepik provinces) are currently undergoing feasibility studies, and the Autonomous Bougainville Government has plans to reopen the Panguna mine if it can mobilise popular support for this to happen.

The large-scale mining industry in NC produces only one commodity, nickel, which means that the country's economy is especially vulnerable to fluctuations in its price. The main island (Grand Terre) accounts for roughly 9 per cent of global nickel production, and still contains between 10 and 30 per cent of the world's known reserves of this mineral, despite the fact that it has been mined since 1873. Since the 1930s, NC's mining landscape has been dominated by a company called Société Le Nickel (SLN), which was founded in 1880 and controlled by the Rothschild Bank from 1888 to 1974. Following a bailout by the French Government, it came under the control of a newly established French company called Eramet in 1985. The core of this operation is the Doniambo processing plant in Nouméa, which derives its raw material from four different mine sites. In recent years, the SLN complex has been joined by two other large-scale mining and processing operations—the Koniambo project in North Province and the Goro project in South Province.[3]

---

2   It has also hosted a number of medium-scale projects with lower production values, most of which have operated for only a few years.

3   The nickel mining industry in NC also contains a number of small- and medium-scale mining companies, known as *petits mineurs*, which are not controlled by one of the three big operators, and which tend to move in and out of the supply chain, or raise and lower the volume of their output, with fluctuations in the nickel price (Freyss 1995). These companies have been the focus of recent political debate about the grant of permits for the export of nickel ore to be refined in China.

The PNG Government reserves the right to purchase anything up to 30 per cent of the equity in any mining project for which it grants a development licence, but has generally ended up with a smaller stake or no stake at all. The balance of the shares in all large-scale mining projects has normally been held by foreign investors, including the companies that operate them, but the Ok Tedi project is now an exception to this rule because the former operator (BHP Billiton) decided that it was a liability rather than an asset. When the government has purchased a stake in a large-scale mining project, all or part of it has commonly been held in trust for the provincial government, local-level government or landowning community that hosts the project. These other entities are allowed to choose whether they wish the national government to exercise this option on their behalf.

The PNG Government currently holds a 5 per cent stake in the Porgera project, which is operated by the Canadian company Barrick Gold, and this stake is held in trust for the Enga Provincial Government and the local landowners. A comparable stake in the Lihir project, which is operated by the Australian company Newcrest, was acquired and then sold by the local landowners. The government did not exercise the option to purchase shares in either the Hidden Valley project, which is also operated by Newcrest, or in the Ramu project, which is operated by the China Metallurgical Group Corporation. BHP Billiton bequeathed its shares in the Ok Tedi project to a charitable trust known as the PNG Sustainable Development Program, but this entity was nationalised in 2013, so this project is now wholly owned by the government.

State participation in NC's three large-scale nickel mining operations is organised through the three provinces.[4] Together they own 34 per cent of the SLN complex and 5 per cent of the Goro project, which is operated by the Brazilian company Vale. The Koniambo project has a very different complexion because it is the result of a political initiative that was part of the process of 'negotiated decolonisation' initiated by the Nouméa Agreement in 1998. This project is operated by the Swiss company Glencore (formerly Xstrata), but 51 per cent of the equity is controlled by the Kanak-dominated North Province through its majority of shares

---

4   While each province has its own elected assembly and its own administration, the word 'government' (*gouvernement*) is not applied to this level of political organisation. Although such entities would be recognised as 'provincial governments' in PNG, we shall here follow the French practice of assigning political agency to 'provinces' and 'provincial authorities'.

in a company called Société Minière du Sud Pacifique (SMSP), which was formerly part of the SLN complex. SMSP also holds a 51 per cent stake in a nickel refinery in South Korea, and is planning a similar investment in a second refinery in China. The Korean refinery derives some of its raw material from excavation sites owned and operated by SMSP, so the SMSP complex currently consists of two refineries and their suppliers.

Development Agreements

The development of a large-scale mining project in PNG is framed by three different types of agreement that are normally negotiated in the following order. First, there is a compensation agreement between the holder of an exploration licence and the customary owners of the land covered by that licence. Since 97 per cent of the land in PNG is generally held to be customary land, there is no such thing as an exploration licence without customary landowners. Second, there is a development agreement (or mining development contract) between the national government and the project proponent based on feasibility studies that the proponent provides to the government. This type of agreement is linked to the production of an environmental impact statement that must also be approved by the national government. Third, there is a benefit-sharing agreement between the national government, the provincial and local-level government (or governments) hosting the project and the customary owners of the land required for development purposes, which is negotiated through an institution known as the development forum. The project proponent is also represented in the negotiation of this third type of agreement, but a development licence is not normally granted until it has been finalised. The first two types of agreement have been features of PNG's mineral policy framework since independence in 1975; the third type was added in 1988 in response to political pressure from provincial governments and local community representatives (Filer 2008; Filer and Imbun 2009).

In PNG, development agreements have normally required that national participation be specified in training and localisation plans and business development plans whose implementation is then reported to the national government at regular intervals. These plans are subject to the 'preferred area policy', which has come to inform the negotiation of benefit-sharing agreements. The origins of the preferred area policy can be traced back to a pair of decisions made by the newly independent national government in 1976. The first decision was to repatriate a sum equivalent to the whole of the royalty collected by the national government, in its capacity as the legal owner of subsurface mineral resources, to the province from which those resources were extracted. This decision was made in response to threats of secession from the province that hosted the Panguna mine, but it would have general application under the new system of provincial government that was put in place at the same time. The second decision was to oblige the future developer of the Ok Tedi mine to give preference in training, employment and business development to the people of the area most directly affected by the mining operation. This decision was not initially meant to have general application, nor did it apply to the Panguna mine. It was justified by the observation that the people living around the Ok Tedi mine were exceptionally poor and therefore deserved this form of affirmative action (Filer 2005; Filer and Imbun 2009).

Considerations of poverty and equity have long since disappeared from PNG's preferred area policy. The allocation of royalties and the allocation of entitlements to training, employment and business development opportunities are now included in the range of benefits that are subject to benefit-sharing agreements through the development forum. In effect, the preferred area policy creates concentric rings of entitlement to a range of benefit streams that are subject to such agreements, including entitlements to training, employment and business development opportunities. The innermost ring is occupied by the customary owners of the land covered by development licences, the next by 'project area people' (however these might be defined), the next by the people or government of the host province, and the outermost ring by the population or government of the nation as a whole (Filer 2005). In the period since the development forum was invented in 1988, there has been a steady increase in the proportion of government revenues from each new resource project that is captured by organisations or individuals in the three inner circles of entitlement. Since 1993, the economic privilege bestowed on preferred areas has been compounded by a tax credit scheme for developers who supply social and economic infrastructure to local communities (Filer 2008).

There is no direct equivalent of the development forum in NC. Instead, there is a recent history of agreements between mining companies and indigenous communities or organisations, a separate history of agreements between mining companies and provinces or municipalities, and some instances of tripartite agreements between company, state and community representatives (Le Meur and Mennesson 2012). There is no clear division between 'compensation agreements' and 'benefit-sharing agreements', and most agreements do not conform to the ideal type of a 'community development agreement' (O'Faircheallaigh 2013). If compensation and development agreements in PNG have come to be framed by national government laws and policies, the situation in NC is better described as one in which policies derive from the accumulation of agreements (Le Meur, Horowitz et al. 2013).

The recent history of development agreements and policies in NC's mining sector cannot be understood in isolation from the politics of decolonisation. This was already evident when the Front de Libération Nationale Kanak et Socialiste (FLNKS) identified the Koniambo project as the main vehicle for the redistribution of economic power that was envisaged by the Matignon-Oudinot Agreements of 1988. Before the Nouméa Agreement was signed in 1998, the FLNKS insisted on a separate agreement, known as the Bercy Agreement, which established the feasibility of the Koniambo project by means of an 'exchange' of ore bodies between SLN and SMSP.[5] In 2002, the government agency in charge of land reform identified the 'areas of influence' of the clans with customary land rights around the future Koniambo project plant site at Vavouto. The result of this inquiry was an agreement under customary law that was designed to settle a dispute that had already caused the plant site to be relocated (Horowitz 2002), but also aimed to forestall any future disputes about the distribution of mining revenues, especially in the form of business opportunities for local enterprises. In this second respect, it seems to have set the precedent for a similar agreement made between Kanak leaders and SLN in 2004. Since 2007, SLN has been party to tripartite agreements that also involve the provincial and municipal authorities and thus enhance their regulatory powers over the mining industry. However, the provincial and municipal authorities were not party to the 'Pact for the Sustainable Development of the Great South' that Vale made with the representatives of local customary authorities and indigenous organisations in 2008, and which is widely regarded as another significant landmark in the mineral policy domain. Instead, Vale signed a separate 'Convention for Biodiversity Conservation' with South Province in 2009. This might be regarded as a sort of compensation

---

5   The Bercy Agreement was signed by representatives of the metropolitan and territorial governments as well as the companies involved in the transaction. The role of the French Government was critical in ensuring that this agreement survived the economic conditions that SLN sought to impose on it.

agreement but, unlike compensation agreements in PNG, it was not an agreement made with the holders of customary land rights in the area affected by the Goro project.

Although questions of land tenure are linked to the participation of Kanak community leaders in development agreements with mining companies, they are not simply articulated in terms of an 'ideology of landownership' (Filer 1997), mainly because mining projects have been developed on public land. However, there has been a general shift from demands for the recognition of customary land rights to demands for the recognition of a broader range of indigenous political rights, which include the right to make this type of agreement without making land claims (Le Meur 2010). In this respect, the situation of Kanak communities in NC resembles that of Aboriginal communities in Australia more than it resembles that of the 'customary landowners' who dominate PNG's political system. But as customary institutions have become increasingly central to economic arrangements within (and beyond) NC's mining sector, they have acquired greater complexity and have been subject to greater regulation. This is why development agreements with local or indigenous communities are not just compensation agreements or benefit-sharing agreements, but also political agreements. And while the development forum could in some ways be seen as the central institution in PNG's current mineral policy framework, development agreements in NC's mining sector have not played the same pivotal role in the formulation of mineral policy.

Local-Level Politics

There is no simple answer to the question of what actually constitutes the 'local' level at which large-scale mining operations become the subject of local-level politics. There is no necessary correspondence between the boundaries of a mine-affected area and those of a set of political or administrative units, even if each mine-affected area is conceived as a unique geographical space surrounding a large-scale mine. The question of what actually constitutes a mine-affected area is itself a political question, especially if people's inclusion in such an area entitles them to special consideration in the payment of compensation or the distribution of project-related benefits. Project operators have an obvious interest in limiting the size of such an area in order to treat broader social and environmental impacts as externalities for which they cannot be held accountable. On the other hand, there may be some circumstances in which the boundaries of the area affected by one project overlap those of the area affected by another project, and it is also possible to represent the areas affected by several different projects as part of a larger region which experiences the cumulative social and environmental impacts of all of them. As we have seen, the presence of three large-scale projects or complexes in NC's nickel mining industry does not mean that there are only three geographical areas that count as areas affected by their operation.

Since the population and the land mass of NC are both smaller than those of the average province in PNG, the gap between the practice of politics at different levels of political organisation is also bound to be much smaller, especially when it concerns the costs and benefits of three large-scale mining projects that make such a big difference to the national economy.[6] But the internal political divisions of the country are not without their own significance. South Province contains 75 per cent of the total population, as well as two of the three nickel refineries; North Province contains 18 per cent of the population, one refinery and several different mine sites;[7] while Loyalty Islands Province contains the remaining 7 per cent of the population and no mining operations at all. The whole country is divided into 33 municipalities or communes, in the normal French manner, but the indigenous Kanak population is also divided between eight 'customary areas', each of which has two elected representatives in the country's Customary Senate, and 57 traditional 'districts', each of which has a 'big chief'.[8] These political spaces can be distinguished from each other by the form and extent of their engagement with the politics of the large-scale mining industry.

Given that Kanaks account for roughly three-quarters of the population of North Province, but only one-quarter of the population of South Province, it is understandable that the large-scale mining industry has become the focus of a specific form of local-level politics that is part of the process of 'negotiated decolonisation' initiated by the Nouméa Agreement.

6   The same point would apply to the Autonomous Region of Bougainville if it were to become an independent country.

7   The SLN refinery (at Doniambo) draws its raw material from several different mine sites. Only one of these (Thio) is in South Province, while the others are in North Province.

8   When a municipality contains more than one traditional district, the name of the municipality may be the same as the name of one of its component districts. All but one of the 33 municipalities belong to one of the three provinces, and all but one of the 57 districts belong to one of the eight customary areas.

Development of the Koniambo project in North Province is seen as a way of 'rebalancing' the distribution of economic power between the north and the south in order to compensate for the colonial marginalisation of the indigenous population by enabling them to develop their own economic policies and public investment strategies. On the other hand, the development of the Goro project in South Province has offered new scope for political parties opposed to independence to argue that further decolonisation could discourage foreign investment,[9] while Kanak people living in the area affected by this project have mobilised a discourse of indigenous rights that looks like a 'weapon of the weak' (Scott 1985) in a context where they have not had the same amount of bargaining power.

Since the French Government's contribution to public expenditure in NC accounts for 15 per cent of the country's gross domestic product, the trade-off between this 'administrative rent' and revenues from the mining industry has become a key issue in both national and local-level politics. From an economic point of view, the administrative rent compensates for shortfalls in mineral revenues during periods when nickel prices are abnormally low. From a political point of view, there is a complex game in which both sides try to demonstrate their superiority in managing the trade-off. The loyalist parties claim that the NC Government's ability to manage the mining industry is an indirect function of the subsidies it continues to receive from the metropolitan government, since these contribute to the country's high standard of living. On the other side, the pro-independence parties claim that a Kanak-dominated province has shown a superior capacity to manage the mining industry in ways that reduce the need for metropolitan subsidies and hence pave the way for economic, as well as political, independence. The pro-independence parties also advocate a form of state capitalism in opposition to the economic liberalism that is generally favoured by the loyalist parties, and that in turn entails a difference in their understanding of the relationship between local-level politics and economic policy.

As befits a much larger country, PNG has four tiers or levels of political organisation. The second tier consists of the National Capital District, 20 provinces and one autonomous region (Bougainville) that used to be a province. Each of these 22 entities has its own elected representative in the national parliament and 21 of these representatives are known as

9   The anti-independence bloc contains divergent views about the desirable level of public participation in the SLN complex.

governors.[10] At the next level down, there are 89 'open electorates', also with their own elected representatives in the national parliament, three of which are subdivisions of the national capital, while the rest are known as districts in their own right. The fourth tier comprises 332 local-level governments, each with its own directly elected president who participates in decisions made at the district and provincial levels, but not at the national level. Twenty-nine of these local governments represent towns, most of which are provincial capitals, while the rest represent rural areas.[11]

Since most of PNG's provinces do not host a large-scale mining project, nor any of the facilities associated with the export of oil and gas, there is an ongoing political debate about the way that national revenues from the extractive industry sector should be distributed between the minority of project-hosting provinces and districts and the majority that cannot currently claim ownership of such a project. The point at issue here is the so-called 'derivation principle', which says that a national government should transfer some of the revenue it derives from any economic activity to the lower levels of government responsible for the area where the activity occurs so that these lower levels of government will have an incentive to support this activity. This is a national issue, not a local one, and is only indirectly connected to the distribution of revenues from foreign aid. Foreign aid now accounts for only 3 per cent of PNG's gross domestic product, and its distribution across the country could not possibly compensate for the effects of the preferred area policy, even if policymakers thought this would be a good idea.

In the context of the mining industry in PNG, local-level politics seems to have a quite distinctive focus on the negotiation and implementation of promises to provide compensation and benefit packages to the customary owners of land in mine-affected areas, but also on the distribution of the contents of these packages among the people who are entitled to a share of them. The size of the enclaves that are demarcated for this purpose varies from one project to another, but the boundaries may also change during the course of the project cycle. At one extreme is the Lihir group of islands, which is comparable in size and population to Loyalty Islands Province in NC, but accounts for just one of the ten local government areas in PNG's

---

10   The Autonomous Bougainville Government has an elected president whose office is distinct from that of the regional member of the national parliament, so the latter is not known as the governor of the autonomous region.

11   There are no local governments in the National Capital District, but there is an 'assembly' that represents the district's indigenous population.

New Ireland Province. At the other extreme is the area now recognised as the one affected by the Ok Tedi mine, which has grown to include all or part of ten out of the 14 local government areas in Western Province, and whose physical extent is greater than the whole of NC's land mass. The larger the scale of a mine-affected area, the greater the scope for it to be internally divided into separate geographical zones, each of which may then become the site of a distinct set of political activities related to the mining project that affects it.

A mining enclave or mine-affected area cannot be neatly demarcated by lines on a map in the same way that a mine site or plant site is bounded and guarded by fences, gates and signposts. It is also constituted in an institutional and ideological sense, by means of its connections with different political and administrative levels and spaces. These connections can be made of legal rules or administrative norms, the distribution of shareholdings in different companies, networks of patronage or clientelism, or the positions assigned to the practice or promise of mining in different political imaginations (Le Meur, Ballard et al. 2013). The multidimensional nature of the mining enclave or mine-affected area therefore means that local-level politics is not simply politics conducted within a particular physical space, or even at one level of political organisation.

Stakeholders, Actors and Characters

Politics is not like football. The people who 'play politics' around the mining industry, at any level of political organisation, are not themselves organised into teams that simply pass a ball of power between them until they score a goal against the other side. World Bank functionaries may like to represent themselves as referees in this sort of contest (Szablowski 2007), but sporting metaphors cannot do justice to the social and economic context in which different groups of political actors take to a political field, nor to the way in which their relationships develop once they get there.

The Triangular Model

During the 1990s, the politics of the large-scale mining industry came to be regarded, in several quarters, as a triangular contest between the representatives of mining companies, government agencies and local communities (Ballard and Banks 2003). In the case of developing countries, this triangular model of political relationships could be seen as a substitute for the previous binary or bipolar model in which the representatives of national governments contested the terms of large-scale resource development with the representatives of multinational corporations (Girvan 1976; Bosson and Varon 1977; Cobbe 1979; Faber and Brown 1980; O'Faircheallaigh 1984). But in the case of capitalist countries, the new triangle replaced an older triangle in which local communities were represented as communities of mineworkers engaged in a class struggle with their employers, while governments were represented as umpires in the contest or more often as supporters of the ruling class (Dennis et al. 1969; Burawoy 1972; Bulmer 1975; Gordon 1977; Nash 1979; Williams 1981; Robinson 1986; Moodie with Ndatshe 1994; Finn 1998). So what had really changed if there were still three parties to the contest?

The end of the Cold War is only one part of the answer to this question. During the 1970s and 1980s, the mining industry underwent a process of modernisation that had several different aspects, each of which had its own political repercussions. The process did not begin and end in the space of 20 years in every single part of the global mineral economy, but it went far enough to change the political complexion of the industry in those regions and countries where new mining projects were being developed.

The new mines were typically open-cut (or open-cast) mines in which huge machines were applied to the excavation of very large holes in the ground, and the material excavated from these holes was then subjected to a complex sequence of mechanical and chemical operations in order to separate specific mineral commodities from much larger volumes of waste material. The mechanisation and partial automation of different aspects of the process of extraction was accompanied by an increase in the capital cost of constructing new mines at any particular scale, which added to the cost and complexity of feasibility studies undertaken in advance of mine construction, and made mining companies more dependent on capital markets to finance the cost of construction. But an increase in the rate of extraction or 'throughput' also had the effect of shortening the overall period of time that it would take to exhaust an ore body of any particular size, and thus gave rise to a form of planned obsolescence in which the lifetime of a new mine had to be calculated before it could be financed and constructed.

Since many of the new mines were developed in regions remote from existing centres of population and industry, where the costs of exploration were correspondingly high, mining companies sought ways to reduce or avoid the additional cost of building and operating a new township to accommodate their workforce, mindful of the fact that such a town might serve no useful purpose at the end of an operation abbreviated by the increased scale and rate of extraction. As the cost of air transport fell relative to the other costs of exploration, construction and operation, mining companies developed a preference for what came to be known as a 'fly-in/fly-out' form of employment, in which most of the workers are flown from their point of recruitment to the mine site at intervals of less than one month, work on the mine for several days at a stretch while sleeping in dormitories, and are then flown out for 'field breaks' in their normal places of residence (Jackson 1987; Brealey et al. 1988; Storey and Shrimpton 1988; Houghton 1993; Storey 2001; Markey 2010).

This last feature of the process of modernisation is clearly evident in PNG, where the Ok Tedi project was the last one to incorporate plans for a conventional mining town when construction started in 1982. It is much less evident in NC, where the 'fly-in/fly-out' form of employment has only been characteristic of the construction phase of the Koniambo and Goro projects.[12] Most of the workers employed in the operation of the nickel mining industry are local residents who commute to work on a daily basis.[13] This can readily be explained by the number of production sites scattered around a very small country, and by the quality of the road transport network, which is much better than that encountered in any province of PNG. In this respect, the large-scale mining industry in PNG seems rather more 'modern' than its counterpart in NC for the somewhat paradoxical reason that PNG is a less developed country. Furthermore, the SLN complex, in contrast to the Koniambo and Goro projects, has a rather 'traditional' mode of operation because it has been operating for such a long time.

Regardless of these differences, we still need to consider the political repercussions of this process of modernisation in those places where one or more of its features have been in evidence.

---

12 About 7,000 workers were employed in the construction of each of these projects. Most were recruited from Asian countries (Philippines, China, South Korea and Thailand) under special (and some would say unfair) labour regulations.

13 The Koniambo and Goro construction camps are still partially occupied by local workers who work and sleep on site for four or five days at a time and then have two or three days at home.

Since the modernisation of the mining industry was accompanied by the globalisation of all sorts of capital, the new generation of mining projects entailed a surge in development agreements between foreign mining companies and national governments claiming ownership of the mineral resources that were the targets of their investment (Bridge 2004). In the negotiation of these agreements, national government representatives generally tried to maximise the national share of the rents or benefits without creating too much of a dent in the surplus left over for distribution to foreign shareholders and thus making other countries more attractive as investment targets (Daniels et al. 2010).

At the same time, the growth in the scale and rate of extraction made the new generation of mining projects a sitting target for a new generation of environmental activists, and thus put additional pressure on government and company representatives to negotiate a second trade-off between the distribution of economic benefits and environmental costs (Keck and Sikkink 1998). The institutionalisation of environmental impact assessment was one aspect of this second form of negotiation (Paehlke and Torgerson 1990; Petts 1999), but this did not happen overnight, so the new generation of mining projects caused a great deal of environmental damage (and political conflict) before they were subjected to more effective forms of environmental regulation.

In those countries where national governments began to attach environmental conditions to development agreements with foreign mining companies, the prior process of environmental impact assessment sometimes involved a parallel process of social impact assessment and a form of consultation with the representatives of mine-affected communities (Little and Krannich 1988; Howitt 1989; Joyce and MacFarlane 2001). When these communities were identified as groups of people who would suffer most of the negative impacts of a new mining project, efforts could be made to compensate them for the damage, or even to provide them with a special share of the national economic benefit derived from the project. But considerations of environmental justice were rarely sufficient to grant the representatives of such communities a seat at the high table of project negotiations unless they could also be represented as *indigenous* communities with a special attachment to the land and other things that would be damaged or destroyed (Geisler et al. 1982; Hobart 1984; Howard 1988; Chase 1990; O'Faircheallaigh 1991, 2002; Howitt 2001; Downing et al. 2002). The third point in the new triangle was thus assembled from a combination of demands for environmental justice and for recognition of the rights of indigenous peoples (Horowitz 2011). However, it is worth noting that Kanak political parties have been wary of the idea that Kanaks belong to separate 'indigenous communities', whether or not they be mine-affected communities, in case this weakens the political case for national independence.

It should now be clear that this type of community bears little resemblance to the traditional community of mineworkers. It is true that the compensation and benefit package delivered to members of a mine-affected community might include a provision for them to be granted special access to employment on the mining project, but this would not establish much in the way of a common interest with members of the workforce recruited from other places, especially if those other workers were long-distance commuters who had their own families or communities somewhere else. In effect, mine-affected communities had now come to be defined as communities united by the social relations of compensation, not by the social relations of employment, even if jobs are still understood to be part and parcel of a compensation package.

This does not mean that the political repercussions of the process of modernisation were self-evident to all the relevant actors at each stage in this process. By 1990, some actors in the mineral policy process were starting to think along these lines because of their engagement with the interests and concerns of mine-affected communities. But there had to be moments of political crisis before the new model of political relationships could be firmly established as a feature of that process at all levels of political organisation. One such moment was the 'community rebellion' that forced the closure of the Panguna copper mine in 1989.

The Lessons of Panguna

The construction, operation and closure of the Panguna copper mine on the island of Bougainville, in the period from 1967 to 1989, provides a clear illustration of the process of modernisation and its political repercussions. But, in this case, the illustration also served to inform the new model of stakeholder politics that was applied to the large-scale mining industry at a global scale.

The Panguna mine exemplified the technical innovations that were typical of the new generation of open-cut mining projects. However, because it was commissioned at the very start of the process of modernisation, no environmental conditions were attached to the licences granted by the Australian colonial administration, so the mine was designed to discharge its waste material directly into the Jaba River. The construction of this most modern mine was also accompanied by the construction of a modern mining town to accommodate the families of a workforce with multiple technical skills. Long-distance commuting was not yet thought to be an economic option for the employment of this type of workforce.

In the short period of time between self-government in 1973 and full independence from Australia in 1975, representatives of the new nation-state renegotiated the Bougainville Copper Agreement that it inherited from the colonial administration and, in so doing, made PNG look like a very smart young Third World country (O'Faircheallaigh 1984). But the resulting increase in the national share of the incomes generated by the new mining project was just one aspect of a longer policy process that also established a set of rules to govern the distribution of these incomes between different groups of national stakeholders. These groups included national members of the mining workforce, the customary owners of land leased to the mining company and the people of Bougainville (or North Solomons) Province.[14] Representatives of the national government and the mining company thought that this process had been completed by 1980.

The outbreak of the Bougainville rebellion in 1989 was the start of an entirely new policy process that was not confined by national borders. It was the first in a sequence of 'bad events' that eventually forced the captains of the global mining industry to articulate new standards of corporate social responsibility (MMSD 2002; Filer et al. 2008). But the lessons of the rebellion were not easy to learn, since it was not clear whether and how it might have been caused by the negligence of Bougainville Copper Ltd (BCL). Some observers blamed the Australian colonial administration for approving the development of the mine without regard for what Bougainvilleans wanted. Others blamed the national government for failing to share enough of the benefits of the renegotiated mining agreement with the provincial government or local landowners. Since the first phase of the rebellion was organised by a local landowner association, much attention was paid to the grievances that its

---

14   The name 'North Solomons' was adopted by the local advocates of secession when PNG became independent in 1975, and was recognised by the PNG Government when the North Solomons Provincial Government was established in 1976. The previous name, 'Bougainville', has been restored as part of the peace process that has led to the grant of regional autonomy.

leaders held against the mine itself. Were they upset about the influx of outsiders taking jobs or doing business with the mine; about the damage being caused to their own land or physical environment; or about the inability of their own social institutions to cope with a rapid process of social, economic and environmental change?

The reason that these questions are so hard to answer, even with the benefit of hindsight, is that the rebellion was not a single 'event' with a discrete set of causes. It was a local and provincial political process with deep historical roots that continued to evolve for years after it had triggered a set of national and global policy responses within the mining sector. If all the local and provincial actors who were involved in this process at one time or another had shared a common set of interests and motivations, then the process would not have lasted for as long as it did (Regan 2003).

So what lessons were to be drawn from the events that forced the closure of the mine if the allocation of responsibility or blame is still contestable? The first lesson was that a disaffected mine-affected community had the power to close down a large-scale mining project if government forces were too weak to stop this from happening. The second was that a mining company could not always rely on a Third World government to police and enforce the compensation and benefit-sharing agreements under which it operated. From which followed a third lesson, that mining companies operating in fragile or chaotic political environments had to develop new strategies to manage relationships between all three groups of stakeholders in the triangle.

There may have been a fourth lesson, but it was not quite so obvious to external observers. It could be drawn from the fact that Francis Ona, the rebel leader, emerged from the ranks of BCL's own workforce. He and some of his fellow mineworkers were not only members of the landowner association that demanded huge amounts of monetary compensation from the company in 1988; they were also responsible for escalating acts of sabotage against the infrastructure of the mine that culminated in the outbreak of armed conflict at the end of that year (Filer 1992). In this respect, the first phase of the rebellion was an 'inside job'. In the second phase, the company was forced to close the mine because it could not guarantee the safety of its workforce; however, by that time, the rebellious landowners in the workforce were already on the other side of the fence. Nearly all of the workers who lost their jobs in 1989, including many Bougainvilleans, left Bougainville to look for work elsewhere. The rebellion did not serve their economic interests at all, but nor was it inspired by rational economic calculations on the part of the rebels, and that is one reason why the government and the company were unable to contain it. So the lesson would be that recruitment of more workers from a mine-affected community is not necessarily sufficient to compensate for the negative impact of a mining project on their society and their environment.

The Fourth Corner

The most obvious lessons of Panguna helped to inform the efforts of the World Bank to persuade mining companies and national governments to form a new kind of 'public–private partnership' to deal with the social and environmental impacts of large-scale mining projects on local communities. To this end, the bank's own mining department sponsored a series of regional 'mining and community' conferences, one of which was held in PNG in 1998. The bank's interest in this matter was not the direct result of its previous engagement with the mining sector in PNG or in other developing countries, but was part of a wider effort to show how much it cared about 'sustainable development'. In this respect, the bank was responding to public criticism of the social and environmental impacts of a sequence of major infrastructure projects that it had funded during the 1980s (Rich 1994; Wade 1997; Fox and Brown 1998; Clark et al. 2003). It was now in the business of sharing the lessons of that experience with an audience of investors and regulators whose complacency had been shaken by a set of scandals and disasters in their own backyards.

The paradox in the World Bank's embrace of the triangular model of stakeholder relationships is that the bank's own stake does not clearly belong in any of the three corners. And the bank is not alone. A range of other stakeholders from outside the triangle got mixed up in the politics of the large-scale mining industry during the course of the 1990s. Ballard and Banks (2003: 304) describe this as a 'fourth estate' comprising 'a wide variety of NGOs, financial intermediaries, lawyers, business partners, and consultants'. So why not simply replace the triangle with a rectangle, call the fourth corner 'society', and let the World Bank be part of it?

There seem to be two main reasons why academic observers have hesitated to take this step. The first is the risk of confusion between 'society' and 'civil society'. If we think of civil society in the old-fashioned way, as the set of institutions or organisations that do not belong to the state or the church, then mining companies could be part of it, and so could landowner associations, perhaps even the World Bank. But when people now talk about the relationship between civil society and the mining industry, or civil society and the World Bank, this is clearly not the usage that they have in mind. In their usage, 'civil society' simply refers to a group of critics or enemies of whatever institutions or organisations are on the other end of the relationship. In its most inclusive form, the phrase has thus come to stand for all those stakeholders who are opposed to the power of capital in all its manifestations and therefore do not wish to be seen as 'stakeholders' at all.[15]

Those self-proclaimed members of 'civil society' who make it their business to attack the mining industry or the World Bank may claim that what they are doing is itself an essential feature of 'neoliberal capitalism' (Kirsch 2014). This may well be the case, but it does not mean that civil society, in this narrow sense, is the entity from which the targets of their attack have now decided to seek what they call their 'social licence to operate'. Some captains of the mining industry discovered their need for such a licence at the same time that some World Bank officials began to sponsor 'mining and community' conferences. But there is an important sense in which the World Bank and other financial institutions were more significant members of the 'society' from which this licence was being sought than were the members of mine-affected communities or the members of 'civil society' who acted as the allies or advocates of those same communities.[16] It is not necessary to believe that mining companies *deserve* a social licence to operate in order to appreciate that there is a large group of actors in the global mineral policy process whose values and opinions are now just as critical to the operation of the mining industry as those of national government or local community representatives (Dashwood 2013; Owen and Kemp 2013).[17]

15   Which is not to deny that the World Bank may have its own reasons for not wishing to be seen as a 'stakeholder'.

16   Some scholars would regard the World Bank's new-found enthusiasm for 'community development' as a straightforward manifestation of neoliberal governmentality (Rose 1999; Li 2007). We have taken a somewhat different position.

17   Recent literature on this subject has stressed the significance of local community support in corporate constructions of the 'social licence' (Parsons et al. 2014). Our argument would be that what matters here is the demonstration of community support to a wider social audience, since there is little point in demonstrating its existence to the very people from whom it has been acquired.

The second reason why academic observers have hesitated to adopt 'society' as the name of the 'fourth estate' is that it may be taken to imply that this is a group of actors or characters with a set of common interests that contrast with those of the other three groups. Each of the other three groups contains a set of organisations of the same general type, whose interests are represented in the way that they are organised. The fourth corner is not occupied by organisations of the same type, unless they are all assigned to the residual category of 'non-governmental organisations', so their interests may be as diverse as those of all the organisations in the first three corners. From the windows of the World Bank, it looks as if the global mining sector consists of a number of national governments, each with an agency dedicated to the regulation of the mining industry, a greater number of mining companies large and small, and an even greater number of mine-affected communities, one for each of the mines that affect them. But there is no equivalent plurality of 'societies', because national (or subnational) societies are not the authorities from which mining companies seek to obtain their 'social licence'.[18] This is probably why the World Bank prefers to represent 'civil society' as a fuzzy set of actors who could be acting at any level of political organisation (Barma et al. 2012).

But even at the local level of political organisation that is invoked by the title of this volume, there is no large-scale mining project that involves just one company, one government agency or even one mine-affected community. Instead, there is a corporate space, a government space and a community space, each of which is liable to be occupied by a number of organisations or agencies in different relationships with each other, some of which count as political relationships, and within each of these organisations or agencies there are more political relationships. The 'social space' may be a residual space, but it is hardly any different in this regard, unless we assume or discover that all the people who enter it only do so as allies or supporters of people who occupy one of the other three spaces, in which case the fourth space virtually disappears.

The basic problem here is to justify the claim that the space occupied by the politics of the large-scale mining industry, at any level of political organisation, has any specific shape at all. If politics goes on inside each of the organisations that occupy this space, as well as between them,

---

18   For example, no such licence can be obtained from the 'societies' of PNG, NC, France, Melanesia.

then it is surely possible for groups of political actors to be assembled in many different ways in relationship to many different issues. In that case, why should the number of such groups not vary from place to place and time to time? Our argument would be that a rectangular representation of stakeholder relationships in the large-scale mining industry is not in itself an essential and permanent feature of 'neoliberal capitalism', but is just a way of representing the conventional wisdom that currently reflects a specific process of economic and technological change at multiple levels of political organisation. That said, we still need to think of more specific ways to represent what political actors actually do with their political relationships, and how these activities vary from one mine site to another.

The Problem of Collective Agency

In their recent studies of local-level politics at two mine sites in PNG and Indonesia, Alex Golub and Marina Welker both pose the problem of collective agency by asking how individual actors get to 'enact a mining company' or 'personate a mine' (Golub 2014; Welker 2014). Both clearly recognise that the internal constitution of a mining company is not sufficient to explain the political behaviour of its employees, either in their dealings with each other or in political relationships with external actors who are enacting, impersonating or representing government agencies, landowner associations or tribal communities. All such collective entities, organised or otherwise, are subject to an ongoing social process of construction, deconstruction and reconstruction, from within and without.

Both authors also recognise that the boundaries of such entities are not always clearly demarcated, in the sense that some actors are able to represent both sides of the relationship between two such entities in two different corners of our rectangular model. One obvious and common example of such a 'conflict of interest' would be the community affairs manager recruited by a mining company from the community whose affairs he is supposed to manage. Francis Ona is another case in point. He was not a community affairs manager. Indeed, he elected to work night shifts so that he could spend part of the day harassing the occupants of BCL's Village Relations Office. Some might say that he was only a political actor in his capacity as secretary of the Panguna Landowners Association, but that would miss the point. BCL kept him on the payroll precisely because of his leadership role in the mine-affected community, but this just gave him more power to disrupt the company's operations.

Golub and Welker both see their case studies as applications of actor-network theory (Callon and Latour 1981; Latour 2004), but they also echo the agent-based or actor-oriented form of political anthropology that was exemplified in the work of scholars like Turner (1957), Epstein (1958), Barth (1959) and Bailey (1969). Turner's conception of political process as social drama is especially pertinent here because it entails a clear distinction between actors and the roles or characters or masks that they adopt. In this way, we avoid the semantic confusion implicit in the concept of a 'state actor' or a 'company actor'. The 'stakeholders' who inhabit the four corners of our rectangular model are not groups of actors; they are groups of roles or characters that constitute the structure of a political domain. Actors move between these positions at intervals that can vary from an hour to a decade and, in so doing, can function as brokers or gatekeepers at numerous points of intersection between the four corners (Boissevain 1974; Long 1989; Bierschenk et al. 2000). The frequency of such movement in the domain occupied by the large-scale mining industry has accelerated with the process of modernisation and globalisation. And that is one reason why the political process or drama in which the actors are engaged has accelerated the transformation of the political structure that is the stage on which they act. This is how the politics of the large-scale mining industry are subject to the process known as 'structuration' (Giddens 1979). The Bougainville rebellion was a perfect illustration of it.

From this point of view, we should be wary of the idea that local-level politics is simply part of the impact that large-scale mining projects have on people living in mine-affected areas. It may be true that the presence of a mining project intensifies relations of competition and conflict between the members of one or more local communities, and causes new kinds of argument about who counts as a community member or representative. But one of the lessons of Panguna is that this sort of impact need not reduce the capacity of such people to make life difficult for other stakeholders or transform the political structure of the industry at a level beyond that of the locality in which they are situated.

The destabilisation of this political structure adds a new significance to the point made many years ago, that local-*level* politics, as opposed to 'purely local' politics, is by definition a form of politics in which some of the actors also act at other levels of political organisation (Swartz 1968). As actors in the world of mining politics have changed their characters or hats with ever greater frequency, so has there been a growth in the number of such actors who play different parts at different levels of political organisation, in what Kirsch (2014) calls the 'politics of scale'. This is largely a function of rapid change in the technology of communication, not in the technology or organisation of the mining industry. But it does have the effect of reinforcing the position of the 'mine-affected community' as the key point of reference in the definition of what counts as 'local-level politics', even when representatives of that community engage in political relationships with other stakeholders outside the area where the mine site or the plant site is located (Bebbington 2013).

Politics and Power

If we have so far made an argument about the way that political relationships around the large-scale mining industry have been transformed by a process of economic and technological change, we still need to say more about what actually constitutes political activity within this set of relationships. We have suggested that politics necessarily involves conflict, and seems to involve the exercise of power, but we have not clearly stated what the conflict is about, except to say that it often features demands for greater local control over large-scale mining projects, and the only thing we have said about power is that some of it 'belongs' to mine-affected communities.

Two Big Issues

The main reason why the political structure of the large-scale mining industry has a well-defined shape, even while it has been destabilised, is that stakeholders are not contesting an indeterminate number of political issues. There are only two big issues at stake here. The first involves the distribution of the economic, social and environmental costs and benefits of a large-scale mining project. The second involves the problem of representation, impersonation or structuration.

The distributional issue encompasses the social relations of compensation, since that is the way that it typically appears to people representing the interests of a mine-affected community. It does not encompass workplace relations or the social relations of production, except insofar as the wages paid by mining companies or their contractors count as part of the larger package of costs and benefits that is being contested. The modernisation of the industry has split the wage-earners between two groups. On one hand there are those, like Francis Ona, who move back and forth between the community and the company, and can become political actors in one or both of these spaces. On the other hand, there are those who do not count as members of a mine-affected community, and therefore cannot claim to represent it. The terms and conditions of employment at a modern mine site are rarely such as to unite workers in both camps in a political contest with company managers, unless we allow that a 'modern' mining operation can have lasted long enough to unite them.

The prominence of the distributional issue as a political issue explains why the process of environmental (and socioeconomic) impact assessment is necessarily a political process, even if it is carried out in a technical or superficial manner (O'Faircheallaigh 1992). In most countries, this process is a legal hurdle that a mining company has to jump over in order to extract a development licence from a national government, and in many countries, the government has little capacity to hold a company to account for the impact of its operations once the licence has been granted.[19] However, there are other institutions that have been created to accommodate the distributional contest at different levels of political organisation, and some of them remain active throughout the mining project cycle.

The representational issue is really just one aspect of the distributional issue, since it revolves around the question of how individuals get themselves or each other into roles in which they can participate in the contest over the distribution of costs and benefits as the representatives of whole groups of other people. For reasons already explained, this issue takes on a life of its own because of the frequency with which political actors move between representative roles, and the consequent appearance of multiple 'conflicts

---

19   NC's new mining code (issued in 2009) constitutes an attempt to overcome this limitation by demanding that mining companies produce long-term action plans covering the entire process of mine closure and its aftermath, but the PNG Government has been making similar demands for more than three decades while still losing the capacity to hold companies to account for their implementation.

of interest'. The representational issue often raises questions of authority or legitimacy, but these questions are best addressed in an empirical analysis of the political process rather than by means of abstract definitions of such concepts. An empirical approach can reveal the extent to which distribution and representation are mutually constitutive issues.[20]

Occupants of the fourth (social) corner in our rectangular model of stakeholder politics do not normally get involved in the distributional issue as people claiming a share of project benefits for themselves. They may obtain such benefits from people who occupy one of the other three corners, may represent those people in a contest over this issue, and may claim 'credit' from this engagement that can be invested in other fields of action. But they may also enter the political process as purveyors of ideas about things like environmental justice or the rights of indigenous people without making any such claim or commitment. World Bank functionaries see their roles this way, even if some members of 'civil society' would prefer to place them squarely in the company corner. The presence of this fourth group of stakeholders therefore adds to the complexity of the representational issue.

Political Actors and Roles

Given the complexity of the representational issue, it is no easy matter to determine who counts as a political actor and what counts as a political role. Many political actors are not keen to be known as such, and are therefore quite happy to go along with the old-fashioned idea that a political role is equivalent to a public office, and all such roles must therefore be part of the state. Even public servants will say that they are not political actors if they have not been elected to the offices they hold; they merely serve the public interest by acting on the orders of elected politicians.

Elected politicians in PNG turn this argument on its head by constantly accusing each other of 'playing politics', which is surely what they are supposed to do, and then making the same accusation against all sorts of other people as well, while those people make the same accusation against each other. The net result, as in many other countries, is that 'politics' appears to be an extremely widespread but very unpopular sort of activity. In PNG this virtual paradox is carried to extreme lengths because of the

---

20   This is the way that Lund (2002) has previously dealt with the relationship between property and authority, albeit in another context.

sheer numbers of people who compete to be elected as members of the national parliament or as presidents of local councils (May et al. 2013; Schwarz 2013). A member of parliament has every reason to fear that lots of people in his constituency, including local councillors and public servants, are 'playing politics', because dozens of them could be planning to stand against him (or very occasionally her) at the next national election.

This form of political competition is no more intense in the minority of electorates that host large-scale mining projects than it is in the majority that do not. However, there is another form of political competition that is especially prevalent within mine-affected communities, both in PNG and in NC, which is the competition to hold office in landowner associations or so-called landowner companies (which are known in NC as *sociétés par actions simplifiées*—'simplified joint-stock companies'). The existence of such entities is authorised by the state, but they are obviously meant to represent the interests of people who belong to the local community. These roles are political because they are an integral part of the social relations of compensation. The proliferation of such roles in the vicinity of large-scale mining projects reflects the level of popular concern with the distribution of project-related costs and benefits (Bainton and Macintyre 2013; Grochain 2013; Golub 2014).

This intensification of local-level politics also entails the reconfiguration of customary leadership in response to the demands of 'development' or the demands of legislation. This underscores the capacity of actors to create new roles for themselves, instead of simply acting within the confines of a political structure that is already given as a fact of life. But it also tends to obscure the distinction between public and private domains of political action, because 'state actors' and 'company actors' get tangled up in a contest in which 'community representatives' dominate the distributional issue by sheer weight of numbers, and everyone accuses everyone else of attempting to manipulate the distribution of project-related costs and benefits to their own personal advantage (Filer 1998; Bainton 2009). The boundaries between 'politics' and 'economics' are also blurred as political orientations encompass the activities of local entrepreneurs (Le Meur et al. 2012; Grochain 2013).

A focus on the distributional and representational issues does have one analytical advantage, for it enables us to ignore forms of political behaviour that are disconnected from these issues. The practice of office politics is no less real than the occupation of political office, but some of the political relationships that are internal to the organisations that occupy each of the four corners in our rectangular model do not have any bearing on the contest at the centre of the square. The contest creates a number of roles for people representing the state, the company, or even 'society', and that number also tends to expand as the contest intensifies. But here again, it is not possible to deduce the content of their political relationships from the way in which these roles are constructed.

Metaphors of Power

It is hardly possible to comprehend the content of political relationships without reference to the concept of power. There is a school of thought that actually defines politics as the pursuit and exercise of a certain type of *personal* or *interpersonal* power, which is the power that some people exercise over other people. Personal power is thus contrasted with the *impersonal* power that people may exercise over things or that things may exercise over people. In this school of thought, person A is understood to exercise power over person B if A can get B to do what A wants B to do. This statement of their mutual relationship may be qualified in various ways by reference to the intentions or desires of the two parties, or by reference to what actually gets done, or to the ways and means by which compliance is secured (Lukes 2005; Searle 2009). Max Weber is seen as the founder of this school of thought within the discipline of sociology, when he defined power as the probability of individuals getting what they want in the face of resistance from other individuals. But this was also Machiavelli's way of understanding politics.

Machiavelli's version of methodological individualism was a self-conscious departure from the Aristotelian tradition in which politics is understood as the collective pursuit of specific moral goals by the citizens who are members of a political community. Insofar as this activity involves the exercise of power, it is the power to determine what we would nowadays describe as the public interest, which is a form of impersonal power that people exercise over things. Furthermore, it is a form of power that people exercise by virtue of their occupation of specific positions or roles in a political system that might or might not be a nation-state (Easton 1990). We do not have to subscribe to functionalist forms of social theory to see the potential value of this alternative conception of political practice, especially when we are dealing with a situation in which many of the actors appear to prefer the Machiavellian version of politics as an interpersonal power game.

Since we have defined the game in question as a contest over the distribution of costs and benefits,[21] it is all too easy to assume that power is a form of currency that people use to purchase a bigger share of the benefits, or to bear a smaller share of the costs, either for themselves or for the groups or entities they are supposed to represent. If the outcome of the contest, which is the actual distribution of costs and benefits at some particular moment in time, is then calculated according to a single standard of measurement, like money, then we might be tempted to think that we have measured the balance of power between competing economic interests. But all we should have done is to represent political relationships as if they were market relationships between actors who possess the same kinds of assets, the same kinds of values, and the same kinds of goals.

This is clearly not the case. Just because we have defined our political field as a contest, this does not mean that all the political relationships in this field are relationships of competition or conflict that entail the exercise of just one kind of power. The different actors involved in this contest may have all sorts of different ideas about the definition of economic, social and environmental costs and benefits, just as they may have different ideas about what is public and what is private, what is right and what is wrong, who has the power to do what, or who has power over whom.

Participation, Exclusion, Domination

Steven Lukes (2005) took issue with the Aristotelian school of thought, as represented in the work of Talcott Parsons and Hannah Arendt, on the grounds that it simply fails to accommodate what people mean when they talk about 'power games' or 'power struggles' as an integral part of any political process. He also tried to avoid the methodological individualism that is commonly associated with an emphasis on interpersonal power as the primary form of political power. Lukes reckons that the exercise of authority or influence by political office-holders does not even count as an exercise of political power if there is no conflict between the interests of different actors in a political relationship. On the other hand, he does not assume that political power can only be exercised by individual actors, nor that its exercise must always be intentional, so it cannot simply be defined as the capacity of individuals to realise their own interests at the expense of other people.

21 Benefits may include political positions or roles, which again links the distributional issue to the representation issue.

Lukes conceives the exercise of political power as something that can take place in three dimensions. The first dimension is the one identified in the work of political scientists like Robert Dahl (1961), who pioneered the study of urban American power games in the 1950s. In their 'pluralistic' conception of local-level politics, power was indeed defined as the capacity of individual actors to influence decisions on issues that were political because they were controversial, and different actors in the political process therefore had different preferences. The second dimension is the one identified in the work of Bachrach and Baratz (1970), who observed that power can also be used by the participants in any political process to exclude some issues and actors from that process, thus constraining the definition of what counts as a legitimate subject of political debate and whose preferences can make a difference to the outcome. Lukes himself adds the third dimension, in which the power exercised by groups or organisations is not reducible to that exercised by their individual members, and the exercise of power can take place without the appearance of political conflict or debate if it is exercised in a way that prevents people from acting in their own interests.

In the first (1974) edition of his book, Lukes was clearly thinking about this third form of power as the sort of power that is exercised in the political relationship between social classes. One of the first attempts to apply the three-dimensional conception of political power to the analysis of local-level politics is of interest here because it was applied to the relationship between capital and labour in what we would call a traditional mining region in the United States (Gaventa 1980). However, the focus on class can be misleading in two ways: first, because it encourages the reduction of all political preferences to economic interests; and second, because it suggests that the first two forms of interpersonal power are much less important than the third one. In that case, it is easy to suppose that the one key question to be addressed in any study of local-level politics is the question of when, where and how the subordinate party in a binary relationship is able to escape the trap of false consciousness and resist the power of the dominant party (Scott 1990).

Even if we put the traditional form of class struggle to one side, and reframe the question of power as a question about the distribution of economic, social and environmental costs and benefits from large-scale mining projects, we may still be tempted to adopt a one-dimensional conception of the political relationship between mining companies and mine-affected communities as a bipolar relationship of persistent domination and occasional resistance in which all other stakeholders are simply allied with one side or the other (Kirsch 2014). Yet Lukes himself now concedes that the third type of political power is not always exercised in a zero-sum game between two groups of people with one type of interest in the result (Lukes 2005: 109).

Nor is there any reason to assume that the parties to relationships of domination and dependency, acquiescence or resistance, must occupy different corners in our rectangular model. Such relationships may just as well be found inside a company, a state or a community. The exercise of interpersonal power through relationships of participation and exclusion may likewise involve actors and roles that belong to one corner or to several corners at any particular moment. The common exclusion of women from the politics of the large-scale mining industry is an obvious case in point (Eftimie et al. 2009).

If there is no objective or authoritative basis for deciding where people's real interests lie, it is obviously more difficult to distinguish different forms of interpersonal power in terms of people's capacity to either advance their own interests or prevent other people from doing so. Political actors make use of a wide range of material and symbolic resources to secure the support, compliance or silence of other actors in any political process, and we might just as well distinguish different forms of interpersonal power by reference to these different forms of 'political capital'.

To take but one example, representatives of communities downstream of the Ok Tedi mine in PNG were able to secure the support of numerous 'civil society' actors in pursuit of their compensation claims against BHP because of the nature and extent of the environmental damage that they were complaining about. The political capital assembled in this way was then deployed against the mining company in the form of a well-publicised law suit that cost the company its 'social licence' to operate the mine, as well as a substantial compensation package, but did not put an end to the environmental damage. The community representatives who gained a partial victory in that legal contest then tried and failed to convert that form of political capital into votes from other people in the mine-affected area when they stood for election to PNG's national parliament (Banks and Ballard 1997; Kirsch 2014). In this instance, we could say that they had three different degrees of success in the exercise of three different forms of interpersonal power, but these are just three of many possible forms.

The Powers of Things

In most of the vernacular languages of Melanesia, the word that most closely approximates the concept of 'power' is the one that signifies a type of dangerous supernatural force that is beyond human control. There is nothing political about this sort of idea, except perhaps in those cases where it is associated with the status of a chief, and thus approximates the well-known Polynesian idea of *mana*, but even then, it represents a supernatural form of traditional authority that cannot be manipulated by its notional owner. In some respects, it is no more political than the power that comes out of a modern power station. Foucault's concept of 'biopower' has similar properties, for it is too pervasive and irresistible to be part of anyone's power games or power struggles, and may only seem to be political if politics is either identified with the practice of government or conceived as the domination of all individual subjects by a singular and mysterious social (or even supernatural) force (Brown 2006; Searle 2009). So how can the power of things over people get more political than this?

In the present context, one obvious candidate is the power of the mine. Golub (2014) writes about the Porgera gold mine as one of several 'leviathans' that are more or less successfully assembled and 'personated' by the individual actors who contest the distribution of its costs and benefits. It is not clear whether this construction of the mine as a person distinct from the mining company or its representatives is just a way of paying homage to actor-network theory (Callon and Latour 1981), by showing how individuals assemble social groups, or whether it reflects the way that people in the mine-affected community actually talk. Anthropologists are certainly not the only people who indulge in this form of reification, by talking about a mine as if it were a person.

There is one obvious way in which a large-scale open-cut mine can seem to exercise a degree of impersonal power over its own workforce that goes beyond the executive or supervisory powers of its own managers. The regimentation of the process of production and consumption within a modern mining compound creates a sort of 'total institution' in which every individual is subject to the same form of discipline or self-discipline, commonly construed in terms of health and safety, regardless of their status in the occupational hierarchy. Yet this is only an approximation of the power of surveillance exercised in Foucault's world of extreme discipline: insofar as it works, it takes the politics out of the organisation. The impersonal power of the mine gets to seem a lot more political when conceived as the power to transform or 'overflow' the physical environment (Letté 2009), since this transformation is one of the main bones of contention in the social relations of compensation, as distinct from the social relations of supervision or surveillance. It makes sense for all political actors contesting the distributional issue to represent the mine as a powerful agent in its own right, since none of them need then accept a personal responsibility for what it does to its environment, but what it does is what they have to deal with in their own political relationships. And once the power of the mine 'overflows' the boundaries of the mining enclave in the form of a violent reaction, then politics presents a second type of limit to governmentality (Li 2007).

Golub would argue that the power of the mine, in this sense, is not something that it acquires as a result of its design, or as a feat of mechanical engineering, but as a work of fiction that, like other 'leviathans', has emerged out of local people's engagement or 'entanglement' with 'wider networks of power and knowledge' (Golub 2014: 83). But he goes on to argue that the mine has acquired more power in the local-level political process through which it has been constituted or 'personated' than the government that supposedly regulates that process. In other words, the balance of impersonal power between the four corners in our rectangular model is decidedly tipped against the state.

It is a commonplace of political science and political philosophy that the power of the state is greater than the sum of the personal powers attached to roles or positions in the state apparatus, in a sense that is not true of companies or communities. This assumption is commonly associated with ideas about the rule of law or the legitimate use of force. France is just the sort of nation-state that is (or was) assumed to wield this sort of power, but the PNG state looks more like a naked emperor or paper tiger, especially in the vicinity of large-scale mining projects. Hence the observation made by one community leader involved in negotiating the development of the Lihir gold mine, that 'the State is only a concept' (Filer 1995: 68). Some social scientists have made similar observations about the 'state as such' (Abrams 1988; Alonso 1994), but even if the state is only a product of some political imagination, some states may have more imaginary power than others.

If the PNG state is unusually 'weak', this does not mean that politicians or public servants have less capacity to use their offices as sources of interpersonal power, even when dealing with mining companies or mine-affected communities. If anything, that sort of power is enhanced when it is not constrained by the rule of law (Dinnen 2001). The weakness of the state must therefore be understood as a function of the incapacity of 'state actors' to act in concert with each other, or as a function of their capacity to move between positions in different corners of the political rectangle (Clements et al. 2007). Golub has a neat example of this type of mobility in the shape of the Porgera district administrator, whose power was a function of his capacity to represent the mine and the mine-affected community, as well as the state, and sometimes to wear all three of these hats or masks at the same time (Golub 2014: 51).

The Power of Ideology

An ideology is a third sort of 'thing', aside from a mining project or a nation-state, that may exercise power over the people involved in a political process by making them think and talk about a political issue in a way that contains a certain sort of bias. The concept of ideology has some of the same drawbacks as the concept of economic interest, since ideologies and interests alike have commonly been seen as the properties of social classes.

If we seek to understand a political process that does not seem to contain any obvious element of class struggle, there may be no reason to think that participants in this process are weighed down with any ideological baggage beyond an understanding of their interests as individuals or as the occupants of specific political roles. The world of Melanesian politics may look like a world without ideology when no distinction can be drawn between the policies of different political parties beyond the self-interest of their leaders (Rich et al. 2006), or when (as in the case of NC) the main point of distinction is the question of how public participation in large-scale mining projects relates to the prospect of national independence. However, those anthropologists who regard all political relationships as variations on a single theme of domination and resistance will say that the 'end of ideology' just means the triumph of a single ideology at all levels of political organisation, regardless of any regional variations in 'political culture'. This is of course the ideology that most of them call neoliberalism, and many associate (a bit hastily) with Foucault's concept of governmentality (Ferguson and Gupta 2002).

Golub (2014) argues that models of 'pervasive governmentality' fail to explain the practice of politics in Porgera precisely because they overestimate the power of the state. But by the same account, a mining company that cannot dominate a mine-affected community, even when it takes on some of the functions of government, can hardly be the vehicle that makes everyone in sight subscribe to a neoliberal ideology when they contest the issues of distribution and representation. This does not mean that ideology has no place in the contest; it only means that the power of ideology is exercised in the construction of political identities and roles, not in the pursuit of class interests.

For example, one of the key things that establishes the identity and the rights of mine-affected communities in PNG is an 'ideology of landownership' which asserts that every automatic (or indigenous) citizen counts as a 'customary landowner' by virtue of his or her membership in one of the multitude of clans that each own one portion of the country's total land area. The emergence of this ideology as a form of national identity reflects the social relations of compensation in a resource-dependent economy, and is therefore closely linked to the form of local-level politics in which 'clans' have been constituted or reconstituted as collective claimants to 'compensation' from large-scale resource projects (Filer 1997). This is what distinguishes the ideology of landownership from ideologies of indigenous or ethnic identity. Golub (2014) obscures this point when he says that members of the mine-affected community in Porgera have come to construe themselves as 'Ipili', that is to say, as members of a partially invented tribal community. Community leaders are more likely to represent themselves as 'landowners' when dealing with other stakeholders in their political contest, and in so doing they take advantage of the fact that most of these other stakeholders subscribe to the ideology of landownership.

The ideology of landownership may perhaps be conceived as a form of popular resistance to the power of neoliberal governmentality, but there is no reason to assume that it constitutes the only form of national or subnational identity that has an impact on the type of political process that we are considering here (Keesing 1992). In NC, the ideology of landownership is constrained by specific legal provisions for the restoration and redistribution of collective customary land rights under policy reforms initiated in 1978. Where big mining projects are at stake, and it is hard to convert indigenous land claims into formal land rights, Kanak people alternate between the assertion of indigenous rights and a more inclusive appeal to municipal community interests. At the same time, these tensions raise questions about corporate boundaries and local citizenship, since mining companies are expected to meet the obligations that arise from being accommodated as a stranger of a special kind within the local community (Le Meur, Ballard et al. 2013; Le Meur 2015). In the PNG case, Golub (2014) notes that Porgeran community leaders could represent themselves or their followers as subsistence farmers or indigenous people in their efforts to secure more compensation from the mining company. Political contests over the distribution of project-related costs and benefits also seem to encourage such leaders to represent themselves as 'chiefs', even in places (like Porgera) where anthropologists would say that no such roles existed in traditional society.

In one sense, this is simply further evidence of the capacity of political actors in such places to invent new roles that carry the impression of additional personal power. But in another sense it could be argued that an ideology of chieftainship has developed alongside the ideologies of landownership, subsistence affluence or indigenous rights, each with the power to create a different form of cultural identity fit for the same political purpose (Rodman 1987). The ideology of chieftainship asserts that people only 'play politics' because foreign forces have corrupted a traditional social system in which chiefs exercised the only legitimate form of personal power, so being or becoming a chief in a local political contest is another way to diminish the power of political roles that do not have this aura of traditional authority (White and Lindstrom 1997). This ploy does not work quite so well in NC, because the roles of great and small chiefs have been officially recognised in the colonial rendition of 'customary authority' since the early years of colonial rule. Nevertheless, the Customary Senate does represent the more recent emergence of neotraditional authority at a higher level of political organisation, and that is where the global discourse of indigenous rights tends to be mobilised.

Ideologies of this kind may be understood as forms of nationalism insofar as they construct the idea of 'the nation' in specific ways (as a nation of customary landowners, subsistence farmers, indigenous people or traditional chiefs), but they do not resemble the nationalisms of European history because they do not treat state institutions (including modern legal codes) as legitimate expressions of this cultural identity (Appadurai 1990; Foster 2002). They therefore have the effect of creating what Ferguson (2005) calls an 'ungovernable space', and in the Melanesian political landscape few spaces are less governable than those which surround a large-scale mining project (Allen 2013). The regimentation of life within a mining compound therefore stands in stark contrast with the unruliness that prevails on the other side of the fence (Golub 2014). In Bourdieu's (1977) terminology, such ideologies can even be counted as forms of symbolic violence against any organisation that threatens the boundary of a neotraditional community.

There is no reason to assume that such ideologies have some sort of monopoly over the terms in which people debate the distribution of costs and benefits from large-scale mines in Melanesia. It is also possible to detect a very different ideology, or set of ideologies, in which the mine itself becomes a symbol of modernity because of its power to deliver what most stakeholders, including local community members, call 'development'. If all actors in a local or national political process agree on a single definition of development, and if it can be shown that this is a 'neoliberal' definition, there seems to be much less scope for any construction of 'culture' to be more than a passing form of 'neopopulist' resistance to the one great power that runs the world. But a closer inspection of what people do mean by 'development' in this particular debate suggests that there is no such general agreement, just as there is no general agreement about the definition and measurement of costs and benefits (Martin 2013). The key point is that 'landowners' may subscribe to an ideology of development in the same way that 'developers' subscribe to an ideology of landownership, not because they agree about the best way to calculate the distribution of costs and benefits, but simply because they want something from the other side. Furthermore, the notion of 'development'—especially of 'sustainable development'—may be deployed as a semantic vehicle to align divergent corporate and community interests in the form of a specific local agreement, as in the case of the Goro project (Horowitz 2012).

Politics and Policy

If the practice of local-level politics in the vicinity of large-scale mines is now to be understood in terms of the relationship between interpersonal and impersonal forms of power, how are we to understand the relationship between the political process that runs alongside the mining project cycle at each major production site and the policy process that creates or transforms the political institutions or 'spaces' in which the distributional and representational issues are negotiated? If a policy process is understood to be one type of political process, we might suppose that it is set apart from other types by virtue of the fact that policies are made by governments and are therefore made at a level of political organisation above the level at which the game of local-level politics is played. However, what we have already said about the power of the state should alert us to the possibility that things are not quite so simple.

To begin with, governments do not have a monopoly on the production of policies. Mining companies have policies too; so does the World Bank, and so do many non-governmental or community-based organisations. Indeed, organisations established to represent the interests of 'local people' in the negotiation of benefit-sharing agreements are liable to manufacture policies and programs precisely in order to demonstrate their moral superiority over the other parties to the negotiation. A notable example of such activity is the long sequence of policy pronouncements made by the Lihir Mining Area Landowners Association since its formation in 1989, which have certainly seemed like acts of symbolic violence to representatives of the mining company at which they are mainly directed (Filer 1995; Bainton 2010). Beyond the proliferation of actors involved in policy production, agreements themselves can also become the building blocks of policies that change the terms in which the distributional and representational issues are resolved in subsequent agreements (Le Meur, Horowitz et al. 2013; O'Faircheallaigh 2013).

The proliferation of actors and negotiation spaces invariably takes place in a specific political and historical context. In NC, the production of mineral policy is inseparable from the politics of decolonisation, as was already evident when the Bercy Agreement became a precondition for negotiation of the Nouméa Agreement in 1998. The Bercy Agreement was not only the result of strong political action by the Kanak and Socialist National Liberation Front; it also served to illustrate the disjuncture between the French metropolitan state and NC's different levels of political organisation in negotiation of both the representational and distributional issues in the mining sector. While the process of 'negotiated decolonisation' creates an active role for the metropolitan state, both as an umpire and a stakeholder in the negotiation of mineral policy, it also fosters state-making processes within NC, where provincial and local agencies are riddled with moving political fault lines.

That sequence of events contrasts with current bilateral negotiations between local Kanak communities and mining companies about the grant of free, prior and informed consent for renewed exploration of the so-called 'forgotten coast', which is the eastern coast of South Province. In this instance, a two-year moratorium has been imposed at the behest of customary authorities at both local and national levels, supported by indigenous organisations and the pro-independence mayors of the two local municipalities (Thio and Yaté). This moratorium has created space for the conduct of environmental baseline studies and the design of a sustainable development strategy for the area. The negotiation now involves customary, municipal and provincial authorities, and could become a sort of showcase for South Province, which favours autonomy for NC but is opposed to outright independence.

In PNG, the capacity of the national government to complete its own policy pronouncements on the questions of distribution and representation in the mining sector has almost evaporated over the past 15 years, despite (or possibly because of) the 'technical assistance' it has secured from the World Bank. The policy *process* has continued, numerous foreign consultants have been engaged to move it forwards or backwards, but no new institutions have emerged as a result. A mineral policy process is most likely to produce a new political institution or settlement when it is undertaken in response to a crisis in the political relationships around a single large-scale mining project. As we have seen, the Bougainville rebellion was the first in a series of such crises that led some of the world's biggest mining companies to the conclusion that they had lost their 'social licence to operate', and hence to the creation of the institution known as the Global Mining Initiative, followed by the Mining, Minerals and Sustainable Development Project, followed by the establishment of the International Council on Mining and Metals (Danielson 2006; Dashwood 2013).

Some observers think that the Bougainville crisis also inspired PNG government officials to invent the institution known as the development forum (Golub 2014: 102), but this is not so. The development forum was invented in 1988 in response to demands by provincial premiers and local community leaders for a bigger share of the benefits to be derived from the Porgera gold mine and a stronger voice in the negotiation of a benefit-sharing agreement (Filer 2008). Once this agreement was finalised in 1989, the prime minister tried to persuade Francis Ona and his followers to negotiate the same sort of agreement, but to no avail. The conflict on Bougainville was eventually resolved by a peace process that was a different sort of policy process, since it bore no direct relationship to political debate about the costs and benefits of the mine that was now closed (Regan 2010). But once that policy process had led to the establishment of the

Autonomous Bougainville Government in 2005, a more specific policy process was instituted to invest this new government with the power to determine the conditions under which the Panguna mine might be reopened or any other mining project might be authorised (Regan 2014).

Although government officials can reasonably claim credit for inventing the development forum in the first place, this institution did not simply become the more or less governable space in which national and provincial government representatives would henceforth negotiate benefit-sharing agreements with the representatives of landowning communities. Each new forum created an opportunity for the participants to turn a political contest into another policy process by making demands that were inconsistent with existing laws and policies. In the Lihir case, for example, the community leaders demanded a share in the ownership of the mine that was greater than the share the government was prepared to purchase on their behalf. They were only persuaded to moderate their demand when the prime minister undertook to raise the rate at which royalties were levied on all large-scale mining projects, and hence to raise the income that the landowners and their community government would receive from their agreed share of the royalties levied on the Lihir mine (Filer 2008).

From such examples it should be evident that a policy process is not best conceived as a process in which one collection of stakeholders operating at a higher level of political organisation determines the rules by which another group of stakeholders operating at a lower level of organisation will sort out their political differences. A political contest over the distribution of costs and benefits derived from one large-scale mining project can turn into a policy process by changing the rules that apply to the same sort of negotiation in other locations (Le Meur, Horowitz and Mennesson 2013). And if governments do not have a monopoly on the production of policies, the transformation of a political process into a policy process can involve the occupants of any number of the four corners in our rectangular model of stakeholder relationships at any particular moment in time. The transformation may therefore involve a form of escalation, in which the number of actors and roles involved in the process grows larger as an issue gets to be contested in new locations or at larger scales, on different or bigger stages.

When a political process does turn into a policy process, there is no need to assume any change in the way that power is exercised over the outcome. The interpersonal powers of participation, exclusion and domination may still be deployed, but the first two are likely to be more significant with an increase in the number of actors and roles involved in the process, since it is less likely that all of them will belong to one of two camps and that one camp will score a decisive victory over the other. When the design of political institutions is at stake, impersonal forms of political power are liable to be exercised or resisted with greater intensity, as more of the actors engage in acts of physical or symbolic violence. But whatever happens along the way, the outcome of a policy process will almost invariably change the balance of both personal and impersonal power between the different elements in the political landscape, since any political institution contains its own distinctive forms of social inequality. Changes in the balance of power that result from the transformation of political institutions must then be distinguished from those that result from what Kirsch (2014) calls the 'politics of time', in which the capacity of different actors to influence the distribution of costs and benefits from a large-scale mining project diminishes in different degrees as the project moves from the point of being designed to the point of being finally closed.

Case Studies

The papers collected in this volume were not originally meant to address a single theoretical question. Most of them are derived from a conference that dealt with the broader topic of 'mining and mineral policy in the Pacific region'. We have taken the relationship between large-scale mining projects and local-level politics in two specific jurisdictions—the independent nation of PNG and the French territory of NC—to be one specific aspect of this broader topic. The authors of conference papers relating to this more specific topic were therefore invited to submit chapters to the present volume, while some additional papers were commissioned in order to make the collection more complete. This introduction has been written in response to their submissions, so the ideas that it contains have not necessarily informed any of the other contributions. In this final section of our introduction, we summarise the key messages of each chapter in the light of our conceptual framework.

New Caledonia

Jean-Michel Sourisseau and his colleagues examine the institutional innovations adopted by a set of local actors, claiming at once a customary and entrepreneurial legitimacy, to deal with, and profit from, the structural effects of the Koniambo project in North Province. They show how these different actors got involved, directly or indirectly, in construction of the nickel processing plant, and compare this practical experience of project management with the expectations, hopes and fears expressed by the same actors before the start of the construction process. Their main focus is on the distribution of employment, the structure of the local economy, and the relationship between economic development and social cohesion. They highlight the appearance of an active learning process on the part of these actors, albeit one that requires a certain level of institutional support to reduce the risk of increasing economic inequality and social disintegration in the project-affected area.

Matthias Kowasch examines the Koniambo project from a different standpoint, by looking at the spatial distribution of the social, economic and environmental changes experienced by local communities—especially Kanak communities—since the project's inception. Three elements are central for the interpretation of these dramatic changes. First, he shows that these communities have been active drivers of these changes and see the Koniambo project as 'theirs', even though the provincial authorities have played a decisive role in its implementation. Furthermore, he shows how the creation of a 'simplified shareholding company' has played a critical role in mediating the participation of Kanak communities in the distribution of project benefits by managing the distribution of contracts to local entrepreneurs. Finally, he warns that these positive aspects of the relationship between the project developers, the provincial authorities and local communities should not lead us to underestimate the negative impacts of the project in the form of escalating land disputes, rising inequalities, economic exclusion or social disruption.

Christine Demmer explores the origins and consequences of the political conflict that resulted in the closure of the Boakaine mine (in North Province) in 2002, after ten years of operation by the Société Minière du Sud Pacifique, and the recent political debate about the possibility of reopening it. Through this case study, the author raises questions about the identity and authority of the different Kanak actors competing over the management and distribution of mineral revenues in the municipality and district of Canala.[22] At stake here is a localised notion of sovereignty in which the struggle for political and economic independence has also been a struggle for control of the 'modern' municipality and 'customary' district in which the mine is located and a struggle for the recognition of indigenous land rights and traditional chiefly authority. While this study reveals the persistence of the segmentary logic of 'traditional' Kanak society in the practice of contemporary politics, it also shows how Kanak demands for participation in the modern mining industry connect with the practice of politics at different levels of political organisation.

Pierre-Yves Le Meur examines another case of political conflict around Thio, one of NC's oldest mining locations, in 1996. A two-week blockade of the two main mines and the wharf by local Kanak residents resulted in a new agreement between the operating company (Société Le Nickel), the local customary authorities, the municipality and the province. This was a wide-ranging agreement that could be described as an 'impact–benefit agreement' because it covered a mixture of social, economic and environmental issues, including customary land rights, access to employment and the prospect of opening a new mine under Kanak ownership and control. The reconfiguration of the local political arena that was prompted by the eruption and resolution of this conflict is analysed from the different local perspectives represented in the substance of the agreement, but also placed in the broader historical context established by the Matignon-Oudinot Agreements of 1988 and the Nouméa Agreement that was signed ten years later. In this broader perspective, the conflict of 1996 can now be seen as the starting point or harbinger of a major shift in the political complexion of NC's mining industry, from both a discursive or ideological point of view and in terms of the relationships between different political actors and their roles.

Claire Levacher provides another side to this story of political change, and a contrast to the story of the Koniambo project, in her account of the way that discourses of environmental justice, indigenous rights and sustainable development were mobilised by Kanak representatives in their negotiation of the 'Pact for the Sustainable Development of the Great South' that was made with the developers of the Goro project in 2008, ten years after the Nouméa Agreement. This agreement was not based on indigenous land claims, but it did coincide with an election that brought

---

22   Canala is the name of the municipality and one of the two districts that it contains.

the municipality of Yaté, where the project is located, under the control of an indigenous political association known as Rhéébu Nùù ('Eye of the Country'). While negotiation of the agreement revealed significant differences in the conceptions of natural and cultural heritage espoused by the members of indigenous and environmental groups, the subsequent alliance between the mining company and indigenous political leaders has been forged at the expense of both a radical environmental ideology and the local influence of the provincial authorities.

Papua New Guinea

Glenn Banks and his colleagues ask why and how mining companies in PNG have delivered community development programs to mine-affected communities. They observe that company motivations vary along a continuum that ranges from contractual obligations to corporate philanthropy, with considerations of social responsibility and social licence somewhere in the middle. But regardless of the motivations, they see all such programs as ways of countering the 'unruliness' of local-level politics at the same time as they are meant to mitigate negative social impacts, and in this sense they are inherently conservative forms of intervention. Although such programs can make mining companies look like aid agencies, the resemblance is only superficial, because mining companies rarely even pretend to engage local communities in the design and implementation of different projects, while their lack of interest in the monitoring and evaluation of development outcomes reflects a fundamental lack of accountability to any public audience.

Colin Filer and Phillipa Jenkins make similar points in their discussion of the way in which the distributional issue has been negotiated between the stakeholders in the Ok Tedi mining project, but they also focus on the question of how the distribution of power, as well as the distribution of costs and benefits, between the different stakeholders has been modified through the lengthy political and policy process associated with the timing of mine closure. While the Ok Tedi mine is rightly renowned for the extent of the environmental damage it has caused, less attention has been paid to the scale and complexity of the institutional superstructure that has evolved out of a sequence of compensation and benefit-sharing agreements between community representatives and other stakeholders. While the authors question the direction of the causal relationship between the transformation of the mining company into a 'proxy state' and the apparent weakness of political institutions in the mine-affected area, there is no doubt that the political life of this particular mine represents an extreme form of the contradiction between resource dependency and sustainable development at a number of spatial and temporal scales.

John Burton and Joyce Onguglo question the extent to which mining companies in PNG have earned their social licence to operate by means of compliance with the various international standards of good practice to which they have made a public commitment. They also discuss the reluctance of some of these companies to acknowledge that mine-affected communities in PNG are also indigenous communities, whose rights and interests therefore demand special consideration. In some respects, this reluctance may be seen as another type of corporate response to the 'unruliness' of these communities, especially those whose leaders look more like warlords than landlords. The PNG Government is also taken to task, not only for its tolerance of corporate hypocrisy, but also because it espouses progressive social policies which it then fails to apply to the benefit-sharing agreements for which the companies are not directly responsible.

Susan Hemer approaches the gender equity issue from a rather different angle, by questioning the effectiveness of the different strategies adopted by the representatives of two women's associations in their efforts to secure a bigger share of the benefits, or a smaller share of the costs, associated with the development of the Lihir gold mine. While a few of these women have been able to air their grievances on a national, or even an international, stage, and thus secure some kind of support from stakeholders outside their own community, this has notably failed to enhance their status or authority within the male-dominated political life of the mine-affected community. Insofar as male and female members of this community remain committed to a defence of their own 'customary' values against the impact of a large-scale mining project, there seems to be very little scope for local women to be 'empowered' in ways that would be recognised and endorsed by members of a national or international audience, since they can only gain 'respect' for doing things that only women do.

Nick Bainton deals with another form of inequality and marginality in the Lihir community by asking how the leaders of that community can simultaneously patronise and demonise the migrants from other parts of PNG who are not directly employed by the mining company yet still seek to benefit from the economic opportunities that it offers. In some respects, the growth of this population of strangers underscores the distinction between the two inner circles of entitlement prescribed by the preferred area policy, since they have settled on land in close proximity to the leases held by the mining company, and the customary owners of this land are also first in line to receive royalty and compensation payments in respect of these leases. Individual members of the 'local political elite' have recruited individual migrants as their clients, either as tenants or employees, but each individual patron has more to fear from the clients of other landowners than he can hope to gain from the support of his own clientele, so the growth of migrant numbers has induced a widespread sense of 'civic insecurity' that now verges on a 'climate of fear'. So long as the migrants all have patrons, the local-level government cannot respond to calls for their wholesale eviction, so these calls are redirected to the mining company, accompanied by the tacit threat of violent conflict between the insiders and the outsiders, which could escalate to the point at which it endangers the company's social licence. The same problem is far more acute at Porgera, where the migrant population has long outnumbered the population of traditional landowners, but the recent history of violence at Porgera has only served to heighten the sense of insecurity at Lihir.

Anthony Regan concludes the discussion of large-scale mines and local-level politics in PNG by revisiting the sequence of events that led to the forced closure of the Panguna copper mine in 1989 and put Bougainville squarely at the centre of a new global debate about the social and political impact of large-scale mining operations on local or indigenous communities. No one is better qualified for this task, since the author has communicated directly and extensively with all of the main actors in that social drama, as well as compiling a voluminous record of all the documentary evidence relating to their actions at the time. From this evidence, he argues that closure of the mine was not the primary aim of most of these actors, but he also argues that there were many different groups of actors, with different interests and goals, who played some part in a process that none of them was able to control. The point of this argument is to challenge the conventional portrait of the process as one that was inspired by the grievances of a single group of 'young landowners' from the mine lease areas, led by Francis Ona, who simply wanted to expel the mining company from their territory. The question addressed here is not just a matter of historical interest, since the constellation of political forces that existed on Bougainville in the 1980s was not all that different from the one that still exists within the jurisdiction of the Autonomous Bougainville Government. And that is why Bougainvilleans still hold a wide range of views on the question of whether the Panguna mine should be reopened, or whether another large-scale mine should be developed in the region and, if so, under what conditions.

References

Abrams, P., 1988. 'Notes on the Difficulty of Studying the State (1977).' *Journal of Historical Sociology* 1: 58–89. doi.org/10.1111/j.1467-6443.1988.tb00004.x

Allen, M.G., 2013. 'Melanesia's Violent Environments: Towards a Political Ecology of Conflict in the Western Pacific.' *Geoforum* 44: 152–161. doi.org/10.1016/j.geoforum.2012.09.015

Alonso, A.M., 1994. 'The Politics of Space, Time and Substance: State Formation, Nationalism, and Ethnicity.' *Annual Review of Anthropology* 23: 379–405. doi.org/10.1146/annurev.an.23.100194.002115

Appadurai, A., 1990. 'Disjuncture and Difference in the Global Cultural Economy.' In M. Featherstone (ed.), *Global Culture: Nationalism, Globalization and Modernity*. London: Sage.

Bachrach, P. and M.S. Baratz, 1970. *Power and Poverty: Theory and Practice*. New York: Oxford University Press.

Bailey, F.G., 1969. *Stratagems and Spoils: A Social Anthropology of Politics*. Oxford: Basil Blackwell.

Bainton, N.A., 2009. 'Keeping the Network Out of View: Mining, Distinctions and Exclusion in Melanesia.' *Oceania* 79: 18–33. doi.org/10.1002/j.1834-4461.2009.tb00048.x

——, 2010. *The Lihir Destiny: Cultural Responses to Mining in Melanesia*. Canberra: ANU E Press (Asia-Pacific Environment Monograph 5).

Bainton, N.A. and M. Macintyre, 2013. '"My Land, My Work": Business Development and Large-Scale Mining in Papua New Guinea.' In F. McCormack and K. Barclay (eds), *Engaging with Capitalism: Cases from Oceania*. Bingley (UK): Emerald Group Publishing (Research in Economic Anthropology 33). doi.org/10.1108/s0190-1281(2013)0000033008

Ballard, C. and G. Banks, 2003. 'Resource Wars: The Anthropology of Mining.' *Annual Review of Anthropology* 32: 287–313. doi.org/10.1146/annurev.anthro.32.061002.093116

Banks, G. and C. Ballard (eds), 1997. *The Ok Tedi Settlement: Issues, Outcomes and Implications.* Canberra: The Australian National University, National Centre for Development Studies (Pacific Policy Paper 27).

Barma, N.H., K. Kaiser, T.M. Le and L. Viñuela, 2012. *Rents to Riches? The Political Economy of Natural Resource-Led Development.* Washington (DC): World Bank.

Barth, F., 1959. *Political Leadership among the Swat Pathans.* London: Athlone Press (LSE Monographs on Social Anthropology 19).

Bebbington, A., 2013. 'Natural Resource Extraction and the Possibilities of Inclusive Development: Politics across Space and Time.' Manchester: University of Manchester, Effective States and Inclusive Development Research Centre (Working Paper 21).

Bierschenk, T., J.-P. Chauveau and J.-P. Olivier de Sardan (eds), 2000. *Les Courtiers en Développement: Les Villages Africains en Quête de Projets.* Paris: APAD-Karthala.

Boissevain, J., 1974. *Friends of Friends: Networks, Manipulators and Coalitions.* Oxford: Blackwell.

Bosson, R. and B. Varon, 1977. *The Mining Industry and the Developing Countries.* New York: Oxford University Press for World Bank.

Bourdieu, P., 1977. *Outline of a Theory of Practice* (transl. R. Nice). Cambridge: Cambridge University Press (Cambridge Studies in Social and Cultural Anthropology 16).

Brealey, T.B., C.C. Neil and P.W. Newton (eds), 1988. *Resource Communities: Settlement and Workforce Issues.* Melbourne: CSIRO Division of Building Research.

Bridge, G., 2004. 'Mapping the Bonanza: Geographies of Mining Investment in an Era of Neoliberal Reform.' *Professional Geographer* 56: 406–421.

Brown, W., 2006. 'Power after Foucault.' In J.S. Dryzek, B. Honig and A. Phillips (eds), *The Oxford Handbook of Political Theory*. Oxford: Oxford University Press.

Bulmer, M.I.A., 1975. 'Sociological Models of the Mining Community.' *Sociological Review* 23: 61–92. doi.org/10.1111/j.1467-954X.1975. tb00518.x

Burawoy, M., 1972. *The Colour of Class on the Copper Mines: From African Advancement to Zambianization*. Manchester: Manchester University Press.

Callon, M. and B. Latour, 1981. 'Unscrewing the Big Leviathan: Or How Actors Macrostructure Reality, and How Sociologists Help Them to Do So.' In K.K. Cetina and A. Cicourel (eds), *Advances in Social Theory and Methodology*. London: Routledge and Keegan Paul.

Chase, A., 1990. 'Anthropology and Impact Assessment: Development Pressures and Indigenous Interests in Australia.' *Environmental Impact Assessment Review* 10(1/2): 11–24. doi.org/10.1016/0195-9255(90)90003-I

Clark, D., J. Fox and K. Treakle (eds), 2003. *Demanding Accountability: Civil-Society Claims and the World Bank Inspection Panel*. Lanham (MD): Rowman & Littlefield.

Clements, K., V. Boege, A. Brown, W. Foley and A. Nolan, 2007. 'State Building Reconsidered: The Role of Hybridity in the Formation of Political Order.' *Political Science* 59: 45–56. doi.org/10.1177/003231870705900106

Cobbe, J.H., 1979. *Governments and Mining Companies in Developing Countries*. Boulder (CO): Westview Press.

Dahl, R.A., 1961. *Who Governs? Democracy and Power in an American City*. New Haven (CT): Yale University Press.

Daniels, P., M. Keen and C. McPherson (eds), 2010. *The Taxation of Petroleum and Minerals: Principles, Problems and Practice*. New York: Routledge.

Danielson, L., 2006. 'Architecture for Change: An Account of the Mining, Minerals and Sustainable Development Project.' Berlin: Global Public Policy Institute.

Dashwood, H., 2013. *The Rise of Global Corporate Social Responsibility: Mining and the Spread of Global Norms*. Cambridge: Cambridge University Press.

Dennis, N., F. Henriques and C. Slaughter, 1969. *Coal Is Our Life: An Analysis of a Yorkshire Mining Community* (2nd edition). London: Tavistock.

Dinnen, S., 2001. *Law and Order in a Weak State: Crime and Politics in Papua New Guinea*. Honolulu: University of Hawai'i Press.

Downing, T.E., J. Moles, I. McIntosh and C. Garcia-Downing, 2002. 'Indigenous Peoples and Mining Encounters: Strategies and Tactics.' London: Mining, Minerals and Sustainable Development Project (Working Paper 57).

Easton, D., 1990. *The Analysis of Political Structure*. New York: Routledge.

Eftimie, A., K. Heller and J. Strongman, 2009. 'Gender Dimensions of the Extractive Industries: Mining for Equity.' Washington (DC): World Bank (Extractive Industries and Development 8).

Epstein, A.L., 1958. *Politics in an Urban African Community*. Manchester: Manchester University Press.

Faber, M. and R. Brown, 1980. 'Changing the Rules of the Game: Political Risk, Instability and Fairplay in Mineral Concession Contracts.' *Third World Quarterly* 2: 100–119. doi.org/10.1080/01436598008419480

Ferguson, J., 2005. 'Seeing Like an Oil Company: Space, Security and Global Capital in Neoliberal Africa.' *American Anthropologist* 107: 377–382. doi.org/10.1525/aa.2005.107.3.377

Ferguson, J. and A. Gupta, 2002. 'Spatializing States: Toward an Ethnography of Neoliberal Governmentality.' *American Ethnologist* 29: 981–1002. doi.org/10.1525/ae.2002.29.4.981

Filer, C., 1992. 'The Escalation of Disintegration and the Reinvention of Authority.' In M. Spriggs and D. Denoon (eds), *The Bougainville Crisis: 1991 Update*. Canberra: The Australian National University, Research School of Pacific Studies, Department of Political and Social Change (Monograph 16).

——, 1995. 'Participation, Governance and Social Impact: The Planning of the Lihir Gold Mine.' In D. Denoon, C. Ballard, G. Banks and P. Hancock (eds), *Mining and Mineral Resource Policy Issues in Asia-Pacific: Prospects for the 21st Century.* Canberra: The Australian National University, Research School of Pacific and Asian Studies.

——, 1997. 'Compensation, Rent and Power in Papua New Guinea.' In S. Toft (ed.), *Compensation for Resource Development in Papua New Guinea.* Boroko (PNG): Law Reform Commission (Monograph 6). Canberra: The Australian National University, National Centre for Development Studies (Pacific Policy Paper 24).

——, 1998. 'The Melanesian Way of Menacing the Mining Industry.' In L. Zimmer-Tamakoshi (ed.), *Modern Papua New Guinea.* Kirksville (MO): Thomas Jefferson University Press.

——, 2005. 'The Role of Land-Owning Communities in Papua New Guinea's Mineral Policy Framework.' In E. Bastida, T. Wälde and J. Warden-Fernández (eds), *International and Comparative Mineral Law and Policy: Trends and Prospects.* The Hague: Kluwer Law International.

——, 2008. 'Development Forum in Papua New Guinea: Upsides and Downsides.' *Journal of Energy & Natural Resources Law* 26: 120–150. doi.org/10.1080/02646811.2008.11435180

Filer, C., J. Burton and G. Banks, 2008. 'The Fragmentation of Responsibilities in the Melanesian Mining Sector.' In C. O'Faircheallaigh and S. Ali (eds), *Earth Matters: Indigenous Peoples, the Extractive Industries and Corporate Social Responsibility.* London: Greenleaf Publishing. doi.org/10.9774/GLEAF.978-1-909493-79-7_11

Filer, C. and B.Y. Imbun, 2009. 'A Short History of Mineral Development Policies in Papua New Guinea, 1972–2002.' In R.J. May (ed.), *Policy Making and Implementation: Studies from Papua New Guinea.* Canberra: ANU E Press (Studies in State and Society in the Pacific 5).

Finn, J.L., 1998. *Tracing the Veins: Of Copper, Culture, and Community from Butte to Chuquicamata.* Berkeley: University of California Press. doi.org/10.1525/california/9780520211360.001.0001

Foster, R.J., 2002. *Materializing the Nation: Commodities, Consumption, and Media in Papua New Guinea*. Bloomington: Indiana University Press.

Fox, J. and L.D. Brown (eds), 1998. *The Struggle for Accountability: The World Bank, NGOs and Grassroots Movements*. Cambridge (MA): MIT Press.

Freyss, J., 1995. *Economie Assisteé et Changement Social en Nouvelle-Calédonie*. Paris: Presses Universitaires de France.

Gaventa, J., 1980. *Power and Powerlessness: Quiescence and Rebellion in an Appalachian Valley*. Oxford: Clarendon Press.

Geisler, C.C., R. Stoffle, M. Jake, P. Bunte and M. Evans (eds), 1982. *Indian SIA: The Social Impact Assessment of Rapid Resource Development on Native Peoples*. Ann Arbor: University of Michigan, Natural Resources Sociology Research Lab (Monograph 3).

Giddens, A., 1979. *Central Problems in Social Theory: Action, Structure and Contradiction in Social Analysis*. London: Macmillan. doi.org/10.1007/978-1-349-16161-4

Girvan, N., 1976. *Corporate Imperialism: Conflict and Expropriation*. New York: Monthly Review Press.

Golub, A., 2014. *Leviathans at the Gold Mine: Creating Indigenous and Corporate Actors in Papua New Guinea*. Durham (NC): Duke University Press. doi.org/10.1215/9780822377399

Gordon, R.J., 1977. *Mines, Masters and Migrants: Life in a Namibian Mine Compound*. Johannesburg: Ravan Press.

Grochain, S., 2013. *Les Dynamiques Sociétales du Projet Koniambo*. Nouméa: Editions IAC.

Hobart, C.W., 1984. 'Impact of Resource Development Projects on Indigenous People.' In D.D. Detomasi and J.W. Gartrell (eds), *Resource Communities: A Decade of Disruption*. Boulder (CO): Westview Press.

Horowitz, L.S., 2002. 'Daily, Immediate Conflicts: An Analysis of Villagers' Arguments about a Multinational Nickel Mining Project in New Caledonia.' *Oceania* 73: 35–55. doi.org/10.1002/j.1834-4461.2002.tb02805.x

——, 2011. 'Interpreting Industry's Impacts: Micropolitical Ecologies of Divergent Community Responses.' *Development and Change* 42: 1379–1391. doi.org/10.1111/j.1467-7660.2011.01740.x

——, 2012. 'Translation Alignment: Actor-Network Theory, Resistance, and the Power Dynamics of Alliance in New Caledonia.' *Antipode* 44: 806–827. doi.org/10.1111/j.1467-8330.2011.00926.x

Houghton, D.S., 1993. 'Long-Distance Commuting: A New Approach to Mining in Australia.' *Geographical Journal* 159: 281–290. doi.org/10.2307/3451278

Howard, M.C., 1988. 'The Impact of the International Mining Industry on Native Peoples.' Sydney: University of Sydney, Transnational Corporations Research Project.

Howitt, R., 1989. 'Social Impact Assessment and Resource Development: Issues from the Australian Experience.' *Australian Geographer* 20: 153–166. doi.org/10.1080/00049188908702987

——, 2001. 'Recognition, Respect and Reconciliation: Changing Relations between Aborigines and Mining Interests in Australia.' In R. Howitt (ed.), *Rethinking Resource Management: Justice, Sustainability and Indigenous Peoples*. London and New York: Routledge. doi.org/10.4324/9780203221020

Jackson, R.T., 1987. 'Commuter Mining and the Kidston Gold Mine: Goodbye to Mining Towns?' *Geography* 72: 162–165.

Joyce, S.A. and M. MacFarlane, 2001. 'Social Impact Assessment in the Mining Industry: Current Situation and Future Directions.' London: Mining, Minerals and Sustainable Development Project (Working Paper 46).

Keck, M. and K. Sikkink, 1998. *Activists Beyond Borders: Advocacy Networks in International Politics*. Ithaca (NY): Cornell University Press.

Keesing, R.M., 1992. *Custom and Confrontation: The Kwaio Struggle for Cultural Autonomy*. Chicago: Chicago University Press.

Kirsch, S., 2014. *Mining Capitalism: The Relationship between Corporations and Their Critics*. Oakland (CA): University of California Press.

Latour, B., 2004. *Politics of Nature: How to Bring the Sciences into Democracy* (transl. C. Porter). Cambridge (MA): Harvard University Press.

Le Meur, P.-Y., 2010. 'La Terre en Nouvelle-Calédonie: Pollution, Appartenance et Propriété Intellectuelle.' *Multitudes* 41: 91–98. doi.org/10.3917/mult.041.0091

——, 2015. 'Anthropology and the Mining Arena in New Caledonia: Issues and Positionalities.' *Anthropological Forum* 25: 405–427. doi.org/10.1080/00664677.2015.1073141

Le Meur, P.-Y., C. Ballard, G.A. Banks and J.-M. Sourisseau, 2013. 'Two Islands, Four States: Comparing Resource Governance Regimes in the Southwest Pacific.' In J. Wiertz (ed.), *Proceedings of the 2nd International Conference on Social Responsibility in Mining*. Santiago: Gecamin Digital Publications.

Le Meur, P.-Y., S. Batterbury, S. Grochain and M. Kowasch, 2012. 'Subcontracting as a Social Interface: Rent-Sharing, Control over Resources and Mining Governance in New Caledonia.' Paper presented at the Australian Anthropological Society conference, Brisbane, 26–28 September.

Le Meur, P.-Y., L.S. Horowitz and T. Mennesson, 2013. '"Horizontal" and "Vertical" Diffusion: The Cumulative Influence of Impact and Benefit Agreements (IBAs) on Mining Policy-Production in New Caledonia.' *Resources Policy* 38: 648–656. doi.org/10.1016/j.resourpol.2013.02.004

Le Meur, P.-Y. and T. Mennesson, 2012. 'Accords Locaux, Logique Coutumière et Production des Politiques de Développement en Nouvelle-Calédonie.' *Revue Juridique, Économique et Politique de Nouvelle-Calédonie* 19: 44–51.

Letté, M., 2009. 'Débordements Industriels dans la Cité et Histoire de leurs Conflits aux XIXe et XXe Siècles.' *Documents pour l'Histoire des Techniques* 17: 163–173.

Li, T.M., 2007. *The Will to Improve: Governmentality, Development and the Practice of Politics*. Durham (NC): Duke University Press. doi.org/10.1215/9780822389781

Little, R.L. and R.S. Krannich, 1988. 'A Model for Assessing the Social Impact of Natural Resource Mobilization on Resource Dependent Communities.' *Impact Assessment Bulletin* 6(2): 21–35. doi.org/10.10 80/07349165.1988.9725633

Long, N. (ed.), 1989. *Encounters at the Interface: A Perspective in Social Discontinuities in Rural Development.* Wageningen: Wageningen Agricultural University (Wageningen Studies in Sociology 27).

Lukes, S., 2005. *Power: A Radical View* (2nd edition). Basingstoke: Palgrave Macmillan. doi.org/10.1007/978-0-230-80257-5

Lund, C., 2002. 'Negotiating Property Institutions: On the Symbiosis of Property and Authority in Africa.' In K. Juul and C. Lund (eds), *Negotiating Property in Africa.* Portsmouth: Heinemann.

Markey, S.P., 2010. 'Fly-In, Fly-Out Resource Development: A New Regionalist Perspective on the Next Rural Economy.' In G. Halseth, S.P. Markey and D. Bruce (eds), *The Next Rural Economies: Constructing Rural Place in Global Economies.* Cambridge (MA): CABI.

Martin, K., 2013. *The Death of the Big Men and the Rise of the Big Shots: Custom and Conflict in East New Britain.* New York: Berghahn Books.

May, R.J., R. Anere, N. Haley and K. Wheen (eds), 2013. *Election 2007: The Shift to Limited Preferential Voting in Papua New Guinea.* Canberra: ANU E Press.

MMSD (Mining, Minerals and Sustainable Development Project), 2002. *Breaking New Ground: Mining, Minerals, and Sustainable Development.* London: Earthscan.

Moodie, T.D. with V. Ndatshe, 1994. *Going for Gold: Men, Mines, and Migration.* Berkeley: University of California Press.

Nash, J., 1979. *We Eat the Mines and the Mines Eat Us: Dependency and Exploitation in Bolivian Tin Mines.* New York: Columbia University Press.

O'Faircheallaigh, C., 1984. *Mining and Development: Foreign-Financed Mines in Australia, Ireland, Papua New Guinea and Zambia.* London: Croom Helm.

——, 1991. 'Resource Exploitation and Indigenous People: Towards a General Analytical Framework.' In P. Jull and S. Roberts (eds), *The Challenge of Northern Regions*. Darwin: Australian National University, North Australia Research Unit.

——, 1992. 'The Local Politics of Resource Development in the South Pacific: Towards a General Framework of Analysis.' In S. Henningham and R.J. May (eds), *Resources, Development and Politics in the Pacific Islands*. Bathurst: Crawford House Press.

——, 2002. *A New Approach to Policy Evaluation: Indigenous People and Mining*. Aldershot: Ashgate.

——, 2013. 'Community Development Agreements in the Mining Industry: An Emerging Global Phenomenon.' *Community Development* 44: 222–238. doi.org/10.1080/15575330.2012.705872

Owen, J.R. and D. Kemp, 2013. 'Social Licence and Mining: A Critical Perspective.' *Resources Policy* 38(1): 29–35. doi.org/10.1016/j.resourpol.2012.06.016

Paehlke, R. and D. Torgerson (eds), 1990. *Managing Leviathan: Environmental Politics and the Administrative State*. London: Belhaven Press.

Parsons, R., J. Lacey and K. Moffat, 2014. 'Maintaining Legitimacy of a Contested Practice: How the Minerals Industry Understands Its "Social Licence to Operate".' *Resources Policy* 41: 83–90. doi.org/10.1016/j.resourpol.2014.04.002

Petts, J. (ed.), 1999. *Handbook of Environmental Impact Assessment* (2 volumes). Oxford: Blackwell.

Regan, A.J., 2003. 'The Bougainville Conflict: Political and Economic Agendas.' In K. Ballentine and J. Sherman (eds), *The Political Economy of Armed Conflict: Beyond Greed and Grievance*. Boulder (CO): Lynne Rienner.

——, 2010. *Light Intervention: Lessons from Bougainville*. Washington (DC): United States Institute of Peace.

——, 2014. 'Bougainville: Large-Scale Mining and Risks of Conflict Recurrence.' *Security Challenges* 10(2): 69–94.

Rich, B., 1994. *Mortgaging the Earth: The World Bank, Environmental Impoverishment and the Crisis of Development.* London: Earthscan.

Rich, R., L. Hambly and M. Morgan (eds), 2006. *Political Parties in the Pacific Islands.* Canberra: Pandanus Books.

Robinson, K.M., 1986. *Stepchildren of Progress: The Political Economy of Development in an Indonesian Mining Town.* Albany: State University of New York.

Rodman, M.C., 1987. *Masters of Tradition: Consequences of Customary Land Tenure in Longana, Vanuatu.* Vancouver: University of British Columbia Press.

Rose, N.S., 1999. *Powers of Freedom: Reframing Political Thought.* Cambridge: Cambridge University Press. doi.org/10.1017/CBO9780511488856

Schwarz, N. (ed.), 2013. *The Politics of Give and Take: The 2012 Papua New Guinea National Election.* Goroka: Melanesian Institute for Pastoral & Socio-Economic Service (Point 37).

Scott, J.C., 1985. *Weapons of the Weak: Everyday Forms of Peasant Resistance.* New Haven (CT): Yale University Press.

——, 1990. *Domination and the Arts of Resistance: Hidden Transcripts.* New Haven (CT): Yale University Press.

Searle, J.R., 2009. *Making the Social World: The Structure of Human Civilization.* Oxford: Oxford University Press.

Storey, K., 2001. 'Fly-In/Fly-Out and Fly-Over: Mining and Regional Development in Western Australia.' *Australian Geographer* 32: 133–148. doi.org/10.1080/00049180120066616

Storey, K. and M. Shrimpton, 1988. 'Long Distance Commuting in the Canadian Mining Industry.' Kingston: Queen's University, Centre for Resource Studies (Working Paper 43).

Swartz, M.J., 1968. 'Introduction.' In M.J. Swartz (ed.), *Local-Level Politics: Social and Cultural Perspectives.* Chicago: Aldine.

Szablowski, D., 2007. *Transnational Law and Local Struggles: Mining Communities and the World Bank.* Portland (OR): Hart Publishing.

Turner, V., 1957. *Schism and Continuity in an African Society*. Manchester: Manchester University Press.

UNDP and GoPNG (United Nations Development Programme and Government of Papua New Guinea), 2014. *From Wealth to Wellbeing: Translating Resource Revenue into Sustainable Human Development*. Port Moresby: UNDP.

Wade, R., 1997. 'Greening the Bank: The Struggle over the Environment, 1970–1995.' In D. Kapur and J.P. Lewis (eds), *The World Bank: Its First Half-Century*. Washington (DC): Brookings.

Welker, M., 2014. *Enacting the Corporation: An American Mining Firm in Post-Authoritarian Indonesia*. Berkeley: University of California Press. doi.org/10.1525/california/9780520282308.001.0001

White, G. and L. Lindstrom (eds), 1997. *Chiefs Today: Traditional Pacific Leadership and the Postcolonial State*. Stanford (CA): Stanford University Press.

Williams, C., 1981. *Open Cut: The Working Class in an Australian Mining Town*. Sydney: Allen and Unwin.

2. From Anticipation to Practice: Social and Economic Management of a Nickel Plant's Establishment in New Caledonia's North Province

JEAN-MICHEL SOURISSEAU, SONIA GROCHAIN AND DAVID POITHILY

Introduction

New Caledonia has gone through major changes over the last 40 years. From an economic point of view, four different periods can be identified (Couharde et al. 2016):

- the 'strictly assisted economy' system from 1975 to 1989, characterised by excess savings in relation to investment and an outflow of savings abroad;
- a strong 'rebalancing process' from 1989 to 2005, with gross savings being redirected towards funding projects in New Caledonia;
- an 'industrialisation process' from 2005 to 2013, with foreign direct investment inflows financing the national economy; and
- since 2013, the shaky start of a 'mining and metallurgical economy'.

This pathway to development remains structured around management of the dependency on mining and government revenues, though different options are still open. Depending on the ideological stance, this development trajectory either guarantees that the country and its mineral reserves remain within the realm of the French Republic, or else it can act as a lever for economic, and eventually political, emancipation.

Development choices cannot be separated from political debates. The rise of the independence movement in the 1970s, the polarisation of the political arena between loyalist and pro-independence camps, the violent upheavals of the 1980s, known as *les événements* ('the events'), the return to civil peace with the Matignon-Oudinot Agreements of 1988, and, finally, the period of 'negotiated decolonisation' following the Nouméa Agreement of 1998—all these marked turning points in structural policy for this former French colony. Since 1988, the country has been divided into three provinces that have significant powers. Federalism resulted in the implementation of different economic projects and policies in the pro-independence North Province and Islands Province and in the loyalist South Province.

Ultimately these phases in the political process did not profoundly alter the logic of dependency contained in the economic model, which has proven to be compatible with political emancipation,[1] but structural changes are now under way that could eventually challenge the foundations of this model of dependency. One of these changes—which was initiated by the pro-independence movement in the 1990s—was implemented in the mid-2000s, and should be consolidated in the next decade, is local-level harnessing of mineral revenues as a way to progressively supplant French government transfers.[2]

This chapter proposes to assess the nature of these changes from the perspective of a pro-independence local authority, North Province,[3] focusing on the Koniambo nickel processing plant in which the province is the majority shareholder (Figure 2.1). Having clarified the origins,

---

1   The dependency is self-sustaining, in particular through high salary levels supported by indexation in the administration, which drive consumption prices and justify the permanence of the payments (Freyss 1995). Moreover, it has enabled sustained growth since the late 1980s.

2   Today, after 25 years of a stronger economic autonomy, such a strategy is increasingly taken up by all of the key political and technical actors.

3   As mentioned in Chapter 1, people in New Caledonia do not refer to this entity as a 'provincial government', although it shares many characteristics with what are legally recognised as provincial governments in Papua New Guinea.

trajectory and challenges of this project, we examine the way in which different actors ensured the economic and social management of the plant's construction phase in 2010, and compare their practical experience of the project with the expectations, hopes and fears expressed by the same actors five years earlier. We focus on the topics of employment, the structure of the local economic fabric, and the relationship between economic development and social cohesion. At the end of the chapter, we update the key indicators to give a glimpse into the current state of the province's development strategy since the smelter became operational in 2013. However, we mainly aim to understand the very specific period of the preparation and management of the plant's construction, as a strategic phase in the wider unfolding of a national 'nickel doctrine' based on the public appropriation of New Caledonia's natural resources to achieve the goal of decolonisation.

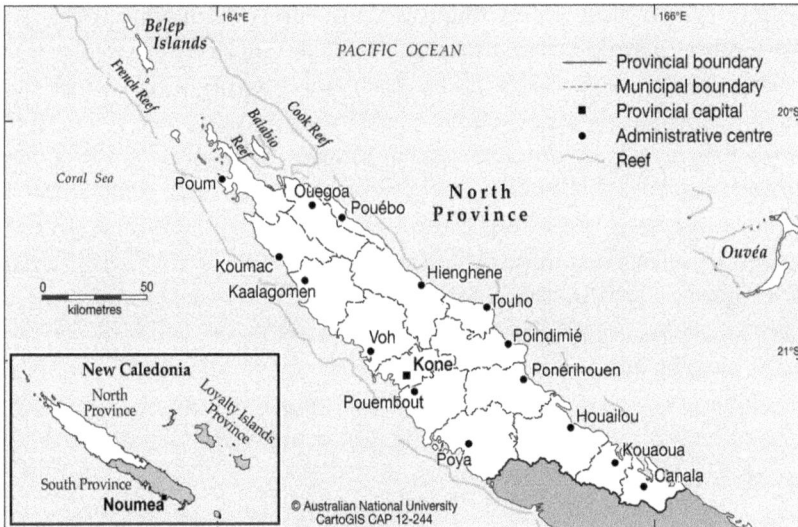

Figure 2.1 North Province, New Caledonia.

Source: CartoGIS, The Australian National University (based on information from Direction des Infrastructures, de la Topographie et des Transports Terrestres, North Province, and Direction des Systèmes d'Information/Service Information et Méthodes, Thierry Rousseau).

This analytical perspective is necessarily partial, since it concerns only North Province, and does not cover all of the national-level changes underway. It can only consider the initial dynamics of a long-term metal production cycle and development process. It does not tackle the height of the construction phase in 2012 and 2013, when there was a massive influx of foreign workers (Blaise et al. 2016: 137; also Chapter 3, this volume).

However, the chapter does enable the contextualisation of questions about the management of mineral revenues by stressing the specificity of previous expectations, which are important to an understanding of prevailing realities and local political stakes. It also enables us to restore the public authorities to the centre of debates about the management strategies applied to mining projects, despite the fact that state actors are often presented as secondary actors, whose role is confined to facilitating or regulating the dialogue between 'companies' and 'communities'.[4]

The Usine du Nord Project and its Ambitions

From the second half of the 1990s, New Caledonian development policies were structured around the enabling of instruments for economic independence, and eventually its political counterpart. The specialisation in mining and metal production was maintained (Néaoutyine 2006), but so was the collection of administrative rent through French Government payments—at least temporarily. Like the political consensus that accompanied it, the latter was perceived by some as evidence of a solid tie with France, and by others as a form of repayment of the colonial debt (Sourisseau et al. 2010). Division of the proceeds from the exploitation of mineral resources was one of the key elements in the negotiation of civil peace. In addition to the purchase in 1990 of the Société Minière du Sud Pacifique (SMSP) by the holding company, Société de Financement et d'Investissement de la Province Nord (SOFINOR), which is owned and controlled by North Province, a *préalable minier* ('mining prerequisite') to the Nouméa Agreement guaranteed that the pro-independence movement would have some control over metal production (Grochain and Poithily 2011; David and Sourisseau 2016).

This issue is important. While the assisted economy model meant that the service sector was still the main contributor to New Caledonia's wealth, the country has actually been dependent on its mineral resources and their strategic nature since it was colonised. Depending on global market conditions, the contribution made by nickel mining and processing to the country's gross domestic product (GDP) has varied from 3 per cent (at the historical low of 1998) to 18 per cent (in 2007, an exceptional year for the

---

4   It is interesting to note that the concepts of 'companies' and 'communities' are rarely clarified; these terms refer to heterogeneous entities that merit better definition as a result of local case studies (see Chapter 1, this volume).

sector). The nickel and non-market sectors combined account for 50 per cent of the growth in added value between 1998 and 2007 (CEROM 2008). Moreover, in the aftermath of the political agreements, thanks to the industrial revolution initiated with the planned operation of three new large-scale smelters, each processing over 50,000 tons of nickel a year, and the expansion in the capacity of the existing Doniambo plant, it was thought possible that nickel's share of GDP would increase to more than 25 per cent (GoNC 2009). Hence, the already significant knock-on effects of the sector were set to become even more extensive. In this context, controlling the income from mining and metal processing was at the heart of the debate surrounding the country's future legal and political status. At stake was the nature of its future development: exogenous or endogenous; part of France or more or less autonomous.

In fact, the development of new nickel smelters had long since come to symbolise this choice. The construction of the processing plant in South Province would appear to have been relatively conventional, with a clear division between the private sector entity responsible for the operation—in this case, the Brazilian company Vale[5]—and the local authorities, who were involved mainly through monitoring the operator's compliance with government regulations and imposing taxes to harness a portion of the value produced,[6] while making allowance for compensation or fees to be paid to local communities. Since the plant was launched in the mid-2000s, the logic has shifted towards compensation of the resident clans and the management of environmental concerns.[7] In response to the difficulties encountered at the beginning of the project, another public stakeholder emerged in the form of local Kanak associations known as 'local groups with special rights', *groupements de droit particulier local* (GDPLs), which collectively purchased a stake in the project's subcontracting businesses (Bouard et al. 2016). However, this move was confined to one local community (Yaté), which had less than 2,000 inhabitants, and ultimately became a sort of enclave (see Chapter 6, this volume).

5   Vale had got involved in the Goro project through its purchase of the Canadian company INCO.

6   Only 5 per cent of the equity in the Usine du Sud (Southern Plant) is held by the state, through the Société de Participation Minière du Sud Calédonien, which was established in 2005. This entity is owned by all three provinces but, in accordance with the logic of 'rebalancing', the distribution of dividends favours the North and Islands provinces.

7   A sustainable development pact was concluded between Vale's subsidiary Goronickel and the customary authorities of Yaté.

While the administration of South Province certainly intervened in such processes, it did so as a facilitator in a process of negotiation focused on the company–community relationship.

Construction of the Usine du Nord (Northern Plant), which was based on the Bercy Agreement, was more clearly rooted in the political bifurcation of the 1980s, and had the task of lending concrete expression to the spatial and economic rebalancing at the root of the political consensus. In 1998, SMSP formed an alliance with the Canadian company Falconbridge to launch the feasibility studies before Falconbridge was bought by the Anglo-Swiss company Xstrata (now Glencore) in 2006.[8] Xstrata officially confirmed its commitment to the project in October 2007, on much the same conditions as those negotiated with Falconbridge. Under this arrangement, confirmed by Glencore, SMSP holds 51 per cent of the capital in the joint venture, Koniambo Nickel SAS (KNS),[9] which was established to implement the project.

The project includes a pyro-metallurgical plant, a power station, a deep water port, an automatic conveyor and various other structures. Costing a total of almost US$5 billion, its construction (like that of the Southern Plant) constituted a major project in global terms. Its completion in 2013 required the employment of a large workforce, which peaked at over 6,000 employees. Around 1,000 jobs are involved in the operation of the plant, which is expected to last for at least 30 years. According to estimates from 2004, the plant was expected to generate 1,700 indirect jobs and attract some 8,000 people to the municipalities near the plant—Voh, Koné and Pouembout (VKP) (North Province 2004; Syndex 2005). At the level of the province and its 45,000 inhabitants, the project embodies the hope of unprecedented economic and social momentum, and therefore justifies the creation of new public infrastructure.

The balanced social management of the expected impacts of the Koniambo project would appear, therefore, to be a prerequisite for its success. The public control of the project aims to boost leadership of the economic development of North Province by the Kanak people and, in some respects, to demonstrate the country's ability to manage its independence

---

8   As a result of various acquisitions and mergers, the New Caledonian nickel landscape is now characterised by the presence of three foreign multinationals—Glencore (Switzerland), Vale (Brazil) and Eramet (France), the latter being the majority shareholder of Société Le Nickel.

9   A *société par actions simplifiée* (SAS) is a 'simplified' company in which the shares are typically held by a number of other companies or associations.

(Néaoutyine 2006). Henceforth, public policy will use the mine and the plant as resources at the service of a territorial strategy. It is intended to reinject the profits into the diversification of the local economy with a view to anticipating shocks in the metals market, preparing for the 'post-nickel' period, and providing protection against a 'Dutch disease' scenario.[10]

From Anticipation to Practice: Questions and Methods

The Usine du Nord project is a high-stakes project from both an economic and political perspective. The possible pitfalls and risks of failure were known to be significant, even at the early stage. Numerous questions existed before the start of the project—and still exist—regarding the public policies and the social and economic structures likely to make the mine and the smelter not merely an economic enterprise, but a real lever for emancipation, rebalancing and the common destiny of the province.

Hence, the question arises as to whether public action was able to respond to the challenges posed by the planning and launch of the project's construction site in the period from 2005 to 2010. How have the issues identified in 2005 been confirmed, invalidated or reformulated in 2010 in light of the experience gained? Which generic lessons can we draw from the management of the beginning of construction in terms of local capacities for intervening or being involved in the process? And which methods or methodological approaches are best suited to answering these questions?

Our hypothesis is that the majority of the fears clearly voiced in 2005 were not realised and that, overall, expectations were satisfied both in 2010 and since then, at least in the VKP area near the plant and at the political centre of the province. We hope to show that, whatever the end of the story may be in future decades, this situation was not only the result of favourable circumstances and sustained economic development throughout the entire country during the period in question, but that it was also the outcome of innovations and adaptations—both individual and collective, public and private—made by the local actors.

10   'Dutch disease' refers to the negative consequences arising from large increases in a country's income. Primarily associated with a natural resource discovery, it can result from an increase in foreign aid or a substantial increase in natural resource prices—anything that changes the structure of the national economy in favour of one export-specific sector. The consequences are a stronger national currency and lower international competitiveness in other parts of the economy.

Our analysis is based mainly on two studies conducted by the Institut Agronomique Néo-Calédonien (IAC) in 2005 and 2010.[11] The first, which was based on investigations carried out in late 2005 and involved almost 250 people throughout North Province, clarifies the nature of the expectations, fears and concerns anticipated by locally elected representatives, economic operators, customary chiefs, representatives of local non-profit and cultural organisations and provincial public servants (Sourisseau et al. 2006). The second is one of the studies produced within the framework of the centre set up by the IAC to monitor the economic and social dynamics of the Koniambo project. It is based on approximately 100 interviews conducted with businessmen from the VKP area who were awarded subcontracts for construction of the plant, and on the analysis of some of the forms of public shareholding set up for the project (Grochain and Poithily 2011). The practices observed in this second study are compared with the forecasts produced in 2005. The comparison is underpinned by the detailed and comprehensive statistical series produced by the Institut de la Statistique et des Études Économiques, the Institut pour le Développement des Compétences en Nouvelle-Calédonie and the provincial Direction des Systèmes d'Information.[12] This evidence has been combined with data from an assessment by the IAC of the financial support distributed by the provincial authorities under their development code, and data from an evaluation produced by the provincial engineering office (Emergences) on the development of customary lands in the directly affected municipalities of Koné and Pouembout (North Province 2008).

Our analysis is focused on three topics that emerged as particularly important in the early stages of the negotiation process and for the actors interviewed in 2005, and which are well documented by the studies conducted in 2010. The three topics are:

1. employment, particularly youth employment;

2. the structuring of a local economic fabric through harnessing a portion of the income from mining and metal production, particularly in light of concerns about competition between local companies and companies from South Province or from overseas; and

---

11 These two studies offer the opportunity for a comparison, but also explain the period chosen for this purpose, even if it would have been relevant to have surveys completed at the very end of the construction phase in 2013.

12 The same sources are used for the update at the end of the chapter.

3. social cohesion, and especially the relationships between customary life and economic development.

It should be noted that our analysis is limited in two respects. First, we propose to recount merely a small part of this experience. The story has already started, so it is not a question of returning to the genesis of the project and the conflicts that emerged from the 'mining prerequisite' of 1998, which have already been extensively documented (Bencivengo 1999; Grochain 2010; Le Meur and Mennesson 2011; David and Sourisseau 2016). And the story is far from over: the period covered by this chapter is primarily the period between 2005, when an irrevocable commitment was made to building the project, and the start of the actual construction phase in 2010. The second limitation is the choice of three particular issues as the focal points for providing answers to a more general set of questions about the development process. In other words, we aim to illuminate a general topic through the adoption of admittedly partial approaches.

The Analytical Context

From an analytical point of view, based on the stated ambitions of the project, an investigation of its impacts on provincial development involves questioning the role of the provincial authorities—considered as a branch of the state—in the management of mining and metal production activities. At the same time, the province acts as a sort of guarantor in the management of the social impacts of a sectoral development project as part of a broader development program, beyond the traditional models of confrontation or negotiation between a company and local communities, whether or not the latter have the support of public authorities or external associations.

In focusing on the public authority, we are attempting to complement the analytical framework developed by Geert van Vliet (1998) to understand mining project cycles and the strategies of their operators (see Figure 2.2). In this framework, the inception of a mining (or petroleum) project is characterised by a strong tendency on the part of the developers to compensate local communities because they are at the beginning of the process and must provide some reassurance, particularly as they have a significant economic stake in the project being accepted. At this juncture, the local people themselves have poor negotiation capacities and are generally reassured by the compensation on offer. Once the project has started, the companies will have replaced the original negotiating team with new managers who are subject to the new demands of construction and operation, with a rapid decline in their disposition to negotiate. Meanwhile, local people gain both experience and expertise, and can better identify the impacts of the project, so their ability to negotiate grows rapidly, often with support from non-governmental organisations. This creates a risky situation, but the windows of opportunity remain open, and the negotiations significantly shape the subsequent stages in the project cycle. At the end of the cycle, especially if a company is not planning to extend its activities in the area, it is much less willing to negotiate because the workforce involved is dedicated to the closure of the operation. By contrast, the negotiating capacity of the local population is at its highest level because the negative impacts of the operation are now well known. This phase of the project cycle has the highest level of risk and conflict.

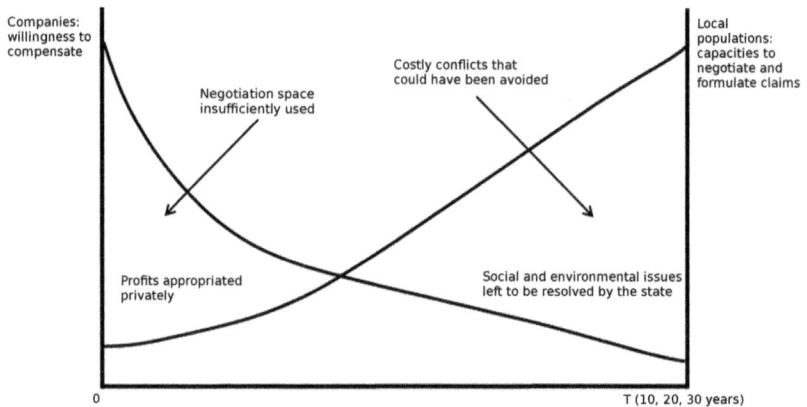

Figure 2.2 The politics of the large-scale mining project cycle.
Source: van Vliet (1998).

Numerous amendments and qualifications can be added to this model in terms of the slope of the curves and the time scales. Having noted that the model is silent on the role of the state and its ability to modify the curves through proactive policies, it is possible to make four additional comments here.

First, the willingness of companies to pay communities at the beginning of the cycle is somewhat questionable. This applies in particular to Chinese companies operating in Africa, or in Papua New Guinea, and surely in many other places, but it could also be the result of new corporate strategies that do not adhere to international standards, but instead involve specific adaptations to particular national contexts.

Second, very severe conflicts are occasionally observed at the beginning of the project cycle, not necessarily between the company and the local communities, but within the communities themselves (see Filer 2006). The question of benefit distribution can generate such tensions if the country or territory does not have a sufficiently clear normative framework, or the framework is not generally agreed upon, or there is limited ability and willingness to implement it. In our case study, we suggest that this type of tension was prevented by the fact that a normative framework was already in place at the start of development process.

Third, the recent rise of discourses of citizen control, and the adjustment of corporate social and environmental responsibility standards, has increased the willingness of companies to pay compensation at the beginning of the production phase,[13] or even towards the end of the project cycle.[14] This tendency has been reinforced by the provision of multilateral assistance for the management of environmental impacts in particular.

Finally, the model does not consider the cumulative experience of populations and companies with a long historical experience of mining operations in a particular area, as is the case in New Caledonia. In our case study, we suggest that, from the outset, the negotiating ability of the local people was greater than is indicated in Figure 2.2. Despite the level of control exercised by the former developer Falconbridge,[15] the project's 'clearing' or approval period from 2000 to 2006 was also a decisive period for the structuring of development policies and public policy in general (Sourisseau et al. 2010; Bouard 2011). This would have enabled the provision of a normative framework that effectively modified the conditions of the dialogue with the company. We support this hypothesis regarding the inflection of the curve with the representation of the province as a shareholder in the project. Beyond the financial aspects, we

---

13   The previously mentioned Grand Sud Pact relating to the Goro project is an illustration of this phenomenon in New Caledonia.

14   A particularly explicit Caledonian example is an agreement reached in Thio in 1996 (Le Meur and Mennesson 2011; see also Chapter 5, this volume).

15   Criticism was expressed to the effect that the voice of SMSP was not sufficiently heard at the information and negotiation meetings (Sourisseau et al. 2006).

establish that the experience acquired through SMSP from 1990, and its concrete participation in the financial and technical investment in the plant, influenced the form of local representation in the project.

The amended conceptual framework hence moves to formalise the lessons learned from the transition from expectation to practice in terms of the management of a project's effects at the beginning of a new industrial cycle.

From Expectation to Practice, 2005–10

The points of view expressed in 2005 bear witness to a quasi-unanimous commitment to the Koniambo project on the part of the local actors. The Usine du Nord was deemed indispensable in giving meaning to the economic rebalancing of the territory and, more broadly, to the general thrust of the political agreements of 1988 and 1998. It also emerges that development of the plant could not, and should not, be considered in isolation; it must be connected with broader sectoral policies and with territorial dynamics as a whole. However, commitment was coupled with enormous focus, which was often expressed as vigilance in relation to the potential negative effects of economic development. Hence the actors appeared to be concerned but also clear: they felt that they were both the potential beneficiaries of the positive impacts of the future plant but were also exposed to its inevitable negative impacts, especially those related to a risky 'submission' to the laws of global markets. They also stressed that 'progress from word to deed' only strengthened the need for vigilance, and that the risk of distrust must not be discounted. In 2005, a common basis existed for the representation of the issues and challenges associated with the Usine du Nord. However, as emerges from a more detailed examination, there were different—and sometimes even contradictory— expectations and fears expressed by different actors.

Disruptions took place around the plant site in 2010, but to a lesser extent than had been expected. The provincial demographic balance was jeopardised by the attractiveness of the site. The loss of 1,500 inhabitants from the east coast highlighted the risk of the emergence of an east–west imbalance in the course of north–south 'rebalancing'. At the same time, construction of the plant was delayed, and the population of the VKP area 'only' grew by 2,000 inhabitants (GoNC 2010)—far fewer than the 5,000 predicted by the development plans in 2004.[16] This was a more gradual demographic transition than the 'tsunami' referred to by several of our informants in 2005.

Overall, it appears that the progress promised in terms of access to health care and education in 2005 was confirmed and reinforced, and the standard of living of local households continued to increase (CEROM 2011). This trend has since been confirmed by the 2014 population census (GoNC 2015).

With the exception of 2008, a year characterised by a general slowdown in New Caledonia, the number of job seekers throughout North Province declined after 2002 and, in 2010, the outlook in this regard was reassuring (GoNC 2011). Assessment of the effects of the provincial Code of Development (CODEV) showed that provincial action had been sustained and diversified, and had yielded convincing results, even from the point of view of people in receipt of government aid. For instance, in 2010, more than two-thirds of the 1,100 projects funded since 2003 were still active; for each 1,000 francs of government subsidy, another 1,500 francs were added through the activity being subsidised, and 56 per cent of the CODEV beneficiaries declared in 2010 that their daily life was better thanks to the project funding (Gaillard et al. 2011). Although this assessment confirms that economic activity followed demographic trends, and that the west had benefited more from the development process, this imbalance was not always perceived negatively, and the returns from the acquired skills and transfers to the community in the east were beginning to take effect.[17] The reorganisation of the Department of Economic Development and the Environment in 2010 was indicative of the additional resources provided for environmental management and the increased specialisation of administrative services within the province. In line with the hopes expressed by interviewees in 2005, the elected representatives and their advisers also insisted on the need to continue their strategy of linking structural projects, with strong government participation, to development of the local area. The interviewees felt empowered by an encouraging assessment of 20 years of decentralisation

---

16 This prediction did not include the foreign construction workers, 5,000 of whom were present in 2011–12, but who mostly stayed inside the construction site. The target of 5,000 local inhabitants was actually achieved in 2014 (GoNC 2015).

17 The expansion of development services within the provincial administration has continued since then (North Province 2015).

or 'provincialisation' and the credibility thus acquired by the provincial authorities, and were generally positive about the progress of the Koniambo project.

However, the public unanimity in favour of the plant appeared less convincing in 2009, particularly because of the visible impact of site clearance on the landscape and the inconvenience caused to some of the local people. Its usefulness in terms of provincial development and its necessity in terms of the north–south rebalancing process were not in question, but vigilance was being maintained and even heightened.

The Local Employment Issue

What the Local Actors were Saying in 2005

In 2005, the locally elected representatives from the VKP area were generally confident about the positive impacts the plant would have for their communities. In their concern 'to be ready on time', however, they deplored the timidity of private initiatives: 'The paradox is that we have high expectations and, at the same time, face a real waiting game'. Outside the VKP area, the elected representatives were hoping for 'some direct employment', but were not counting on it too much because they thought the logic of 'local employment' would to work to their disadvantage. Instead they believed that they would have a role to play in the supply of food (agricultural produce and fish), and in the areas of leisure, weekend tourism and services.

In particular, the elected representatives highlighted the risk of 'disappointment among young people' due to the limited number of direct and indirect jobs that would actually be created, the lack of qualifications that could make many young people 'difficult to employ', and the very narrow conception of 'local' employment. Some mayors, who were generally optimistic in other respects, did not rule out the possibility of such frustrations leading to the formation of 'social movements'.

The risk of an increase in territorial imbalances between the east and west coasts was recognised, particularly by elected representatives from the east coast. They thought the risks were both economic and demographic, with a decrease in the value of grants received by the communities in decline, and some feared that marginalisation would lead to social destabilisation: 'The young people from the east coast are restless and they feel forgotten'.

The elected representatives from the VKP area were more optimistic and believed that 'it is a mistake to think that the east coast will not benefit from the effects of the plant'.

For the customary authorities, the creation of jobs, particularly for young people, was almost always associated with strong fears: 'There will be fewer jobs than the young people are expecting'. They also feared the resulting frustration, and insisted on the need for both training and a provincial focus: 'The mine is not only for Voh, but for the North Province'.

The representatives of economic organisations were also worried by the narrow conception of 'local' employment: 'The politicians invented specific local characteristics and now they no longer know how to manage them'; 'Before the "events" the people were very available, now each one thinks about his tribe, his community, his organisation'. The risk of disappointment on the part of the young people was repeatedly evoked.

Ultimately, the question regarding the jobs that were supposed to be created by the project was sensitive, and a potential source of 'slippage'. All of the local actors pointed out that special efforts had to be made in terms of providing training for the labour force so that the Koniambo project and other development activities would promote the social advancement of disadvantaged people. Accordingly, a greater recognition of other sources of potential employment appeared to be necessary. In addition to the industrial project, interviewees felt that provincial economic development should be based on the other projects supported by SOFINOR (in aquaculture, fishing or tourism), on the structuring of a private sector market, and on the social economy sector, which had been considered for a long time to be the main supplier of self-employment and a form of protection against job insecurity.

What the Research Said in 2010

The employment statistics produced by the Institut pour le Développement des Compétences en Nouvelle-Calédonie showed that the fears expressed in 2005 had only partly materialised. The figures for job offers (Figure 2.3) show the dynamism of the provincial labour market. In 2010, 60 per cent of job offers related to the construction site in the municipality of Voh, but, over the preceding three years, there were more than 1,200 job offers each year relating to work outside the plant site. When compared with the number of applications, this prompted the conclusion that there was a situation of near full employment.

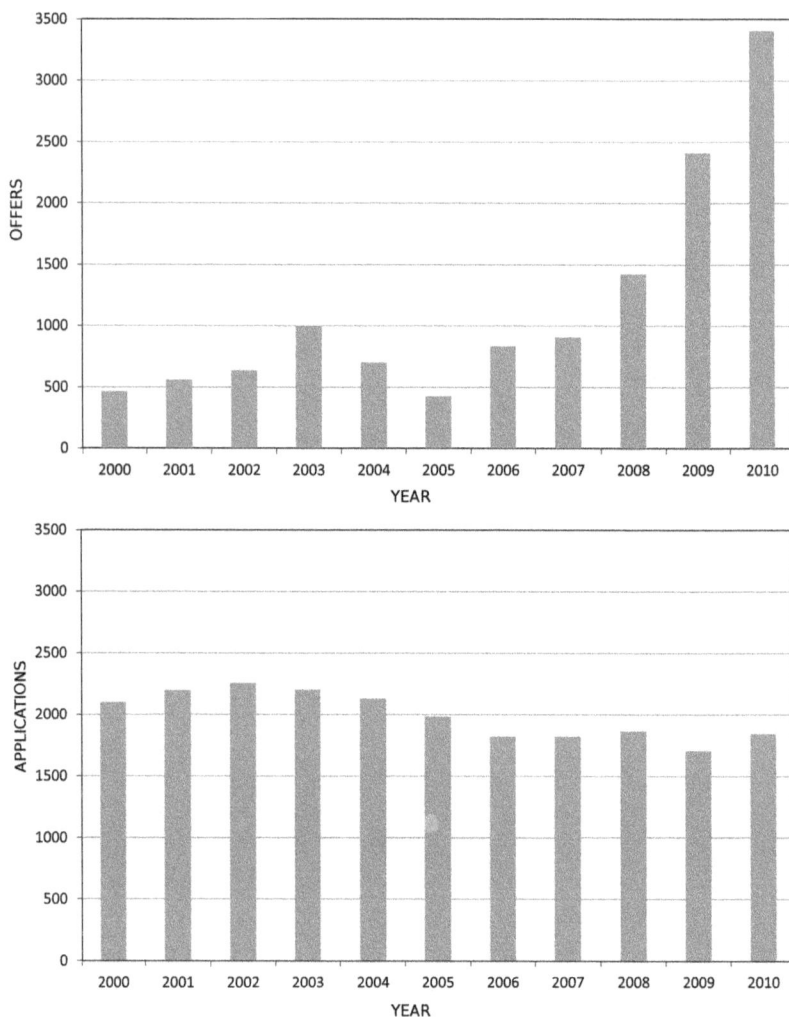

Figure 2.3 Employment offers and applications in North Province, 2000–10.
Source: Institut pour le Développement des Compétences en Nouvelle-Calédonie.

Seventy-seven per cent of the New Caledonian employees working on the Koniambo project itself were natives of North Province; 45 per cent of the local employees were natives of the VKP area, 6 per cent were from other parts of the west coast and 21 per cent were from the east coast. Moreover, 77 per cent of the local employees were Kanaks. Concerns previously expressed about the employment of young people on the Koniambo project had been assuaged, since they constituted a substantial majority of the project's employees: 70 per cent were under thirty-six years of age, and 45 per cent of these were under thirty-one. On the other hand, the position of female employees on the Koniambo project, as in the mining and metal production sector more generally, remained fairly marginal. They accounted for less than 30 per cent of the project's employees, at a time when women accounted for 37 per cent of the total formal workforce in New Caledonia. Hence, the employee profile most commonly found at the Vavouto construction site was that of a young, male, Kanak worker who was a native of North Province (Grochain 2010).

With regard to the position of young people in the North Province job market as a whole, we found that the number of applicants under twenty-six years of age fell from 597 in 2005 to 487 in 2009, but rose again to 582 in 2010. In the end, the proportion of applicants in this age group remained at 31.5 per cent and, like the other age groups, they seem to have benefited from the development of the mining project. Here again, women accounted for nearly two thirds of the unemployment rate. Their representation among the job applicants increased from 58 to 62 per cent, and they benefited least from the new developments. The proportion of permanent contracts among the job offers rose from 5 to 12 per cent between 2005 and 2010, but it peaked at 28 per cent in 2008, before the peak of applications for work on the construction site. Likewise, the share of full-time job offers increased from 77 to 93 per cent between 2005 and 2010. It may therefore be concluded that a trend for longer-term contracts already existed, and this invalidated some of the fears expressed in 2005. However, employment was still very unevenly distributed, with more than 85 per cent of the total job offers being made in the VKP area.

The population census figures indicate that the employment rate among people over fourteen years of age rose from 49 to 53 per cent in North Province as a whole, and from 57 to 67 per cent in the VKP area, between 2004 and 2009. Women in particular benefited from this trend, as their rate of employment rose from 35 to 45 per cent, but men from the VKP area still ended up with the highest rate of employment (78 per cent).

A closer look at the employment of people originating from different tribes enables further clarification of the figures (North Province 2008). Direct surveys of all households from the tribes of Koné and Pouembout show that the number of wage-earners among the working population increased by 178 per cent in 12 years: in 1996, 361 out of 1,180 working people were wage-earners, the others being self-employed in the domestic economy; in 2008, there were 1,005 wage-earners out of a total working population of 1,867, which represents 1.7 salaried employees per dwelling. Among the salaried employees in 2007, almost three-quarters had permanent jobs, which guarantees a certain financial stability for the families in the tribes. In 2008, the tribes that benefited most from the slight improvement were those close to Koné: Noelly, Baco, Tiaoué and Koniambo accounted for two thirds of the salaried employees. We also noted considerable variability in the representation of the different tribes in different economic sectors, but an overall balance between the four major sectors that accounted for 75 per cent of the jobs: administration, food and agriculture, 'other services', and 'other trades or industries'. So the mine and construction of the smelter did not (yet) constitute the dominant sector in the process of job creation.

Within this already positive situation, local businessmen in the subcontracting sector developed strategies to create jobs for their families and social networks. In the businesses themselves, associations of family members were common, and the head of the family typically became the managing director. The company's dividends were generally evenly distributed, but family members without shares in the company could also be expected to contribute to its operation (Grochain and Poithily 2011). The companies were recruiting first and foremost from their family circles, but if the family could not meet the requirements, they looked for trustworthy people in their wider circle of acquaintances. With regard to access to jobs created directly by the construction project, these family and social networks were also mobilised to pass on information and thus facilitate the process of local recruitment.[18]

Finally, efforts to support local companies operating outside the mining sector were being sustained at the provincial level. From 2004 to 2010, implementation of the CODEV program alone created almost 750 jobs, the vast majority of which were intended for local workers, and half of which would be (low-paid) forms of self-employment (Gaillard et al. 2011).

In the period following the preparation and start of the project construction phase, the logics of proximity and provincial assistance enabled the local population to seize the job opportunities offered by the 'mechanical'

---

18   This type of information was also distributed through more formal channels by bodies like the Bureau des Entreprises du Nord, SAS Vavouto, and others.

growth in economic activity. With the exception of concerns about an increase in the intraprovincial east–west imbalance, and the relatively poor integration of women into the labour market, the fears expressed in 2005 had generally not been realised by 2010.

The Business Development Issue

What the Local Actors were Saying in 2005

All the economic operators encountered in 2005 were in favour of any big project that could generate more business activity. A considerable number of them were of the view that 'the mine lights the fuse'. Specifically, the most experienced and well-resourced economic operators thought that 'the long-term prospects are good', even though 'there are a lot of short-term uncertainties' concerning the activities that would actually be promoted by the new local economy and the ability of local companies to respond to the new opportunities. The small companies and traders appeared to be more worried and uncertain because of previous announcements that turned out to be false.

The question of access to landed property was of concern for a number of economic operators who took a positive view of the development of new activity zones, especially on customary land, but were worried about possible delays in establishing new activities, and were undeniably suspicious of the prospects of actually investing in customary land.

However, it was the threat of competition from the businesses of South Province that generated most concern, particularly in the construction phase: 'We must not allow the markets related to the plant to get away from us'. Three explanations were presented: (1) the tendering system was deemed to be disadvantageous to the companies from the north; (2) the elected representatives were perceived as only wanting to deal with 'reliable contractors'; and (3) there was thought to be a lack of relevant skills in some prospective areas of work. The creation of groups of companies for specific tenders was mentioned as a (partial) response to these challenges.

The existing support mechanisms for investment on a significant scale, such as the Institut Calédonien de Participation or the grant of tax exemptions, were considered 'effective and uncomplicated' by those who could benefit from them, even if they had limited knowledge of their operation. CODEV was subject to differing opinions, but was also generally considered to be an 'asset' that 'could be improved on'. A large number of economic operators in the VKP area expected to receive information and support from the Cellule Koniambo, an entity established by the provincial authorities to support the development of businesses related to the Koniambo project. Nevertheless, the provincial economic development and employment situation prompted calls for the strengthening of socio-economic initiatives, which were considered 'very significant in number but still not very well coordinated'. The view was that more should be done to encourage the diversification of business activities so that individuals could really exercise more choice in this field.

Informants agreed that it was necessary to promote the establishment and maintenance of links between the various mechanisms and institutions so as to improve their impacts and benefits. People expressed particular preference for the facilitation and promotion of small business activities, the provision of a personalised advisory service for businesses, a local process for facilitating the emergence and intensive support of individual and collective initiatives, and the establishment of some collective services such as a sorting centre.

What the Research Said in 2010

Between 1995 and 2010, the number of companies in North Province increased by more than 4,100. Nearly two-thirds of these companies were operating in the service sector, while 28 per cent were in the construction sector and 8 per cent in the manufacturing sector. The figures for business start-ups are more significant. In 2010, more than 600 businesses were created in North Province, compared with fewer than 300 in 1995 and 340 in 2005. Taking into account the disappearance of some enterprises, there was a net increase of 120 businesses in the constructions sector and 375 in the service sector over the whole period (Figure 2.4).

The emergence of the VKP urban centre largely explains the increase in the number of start-ups after 2005, and especially in 2010. Thirty-eight per cent of the businesses were created within the VKP area, and over 40 per cent of these start-ups were in the construction sector. On the east coast, the proportion of business start-ups decreased slightly between 1995 and 2008 (from 35 to 30 per cent), and towards the end of the period in question, more than 70 per cent of these start-ups were in the tertiary (service) sector. Moreover, it was possible to observe an increase in the proportion of start-ups in the tertiary sector in all parts of the province after 1995, except in the municipalities bordering the new plant site. Similarly, the manufacturing sector accounted for nearly 10 per cent of the business start-ups on the west coast, compared with only 5 per cent on the east coast.

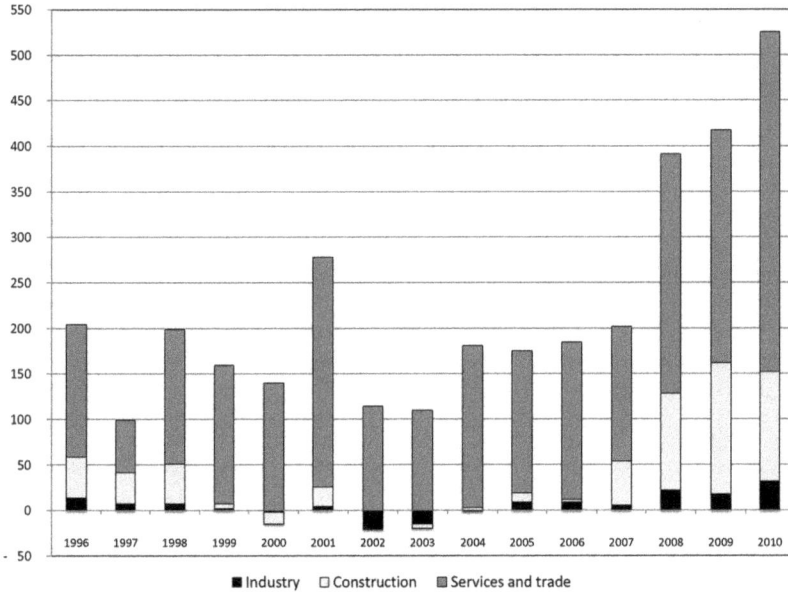

Figure 2.4 Net annual average of semi-annual business start-ups, 1995–2010.

Note: Numbers are negative for some sectors in some years because the number of new businesses created was smaller than the number of existing businesses that closed.

Source: Institut de la Statistique et des Études Économiques, Direction des Systèmes d'Information/Service Information et Méthodes.

Overall, the trends in business creation and development followed those observed in the labour market. Since the early 2000s, it has been possible to observe the initial effects of provincialisation, followed by a noteworthy acceleration with the commencement of work at the Vavouto construction site. In 2010, almost 450 new businesses were established with their headquarters in the VKP area.

In the subcontracting sector, the total value of locally issued on-site orders and contracts from 1998 to 2009 was XPF58 billion, of which XPF38 billion was concentrated in the period from mid-2005 to mid-2009 (Grochain and Poithily 2011). Close to 90 businesses were created over this latter period with the help of the Cellule Koniambo. Seventy per cent of these businesses were headquartered in North Province and operating on the Vavouto site, and provincial policy measures enabled locally owned businesses to capture 60 per cent of the XPF38 billion spent locally through on-site contracts and orders between mid-2005 and mid-2009. Public works accounted for 57 per cent of the value of these spin-offs (22 per cent in earthworks and 35 per cent in construction), followed by transport (23 per cent), marine activities (8 per cent), environmental management (4 per cent) and miscellaneous government contracts (8 per cent). Finally, it should be noted that the new entrepreneurs, both Caledonian and Kanak, often came from modest social backgrounds. These figures indicate that competition from the south did not seriously disadvantage the entrepreneurs of the north, but that was mainly because of provincial policy measures taken to counter that risk.

A few examples will serve to illustrate this point.[19] Although some local businesses were already involved when work first commenced at the Vavouto site, major contracts—especially for earthworks—were awarded to businesses from outside the area. In keeping with the fears previously expressed by local economic operators, businesses in the VKP area were not well structured, were poorly prepared to meet the demand, and thus had trouble asserting themselves against the outsiders. They tried to stress the need for local employment and the political values of 'rebalancing' and 'common destiny' at the expense of market relationships that favoured large companies with prior experience of working with Hatch Technip, which was the joint venture company selected by Falconbridge (in 2006) to oversee the whole construction process.[20] In 2008, the Bureau des Entreprises du Nord (BEN) was established by local entrepreneurs to push for greater local participation, especially in relation to the earthworks contract awarded to the multinational company Vinci. From the outset, BEN united northern business organisations with partnerships, known as *sociétés civile de participation* (SCPs), which represented the local Kanak clans. However, the federal structure of this organisation did not meet the requirements of KNS, and that is why SAS Vavouto was established in late 2008. This new corporate entity was able to raise finance on its own account and take responsibility for issuing subcontracts on behalf of KNS, often by dividing them between local companies that were members

---

19 See Poithily (2010) for more detailed description and analysis of these case studies.
20 This major contract included design, procurement and project management services.

of BEN while underwriting their capacity to meet the contractual requirements. In return, SAS Vavouto received a commission calculated as a percentage of the value of each contract and redistributed the dividends to its shareholders, which were the main clans of the VKP area grouped into SCP partnerships. Although the new company did not secure all of the contracts that it wanted, it did secure enough—including some of the earthworks contracts—to make a significant contribution to the growth in spin-offs for local businesses.

The provincial authorities supported the establishment of another joint venture, the Société Webuihoone Maritime SAS (SOWEMAR), in which a subsidiary of SMSP owns 49 per cent of the shares, while the rest are divided between two clan-based companies from the VKP area. SOWEMAR secured a contract to operate the tugboats that tow the bigger ships entering the port of Vavouto, and this enabled the participating clans to create a long-term and potentially profitable activity in the area. It was not expected that local residents would so quickly capture a share of the market in activities aside from those that have conventionally been assigned to them, like caretaking, cleaning and 'site enhancement'.

The French Government also played an important role through the provision of tax relief. Unlike provincial tax exemptions associated with new projects, which have been criticised by developers for being too small and by contractors for being too slow, national tax relief on vehicle prices facilitated the creation of new businesses in the area.

In light of these examples, it would seem that a variety of local entrepreneurs got more than they had previously hoped to get from the economic development triggered by the project. The strategies adopted, in particular the collective dynamics generated around public shareholdings and embedding of contractual decisions in local social networks, generally proved effective. However, it must be stressed that most of the businesses created were very small in size (80 per cent had no salaried employees), and that provincial economic activity was still highly dependent on the public or semi-public sector as the leading supplier of salaried jobs and provider of financial support for the vast majority of private initiatives. This situation, which the provincial political executive hoped would be a temporary one, demonstrates the inexperience and financial frailty of a network of businesses, the majority of which had only recently been established.

The Social Cohesion Issue

What the Local Actors were Saying in 2005

In 2005, there was a general belief that big projects would accelerate and amplify the changes in local communities, which had already been weakened by the rapid changes taking place, because they would aggravate social inequalities. The role of the customary chiefs in economic development, the institutionalisation of the relationships between the political office holders and customary chiefs, and the need to ensure 'historical continuity' were at the centre of the questions being raised. The need to respect the rights of the clans was highlighted by some customary chiefs, but warnings against opportunistic conduct were also being expressed. The customary chiefs in the tribes closest to the proposed project expressed the greatest fears about the social changes to come: 'People are going to earn money but let's be careful to use the money for our well-being and not to our disadvantage'. The customary authorities were preoccupied with preserving a way of life even though 'life is going to change', particularly as many of them stated that they did not 'know what is really going to happen'.

On a slightly different note, the people's elected representatives were aware that the project would accelerate changes in the local society and thought it was essential 'to prepare to control them'. The influx of people and the growth of economic activities and revenues were considered as positive drivers of a 'co-mingling' between villages and tribes as long as they were controlled. Hence, the 'preservation of a tribal space' that was not excluded from development, but could benefit from it through the adoption of new practices appropriate for customary land, appeared to some of the mayors to be an important way of exercising such control. For a small minority of the mayors, outside the VKP area, the risk of 'bad development' was one that could not be ignored, and they thought it was necessary to remain vigilant against 'the wrong kind of development' that results in a poor distribution of the wealth that is generated.

More broadly, questions remained about the choice of a 'dehumanised' form of economic development, 'favouring individualism'. These identity-based issues reflected the fears of a 'loss of peaceful life in the north' and the risks of growing social ruptures. This sentiment was accompanied, however, by clarity in relation to the differences between European and

Kanak conceptions of development, and the need for corresponding differences in strategy and policy. Finally, there was concern about one demographic aspect of the construction process: 'Many foreign workers will come; they will be single and that can cause many problems'.

What the Research Said in 2010

There are no statistics that provide objective information about the link between development and social cohesion. To assess the differences between expectations and practical experience, we relied on three main sources:

- the CODEV evaluation carried out by the IAC, specifically its assessment of the project's impact on the social integration of households;
- a study of the customary life of the Koné and Pouembout tribes (North Province 2008); and
- a more general study of the solidarity networks in local communities (Grochain and Poithily 2011).

The CODEV evaluation was concerned with the distribution of XPF3.4 billion in support of 3,362 economic projects in North Province. The program was also intended to respect and develop the local Kanak culture, and was therefore not exclusively focused on the production of marketable goods, but was also meant to contribute to an improvement in the social life of the beneficiaries. A noticeable improvement in relationships with the local population was found to have been achieved. Having received assistance, individual CODEV beneficiaries felt that their activities were being recognised and, as a result, they were better integrated into professional networks. For their part, the groups and associations supported by the program were found to enjoy a better social environment as a result of 57 per cent of the projects. For these two types of CODEV beneficiaries, the projects appeared to play an important role in the acquisition of a certain socio-professional recognition. Moreover, 54 per cent of the projects were found to have improved social relationships within tribes, while 46 per cent had this effect within villages. At the same time, however, problems involving jealousy appeared in 28 per cent of the projects that had some social impact (and may have played some role in the 9 per cent that were simply aborted). Overall, less than one-third of the projects were perceived by the beneficiaries as having had a significant impact on their social environment, and many of them were assessed more in terms of how they improved individual living conditions. On the other hand, the beneficiaries perceived a decline in the negative impact of projects on the social environment, which suggests the emergence of a basic trend towards a greater social acceptance of market integration.

The study of the Koné and Pouembout tribes emphasised the rarity of wealth creation in late 2008, and thus corroborated the fears of a growing imbalance in the distribution of benefits from the new project (North Province 2008). The assets of the tribes were found to reside in the social economy, which was often organised around women and the conduct of very small-scale productive activities. The consultancy report was based on an inventory of 48 businesses originating from within the two tribes, which had a combined population of 3,000 residents. Ninety-two per cent of these were fragile micro-businesses that generated 130 jobs between them and focused on activities (such as transport) that were related to the proximity of the plant site. At the same time, the study noted a trend towards a reduction in food-producing activities with the emergence of wage-earning opportunities, the small proportion (only 15 per cent) of micro-businesses that were producing food on customary land, and the disappearance of several livestock farms. The analysis confirmed the risk of social and customary deconstruction and an exodus of villagers to the urban centres. To this scenario was added local people's pessimism regarding the possible impacts of economic development on social relationships, such as the emptying of Kanak villages during the week, the substitution of wage employment for mutual aid, the rise of individualism, a decline in traditional authority, uncertain entrepreneurial capacities, the disappearance of vernacular languages, scholastic failure, family problems, and competition or conflict generated by the flow of regular cash incomes. Nevertheless, the report concluded with a note of optimism in relation to 'the measured and not very alarmist formulation of the problems' and the 'current revitalisation of the councils of clan chiefs who share and act in the context of the customary area' (North Province 2008: 130).

In their analysis of the strategies of the local subcontracting companies, Grochain and Poithily (2011) also reported on new forms of valuation of customary life and relationships. On one hand, these strategies enabled the placement of customary life, as reconstituted on the basis of inter-generational relationships and the social changes already under way, at the core of entrepreneurial practices; on the other hand, they also resulted in new forms of conflict. It was easier to secure contracts because of the developer's acknowledgment of local people's customary connections to the land, but this in itself was far from being sufficient as a basis for the development of a business strategy. Ideally, the business operators should become entrepreneurs while retaining certain fundamental customary elements in their working arrangements and, as previously mentioned, the GDPLs were understood to be bodies that could bridge the gap between the customary and capitalist organisation of work.

The formation of the SCPs, which brought together the clans of the VKP area, was a further illustration of this point. This type of partnership or association, which had already been tested in the context of the Goro project in South Province, was meant to be a way of enabling the customary authorities to be recognised and represented within the project. What was also involved here was an acceptance of the decision by the provincial authorities and KNS not to pay royalties to the customary authorities, but instead to initiate a kind of apprenticeship in business development. The GDPL, the basic organisational unit governed according to customary law, was modified so that it could participate in the SCP, an economic structure managed in accordance with common law that could then invest in other companies. With the support of KNS and the provincial authorities, and following around 100 meetings, the intention was to assemble the various clans in a customary form of capitalism that would offer the best solution for them. In the end, one SCP united the GDPLs (and hence the clans) from the coastal zone with their allies from the plains, another united the GDPLs from the Koniambo Massif, and a third united the GDPLs of the tribes of Baco and Poindah bordering the massif. The SCPs functioned as screening devices that took responsibility for economic issues on behalf of the customary authorities they represented, for example by investing in businesses, or collecting and redistributing the dividends arising from their investments. As we have seen, there was also an overarching structure, SAS Vavouto, which grouped the SCPs with other shareholders for the purpose of facilitating relationships with KNS, and hence formed a second 'screen' between the project and the customary authorities.

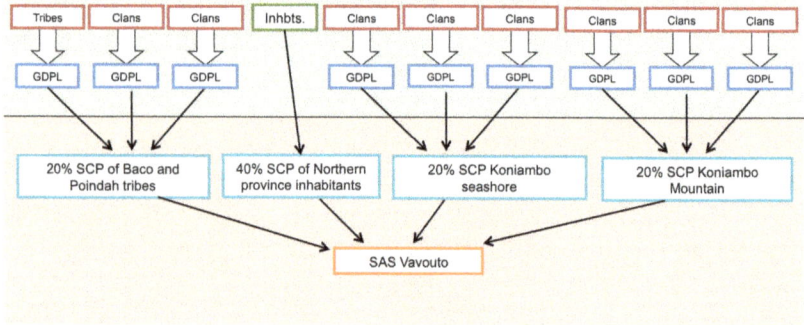

Figure 2.5 Popular participation in SAS Vavouto.
Source: Authors' diagram.

Thus the principle at work here did not involve the multiplication of bilateral relationships between an industrial enterprise and a set of isolated local clans, which might have been the standard practice in the logic of an enclave development, but was scarcely feasible within the local context and territorial ambition of this particular project. The intention was to recognise customary claims through innovative financial instruments, which meant a rethinking of the political relationships within and between the clans. Within this alternative form of organisation, customary relationships continued to facilitate access to markets. Based on their preferential position, particularly in relation to access to information, through their GDPLs and SCPs, the clans actually established a system for exchanging contracts and jobs that was partly free from the logic of economic effectiveness and partly bound to a customary base (Poithily 2010). At the same time, this relational structuring enabled the project's developers to deal with one entity, Vavouto SAS, which was consistent with its management principle of not spreading contracts too thinly between the service providers.

The system that was put in place appeared to operate well enough, and enabled the shielding of local businesses from excessive competition and the linking of customary authorities from the VKP area by contributing to their reorganisation. However, it did not prevent conflicts, which remained frequent even if controlled in their intensity, nor territorial imbalances, which could ultimately become extremely risky. Furthermore, its endurance was not guaranteed because of the weakness of many of the resulting businesses.

With respect to indirect impacts, the integration of the Kanak people into an economic development program for customary land remained a source of ambivalence because of the combination of some successful projects with some internal conflicts. Without turning very violent, the latter have had the effect of 'judicialising' the customary world through the recognition of the clan as a legal entity and the capacity of clan leaders to instigate legal proceedings.

The Situation in 2016

The empirical study documented in this chapter was completed in 2010, so it is now instructive to consider what has transpired since then. So far as local employment is concerned, the figures show the expected decline in the availability of jobs following the end of the project's construction phase, but a relatively stable number of job offers in the operational phase, at a level three times higher than it was before 2005 (Figure 2.6). This is a sign of dynamism in the labour market, largely driven by the efforts to 'rebalance' the supply of infrastructure in North Province, supported by an overall 'densification' of the provincial service sector. The demand for formal employment in North Province has stabilised at around 1,500 job applications per annum in recent years, comparable to the trend in New Caledonia as a whole (GoNC 2016).

The Koniambo smelter employed 999 workers at the end of 2015 (KNS 2016). However, it is worth noting that the local labour market was destabilised in 2016, firstly because of technical problems with the furnaces in 2015, and then because of a fall in the nickel price in 2016.[21] This instability demonstrated the extent to which the local economy was sensitive to the functioning of the smelter. Expatriate positions were the first to be cut, which led to a 15 per cent drop in real estate prices and had a negative impact on economic activity at large. A new wave of redundancies affecting local workers was avoided thanks to the trade union pressure, the reassertion of Glencore's commitments in North Province and slightly better prospects for metal prices.

---

21   Nickel prices have been highly unstable on the London Metal Exchange over the last ten years. They fell from US$55,000/tonne in 2007 to US$10,000/tonne in 2009, rose to US$25,000 in 2010, dropped again to US$11,000 by the end of 2016, but are now rising again.

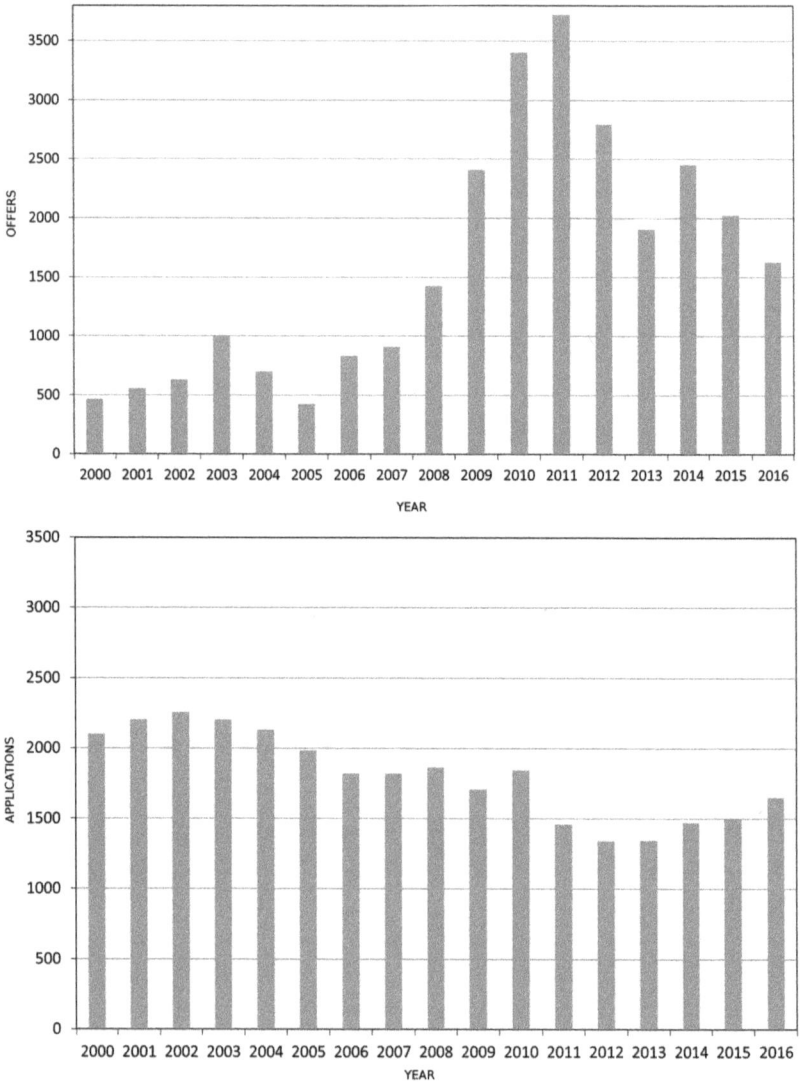

Figure 2.6 Employment offers and applications in North Province, 2000–16.

Source: Institut pour le Développement des Compétences en Nouvelle-Calédonie.

In the matter of business development, there was an understandable peak during the height of the construction phase, and one that was not confined to business activity within the construction sector itself, and business creation outside the construction sector has retained its dynamism in the operational phase (see Figure 2.7). The total value of contracts relating to the construction and operation of the smelter amounted to XPF367 billion by the end of 2015, and businesses based in North Province captured 49 per cent of this value (KNS 2016). The volume of subcontracting activities has understandably decreased, and enterprises have had to close or else reorient their activities, which generated some awkward situations for individual entrepreneurs. However, the departure of enterprises from South Province that were temporarily based in North Province has limited the impact of the post-construction 'demobilisation' on the truly local businesses.

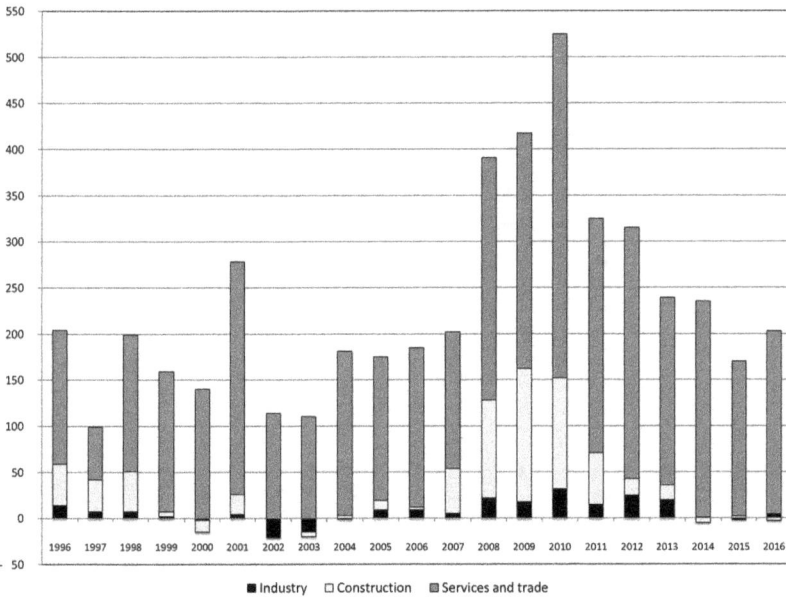

Figure 2.7 Net annual average of semi-annual business start-ups, 1995–2016.

Note: Numbers are negative for some sectors in some years because the number of new businesses created was smaller than the number of existing businesses that closed.

Source: Institut de la Statistique et des Études Économiques, Direction des Systèmes d'Information/Service Information et Méthodes.

It is hard to say whether local entrepreneurs have learned to operate more sustainably as a result of measures implemented before the start of the construction phase. Although the provincial dynamics no longer depend on the direct and indirect impacts of the construction process itself, they are still strongly supported (and subsidised) by the provincial authorities. The Nord Avenir Group was split off from the province's holding company, SOFINOR, as the holder of assets outside the mining and metallurgical sector, in order to promote economic diversification, and now holds shares in more than 80 firms in both North and South provinces. However, it is still dependent on public funding, which could be seen as a sign of frailty.

With respect to the question of social cohesion, the structures designed to bridge the gap between customary and capitalist forms of organisation are still in place (Apithy et al. 2016), but the dynamics at work do not provide clear answers to the concerns formulated in 2005 about the nature and 'quality' of the development triggered by the project. It appears that adjustments are being made on a day-to-day basis, but have still been dictated by the rhythms of the nickel economy, even after the smelter started operating in 2014. The slowdown in the rate of economic growth (IEOM 2016) has not had much effect on the local communities so far, and recent studies suggest that individual members of local tribes have been able to adjust to economic fluctuations through short- or longer-term occupational or geographical mobility (Pestana et al. 2016). These capacities have been enhanced by the level of infrastructural development achieved in the last three decades. However, there is still much uncertainty about the medium- and long-term social impacts that will be experienced through the remainder of the mining project cycle and hence about the opportunities offered by the new industrial regime.

Conclusion

The evidence presented in this chapter partially validates the hypotheses contained in the model of negotiations through the mining project cycle (Figure 2.2), but it also seems necessary to add some new elements into the framework in light of the relatively short period of time that elapsed between the decision to build the smelter and the completion of the construction phase.

First, it seems essential to incorporate the presence of the province, both as the face of the state and as a key actor involved in supporting the project and managing its impacts. Our observations show that its action weighs on the normative framework at the start of the project, with the dual effect of maintaining the company's willingness to pay at a high level and noticeably increasing the negotiating capacity of the local communities. Moreover, far from concentrating its political actions on the one industrial project, the province has broadened of the concept of redistribution and limited social tensions through the implementation of CODEV, through the operation of financial entities like SOFINOR and Nord Avenir, and through the other levers at its disposal, which remain both numerous and well resourced. To understand the effects of the project on a small territory, it would seem essential to shift the focus of the analysis to the capacity of the state (or in this case the province) to connect other dimensions of development to the production of minerals or metals.

Second, it seems that the political dimensions of the project brought about a combination of provincial action with the developer's engineering efforts and with knowledge of the likely impacts of the project on local populations. These conditions contributed to the control of conflict at the beginning of the project cycle. Not only were the rules already generally in place, but the support of the local authorities, occasionally backed up by the developers, created conditions favourable to the emergence of organisational innovations and mechanisms for redistribution of the benefits. The high level of prior awareness of the risks, as demonstrated by the clarity of the anxieties expressed in 2005, is also indicative of a learning process that altered the shape of the curve representing the negotiating capacity of the local actors.

The model cannot, however, account for the different levels of conflict that are likely to extend through the local communities. While the project's construction phase ran quite smoothly, minor conflicts have occurred during the operational phase, which highlight the fact that active power relations are at play, and which could lead to more acute tensions in future. It would appear to be necessary to approach the project cycle in its entirety while being attentive to the discreet changes that emerge in the initial phase and whose effects might only become apparent in a subsequent phase of the cycle.

As we have seen, the weakness of the local economic fabric also presents a major risk. Its disintegration could quickly undermine the durability of the redistribution mechanisms, the assumption of responsibility and the lessons learned through the public shareholding institutions, and the subsequent ability to contain social conflicts.

Generally speaking, due to the far-reaching economic development that has been in train through the construction and operational phases of the project, it is difficult to assess the sources of future tension that may be related to the radical social dynamics at work. We have seen that innovations and proactive policies do not necessarily provide convincing answers or reduce the level of uncertainty about the profound nature of the changes taking place. What may be involved here, again, are the hardly discernible seeds of conflicts to come.

Ultimately, our analysis argues in favour of the ongoing monitoring of changes triggered by the production of minerals and metals throughout the mining project cycle. More comprehensive monitoring would eventually enable the formulation of more detailed recommendations that could in turn enable better anticipation and prevention of conflicts and improvement in the management of social impacts. It would also make sense to develop the analytical framework applied here by improving the integration of the state's capacity for action, focusing on the connections between industrial and broader local development, and taking the heterogeneous nature of local communities into account.

References

Apithy L., S. Bouard, S. Gorohouna, S. Guyard and J.-M. Sourisseau, 2016. 'Dynamiques Économiques et Sociales des Tribus et de la Ruralité: Fragilités et Facultés d'Adaptation.' In S. Bouard, J.-M. Sourisseau, V. Geronimi, S. Blaise and L. Ro'i (eds), 2016. *La Nouvelle-Calédonie Face à Son Destin: Quel Bilan à la Veille de la Consultation sur la Pleine Souveraineté.* Paris: Karthala.

Bencivengo, Y. (ed.), 1999. *La Mine en Nouvelle-Calédonie.* Nouméa: Ile de Lumière.

Blaise, S., J.-M. Sourisseau, O. Hoffer and S. Bouard, 2016. 'Des Mineurs, des Métallurgistes et des Entrepreneurs au Défi de la Concurrence Internationale.' In S. Bouard, J.-M. Sourisseau, V. Geronimi, S. Blaise and L. Ro'i (eds), *La Nouvelle-Calédonie Face à Son Destin: Quel Bilan à la Veille de la Consultation sur la Pleine Souveraineté.* Paris: Karthala.

Bouard, S., 2011. Les Politiques de Développement à l'Épreuve de la Territorialisation: Changements et Stabilités dans une Situation de Décolonisation Négociée, la Province Nord de la Nouvelle-Calédonie. Montpellier: Université Pau Valéry Montpellier III (PhD thesis).

Bouard, S., J.-M. Sourisseau, S. Bellec, O. Hoffer, P.-Y. Le Meur and C. Levacher, 2016. 'Des Stratégies de Développement Local Volontaristes et Différenciées.' In S. Bouard, J.-M. Sourisseau, V. Geronimi, S. Blaise and L. Ro'i (eds), 2016. *La Nouvelle-Calédonie Face à Son Destin: Quel Bilan à la Veille de la Consultation sur la Pleine Souveraineté.* Paris: Karthala.

CEROM (Comptes Economiques Rapides de l'Outre-Mer), 2008. *Les Défis de la Croissance Calédonienne.* Nouméa: CEROM, Institut Economique de l'Outre-Mer (IEOM), Institut de la Statistique et des Etudes Economiques (ISEE), Agence Française de Développement (AFD).

——, 2011. *Les Comptes Économiques Rapides de la Nouvelle-Calédonie en 2009, une Année Mitigée.* Nouméa: CEROM, IEOM, ISEE, AFD.

Couharde, C., V. Geronimi and A. Taranco, 2016. 'La Nouvelle-Calédonie suit-elle Toujours le Modèle de l'Économie Assistée.' In S. Bouard, J.-M. Sourisseau, V. Geronimi, S. Blaise and L. Ro'i (eds), *La Nouvelle-Calédonie Face à Son Destin: Quel Bilan à la Veille de la Consultation sur la Pleine Souveraineté.* Paris: Karthala.

David, C. and J.-M. Sourisseau, 2016. 'De Matignon à la Consultation sur l'Indépendance: Une Trajectoire Politique et Institutionnelle Originale.' In S. Bouard, J.-M. Sourisseau, V. Geronimi, S. Blaise and L. Ro'i (eds), *La Nouvelle-Calédonie Face à Son Destin: Quel Bilan à la Veille de la Consultation sur la Pleine Souveraineté.* Paris: Karthala.

Filer, C., 2006. 'Custom, Law and Ideology in Papua New Guinea.' *Asia Pacific Journal of Anthropology* 7: 65–84. doi.org/10.1080/14442210600554499

Freyss J., 1995. *Economie Assistée et Changement Social en Nouvelle-Calédonie.* Paris: Presses Universitaires de France.

Gaillard C., J.-F. Bélières, P.-M. Bosc, J.-M. Sourisseau and M. Passouant, 2011. 'Evaluation du Code de Développement (CODEV) de la Province Nord de la Nouvelle-Calédonie.' Koné: Institut Agronomique Néo-Calédonien (IAC)/Centre de Coopération Internationale pour la Recherche Agronomique pour le Développement (CIRAD).

GoNC (Government of New Caledonia), 2009. 'Les Rapports des 9 Ateliers du Diagnostic.' Nouméa: Haut-Commissariat de la République en Nouvelle-Calédonie.

——, 2010. 'Recensement Général de la Population Calédonienne.' Nouméa: Institut de la Statistique et des Études Économiques (ISEE).

——, 2011. 'Statistiques de l'Emploi, Année 2010, Bilan Détaillé.' Nouméa: Institut pour le Développement des Compétences en Nouvelle-Calédonie (IDC-NC).

——, 2015. 'Recensement de la Population 2014: Une Démographie Toujours Dynamique.' Nouméa: ISEE (Synthèse°35).

——, 2016. 'Le Marché de l'Emploi en Nouvelle-Calédonie, 2015: Ralentissement du Marché de l'Emploi Calédonien.' Nouméa: IDC-NC.

Grochain S., 2010. 'Les Répercussions Socioéconomiques du Projet Koniambo.' Pouembout: IAC/KNS (working paper).

Grochain S. and D. Poithily, 2011. 'Sous-Traitance Minière en Nouvelle-Calédonie: Le Projet Koniambo.' Nouméa: CNRT Nickel, Programme Gouvernance Minière (Document de Travail 4).

IEOM (Institut d'Emission d'Outre-Mer), 2016. '2015 Nouvelle-Calédonie: Rapport Annuel, Édition 2016.' Nouméa: IEOM.

KNS (Koniambo Nickel SAS), 2016. *Rapport Développement Durable 2015.* Koné: KNS.

Le Meur, P.-Y. and T. Mennesson, 2011. 'Le Cadre Politico-Juridique Minier en Nouvelle-Calédonie: Mise en Perspective Historique.' Nouméa: CNRT Nickel, Programme Gouvernance Minière (Document de Travail 3).

Néaoutyine, P., 2006. *L'Indépendance au Present: Identité Kanak et Destin Commun.* Paris: Syllepse.

North Province, 2004. 'Rapport de Présentation SDAU Voh Koné Pouembout.' Nouméa: Syndicat Intercommunal à Vocation Unique (SIVU), Voh, Koné, Pouembout, Service d'Etat de l'Aviation Civile, Direction de l'Aménagement et du Foncier de la Province Nord, and Empreintes.

———, 2008. 'Diagnostic des Terres Coutumières de Koné—Pouembout, Tome 1: Diagnostic et Analyse des Enjeux.' Koné: Direction du Développement Economique et de l'Environnement de la Province Nord (DDEE-PN), and Emergences.

———, 2015. 'Évaluation de l'Action Provinciale sur la Période 2008–2013.' Koné: Secrétariat Général de la Province Nord.

Pestana G., O. Hoffer and P.-C. Pantz, 2016. 'Mobilités, Dynamiques Territoriales et Urbaines.' In S. Bouard, J.-M. Sourisseau, V. Geronimi, S. Blaise and L. Ro'i (eds), 2016. *La Nouvelle-Calédonie Face à Son Destin: Quel Bilan à la Veille de la Consultation sur la Pleine Souveraineté.* Paris: Karthala.

Poithily D., 2010. Les Stratégies et Motivations des Entrepreneurs VKP. Nouméa: Université de la Nouvelle-Calédonie (Masters thesis).

Sourisseau, J.-M., G. Pestaña, C. Gaillard, S. Bouard and T. Mennesson, 2010. *A la Recherche des Politiques Rurales en Nouvelle-Calédonie: Trajectoires des Institutions et Représentations Locales des Enjeux de Développement (1853–2004).* Nouméa: IAC Éditions/Tabù Éditions.

Sourisseau, J.-M., R. Tyuienon, J.C. Gambey, M. Djama and M.R. Mercoiret, 2006. *Les Sociétés Locales Face aux Défis du Développement Économique: Province Nord de Nouvelle-Calédonie.* Nouméa: IAC/DDEE-PN/CIRAD.

Syndex, 2005. 'Nouvelle-Calédonie Nickel 2010: Une Nouvelle Ère Industrielle.' Colloque International du 7–8 Juillet 2005, Paris.

van Vliet, G., 1998. 'Extractive Industries: Conflict Prevention through Empowering Stakeholders.' PowerPoint presentation to conference on 'Forces for Sustainability', The Hague, 14–15 March. Viewed 21 April 2016 at: www.envirosecurity.org/sustainability/presentations/Kloff.pdf

3. Social and Environmental Transformations in the Neighbourhood of a Nickel Mining Project: A Case Study from Northern New Caledonia

MATTHIAS KOWASCH

Introduction

The irony of 'poverty in the midst of plenty' can be applied to many developing and emerging countries rich in mineral resources. The resource curse thesis 'correlates natural resource abundance to evidence of slow or declining economic growth' (Langton and Mazel 2012: 24; see also Auty 1993; Sachs and Warner 1995; Freudenburg and Wilson 2002; Bebbington et al. 2009). Raw materials such as copper, nickel, gold and uranium are found and widely exploited in regions inhabited by indigenous peoples who live far from the centres of political power. In most cases, the mining sector offers few economic benefits for the indigenous peoples, even if mine operators promise to provide opportunities for social, educational and economic development. In their study of the Pilbara region in Western Australia, Taylor and Scambary (2005: 28) note: 'While the regional labour market has grown in both size and complexity, indigenous participation has remained relatively marginal'. The relationship between mining companies and indigenous communities is largely based on an unequal relationship of power. Even if multinational companies have become increasingly aware of the importance of corporate social responsibility, indigenous communities are in most cases characterised by weakness in the face of these companies. Taking examples from Papua New Guinea, Colin Filer postulates a process of 'social disintegration' within such indigenous communities (Banks 1996: 233).

The French overseas territory of New Caledonia possesses around 25 per cent of worldwide nickel deposits, and mining products represent around 95 per cent of its total exports. Economic development and the nickel industry depend on each other. Historically, indigenous Kanak people were excluded from the mining industry in New Caledonia, while the land was forcibly occupied for farming and other economic activities. Discriminatory regulations were abolished after World War II, and Kanak people progressively became French citizens. They also got jobs in the mining sector, especially as truck drivers or cleaners. The only nickel smelter at that time was the Doniambo plant in Nouméa, which was operated by the French company Société Le Nickel (SLN). This situation changed in the 1990s. The demands of the independence movement for economic development and political emancipation from France resulted in the Koniambo project in North Province.

Figure 3.1 The Koniambo project construction site, September 2012.
Source: Photo by Matthias Kowasch.

North Province, which is governed by the Kanak independence umbrella party Front de Libération Nationale Kanak et Socialiste (Kanak and Socialist National Liberation Front), became the majority shareholder in the project. A majority shareholding means more decision-making power and more influence over the strategy of the mining company or its shareholders. But involvement in the nickel sector will not preclude societal transformations occurring within the Kanak communities living in the neighbourhood of the project. On the contrary, employment, business creation and shareholding lead to social upheaval.

After describing the emergence and background of the Koniambo project in northern New Caledonia, this chapter focuses on societal and environmental transformations within the indigenous communities in the neighbourhood of the project. In 2003, Leah Horowitz completed a PhD thesis about the relationship between the Koniambo project and the closest Kanak village, Oundjo, which is only 2 kilometres from the work site (Horowitz 2003a). Drawing upon this ethnographic study and the papers of other authors working on societal transformations in neighbouring Kanak communities (Emergences 2009a, 2009b; Poithily 2010; Grochain and Poithily 2011; Le Meur et al. 2012; North Province 2013), this chapter is based on more than two years of fieldwork and various research projects conducted at the Institut de Recherche pour le Développement (IRD) between 2007 and 2012.[1] The present chapter sets out the involvement of neighbouring Kanak communities in the Koniambo project, through employment, business creation and subcontracting, and describes conflicts linked to the construction of the nickel smelter. Social and environmental transformations within the communities are analysed in a theoretical framework outlined in the next section.

Method and Theoretical Framework

Fieldwork in post-colonial settings is subject to special rules and presents very particular challenges to the researcher. The French overseas territory of New Caledonia is a country, inscribed since 1986 on the United

1   I would like to thank the IRD branch in Nouméa, especially the research groups UMR Espace-DEV and UMR GRED for logistical support, welcome and encouragement during my PhD and subsequent research projects. In the framework of different research programs (GERSA, Mining Governance), financial grants helped me to conduct the surveys. I received warm welcomes from families in the four villages where I worked. I would like to particularly thank my family in Baco, where I lived for a year.

Nations (UN) list of 'non-self-governing territories' (see article 73 of the UN Charter). The long history of relations between the indigenous Kanak independence movement and the UN has been described by Christine Demmer (2007) and Stéphanie Graff (2012). The civil war in the 1980s, the independence movement and the process of decolonisation have to be considered by researchers who undertake ethnographic studies in Kanak communities. The theft of intellectual property represents a danger that the researcher also has to consider when doing fieldwork. According to Paul Robbins (2006: 313), research in the global South can be a post-colonial and expropriating act. When I started fieldwork in the Kanak village of Baco, clan elders mentioned that it is important to share the results of research studies with the communities themselves. Kanak communities often welcome European researchers who are interested in obtaining information about clan histories, sacred places, and local people's perceptions and ideas—in short, 'intellectual property'.

My fieldwork in northern New Caledonia was part of a PhD thesis in human geography (2006–10), other research projects and a consultancy undertaken for the authorities of North Province in 2012. A quantitative survey of 239 people in four Kanak villages (Netchaot, Baco, Oundjo and Gatope) was carried out from 2008 to 2010 (see Figure 3.2). Young volunteers from the villages were recruited as interviewers for the study. The quantitative approach was complemented by semi-structured interviews with businessmen, customary authorities and politicians. I collaborated with customary authorities to map sacred places and clan housing in order to visualise territorial perceptions in the vicinity of the mining project. We pointed out the places with a global positioning system or on a paper map before exporting the data to a geographical information system. The challenge was to develop a method of spatial analysis to map the heritage values attributed to places by Kanak people and to make this new information available for regional planning programs. A real partnership means having enough space for views to be expressed by both the researcher and the customary representative. Both play the parts of 'learner' and 'teacher' at the same time (Kowasch 2014: 261). Collaboration with customary authorities and representatives is consistent with the conclusion of the political ecologist Piers Blaikie (2012: 237), who suggests that 'the notion of a politicised organigram is sometimes useful to understand what "really goes on" in a relevant part of an administration' (2012: 37). And the regular presentation of research results in Kanak villages underlines the point that researchers can 'give something back' to local populations (Kowasch 2014: 262).

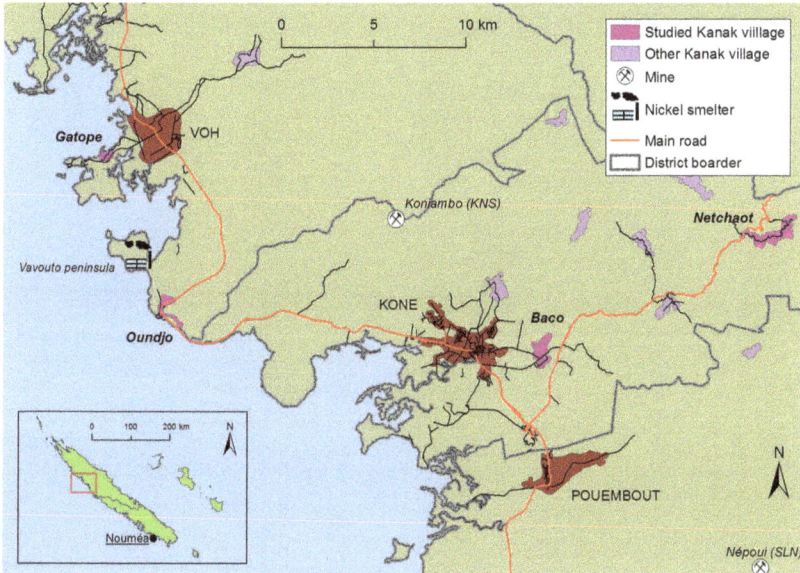

Figure 3.2 Location of the studied Kanak villages in northern
New Caledonia.

Source: Author's map, based on information supplied by the Direction des Infrastructures, de la Topographie et des Transports Terrestres.

The question then is whether the researcher has to be integrated into the indigenous village for a longer period and actively participate in community life. According to Blaikie (2012), social engagement can be useful, and withdrawal into an ivory tower does not permit a deeper understanding of peoples' perceptions and opinions. My research results are based on integration into community life in the village of Baco over a period of six years.

The environment has not only a utilitarian value for Kanak societies but also a cultural heritage value. Research on environmental economics in other countries has attempted to place a monetary value on ecosystems (e.g. Constanza et al. 1997; David et al. 2007). Figure 3.3 shows a simple model to demonstrate the transformation of environmental values.

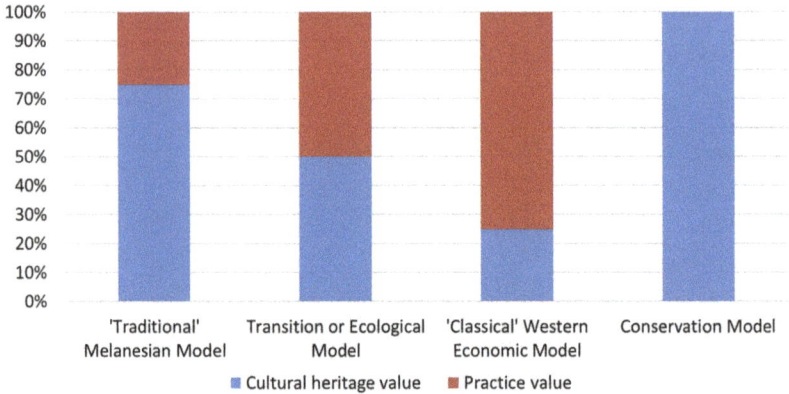

Figure 3.3 Environmental value model.
Source: Author's chart, based on personal communication from Gilbert David and Jean-Brice Herrenschmidt (2010).

Heritage values are transcribed mainly in topographical locations that represent an essential cultural reservoir. Winslow (1995: 14) notes that each clan name is a set of place names—the name of the place where the ancestor of the clan appeared and all other places that the clan occupied in its history. Ancestors are pervasive, and their presence is felt in the passage of a butterfly, in the falling rain or in the swirls of river water. The appropriation of space in a particular place results from naming and recognising the site in one's own language. The real 'owners' of a place are its founders. The specificity of the clan is primarily its history: the place and the circumstances of its appearance, the journey it undertook and the recognised symbolism of place. The social identity and legitimacy of any clan is then presented as a route or as a series of places where the group lived (Naepels 2006: 50). But Kanak property is subject to perpetual updating (Naepels 1998; Le Meur 2010; Kowasch 2012a), because the itinerary of the clan can change, and also because land can be given away. The founding group is therefore not the only one able to claim land rights. Pursuant to a 'host dynamic', the newcomer receives from the founder some ground where he can live and grow crops. It is important to note that the transfer of land represents not the transfer of ownership, but the superposition of different legitimacies: the property is not unique, but rather consists of competing titles. The founder maintains a special relationship to the land. The newcomer in turn can argue that he resided there, made exchanges with the founder and worked there, but he cannot in any way challenge the fact of having received the land of another.

The 'traditional' Melanesian model contains a strong cultural discourse linking places to a myth and/or to a specific history. To these cultural heritage values, which include cultural practices like fishing or subsistence farming, are added economic practices, which are required for people to live and reproduce. In the transitional or ecological model, the discourse seeks to balance the importance of practical values and cultural heritage, describing a changing pattern of Melanesian lifestyle in the cash economy, or otherwise describing the evolution from a free-market economy to a sustainable development model, introducing a notion of balance between the use of the environment and its capacity to retain its heritage features. According to the classical Western economic model, the socioeconomic value of an environment is often reduced to its practical value—the economic goods and services that the environment provides. The conservation model in turn removes the question of practice by promoting the protection of natural areas perceived as part of a common heritage. Only the cultural heritage value exists in this model (see Figure 3.3).

Fishing and subsistence farming have heritage value as cultural practices, because they are used to establish or renew social relationships with other clans and to transmit cultural knowledge to younger generations in the context of customary ceremonies, as when fish are exchanged in a customary wedding ceremony. New incomes from people's involvement in the mining sector not only lead to social transformations within Kanak communities, as when farming products can be replaced by products from the supermarket, but mining projects also result in changes to landscape and territory. Fishing, hunting and farming are practised in what Tolia-Kelly (2006) calls 'territories of belonging'. The exploitation of mineral resources creates conflicts and negotiations over such territories.

Given that cultural heritage sites are also located on mine sites in New Caledonia, and on the customary land where the social and economic facilities required for a metallurgical plant are constructed, would it be true to say that the territories of belonging are currently being transformed by the traditional landowners themselves? Who are the actors in these development projects and what conflicts within the Kanak communities result from the projects? The territorial and social transformations linked to the Koniambo project will be analysed in the rest of this chapter. What are the implications of what Amin (2002: 397) would call this 'politics of place'?

A Short History of the Koniambo Project

The Koniambo project has a strong symbolic value for the Kanak independence party that governs North Province. Shareholders in the project include the local company Société Minière du Sud Pacifique (SMSP), which owns 51 per cent of the shares and is owned by North Province, and the Swiss group Glencore, which holds 49 per cent. The Koniambo nickel smelter has an annual production capacity of 60,000 tonnes of ferronickel. The first pouring of nickel metal was in April 2013, and the plant was fully commissioned in November 2014. However, it failed to operate at full capacity due to technical problems and incidents.

Even before the Koniambo project went into production, North Province was reinvesting its revenues from SMSP nickel mines elsewhere in the country in wider economic development, such as tourism, aquaculture and urbanisation, along with construction of social or cultural facilities. In this case, the nickel sector is a kind of 'instrument' for economic and political independence, and not just for making profits (Winslow 1995; Pitoiset and Wéry 2008; Le Meur and Mennesson 2011; Kowasch 2012a). Investors sometimes mention 'two Koniambo projects': on one hand, the construction of the processing plant and development of the mine; on the other hand, the economic development of the Voh, Koné and Pouembout (VKP) municipalities where the Koniambo project is located.

By taking the initiative in the 1990s to build a nickel smelter in North Province, SMSP sought an industrial partner who could provide the relevant expertise and co-financing (Néaoutyine 2006; Pitoiset and Wéry 2008). The provincial authorities and SMSP initially established a joint venture with the Canadian company Falconbridge.

> SMSP contacted and negotiated with Falconbridge and then convinced the French Government—with assistance of political pressure from independence party leaders—to help it obtain the Koniambo Massif from a French mining company (Horowitz 2004: 290).

SMSP/Falconbridge won the Koniambo Massif in exchange for surrendering the licence to mine another nickel deposit (Poum), in the far north of New Caledonia, to SLN. The deal was fixed in the Bercy Agreement in 1998, signed by SMSP/Falconbridge, the French Government and the French company Eramet, which is SLN's parent company (Pitoiset and Wéry 2008). In 2006, Falconbridge was taken over by the Swiss mining company Xstrata, which thus became the new partner of SMSP (Pitoiset and Wéry 2008; see also Chapter 2, this volume). Xstrata has mainly financed construction of the processing plant on Vavouto Peninsula and exploitation of the nickel deposit, while SMSP holds the mining tenements in the Koniambo Massif and has support from local Kanak communities, even though nickel mining has caused widespread environmental pollution in the country for nearly 150 years. With the recent development of the Koniambo and Goro nickel projects (see Chapter 6, this volume), New Caledonia is expected to triple its nickel production and become the second-largest nickel producer in the world. In October 2012, around 6,000 workers from different countries were working (and mostly living) on the Koniambo construction site. Chinese and Filipino workers, who assembled the components of the nickel smelter and the power plant, travel the world doing such work. Direct employment on the project accounted for between 800 and 1,000 jobs once it came into production.

SMSP already possesses nickel mines in Ouaco and Poya, on the west coast, and Nakety and Kouaoua, on the east coast of Grande Terre, the main island of New Caledonia, all located in North Province. Since 1995, SMSP had become the largest ore exporting company in New Caledonia because SLN, which produces more than 50 per cent of the ores, feeds the majority of its output into its Doniambo processing plant in Nouméa. The primary goal of SMSP was always to become a 'metallurgist' and to transform nickel ores into metal. This milestone was achieved in 2008 with the start-up of a processing plant in Gwangyang in South Korea. The nickel smelter is operated by Société du Nickel de Nouvelle-Calédonie et Corée (SNNC), a joint venture between SMSP, holding 51 per cent of the shares, and its Korean partner POSCO, holding 49 per cent. SMSP and POSCO created a second joint venture, which is simply called Nickel Mining Company (in English), to exploit all of SMSP's nickel deposits in New Caledonia apart from the Koniambo deposit. The revenues from the smelter in Gwangyang were used to co-finance construction of the Koniambo plant. In both cases, the 51/49 per cent model of ownership is remarkable because it has been rolled out to an overseas investor. It is little known globally that an indigenous mining company holds 51 per cent of the shares in a project that is operating in South Korea (with POSCO), as well as another joint venture planned in China, with the Chinese nickel producer Jinchuan. Processing plants are integral to both deals.

The Koniambo project is located in a rural area where the infrastructure required for the new nickel smelter was initially lacking. In 2004, North Province set out to provide a broader development of the VKP area, which only had 8,211 inhabitants at the time (Kowasch 2012c; North Province 2013). In the five years until the next census (in 2009), the local population grew by an average of 18 per cent each year. The resident population of the provincial capital, Koné, where many of the new facilities were built, grew from 5,199 in 2009 to 7,340 in 2014 (GoNC 2016: 31). To welcome the new arrivals, the main priority of the public authorities was to provide a greater supply of rental housing. Anne Pitoiset and Claudine Wéry described the transformation in their biography *Mystère Dang*:

> The villagers discover traffic, real estate speculation, rental housing development and the massive arrival of new peoples. The money flows and no question, this time, people don't want to be dispossessed … For the first time, the nickel boom leaves Nouméa and its effects are felt in *Kanak* country. (Pitoiset and Wéry 2008: 106)

In New Caledonia, it is common to talk of the 'bush' beyond Nouméa in order to highlight the demographic and economic primacy of the capital located in South Province, 270 kilometres south of Koné. Some 40 per cent of New Caledonia's population live in Nouméa, and the capital accounts for roughly 75 per cent of formal employment. The spatial imbalance of the territory, inherited from the colonial period, has continued to widen. A policy of 'territorial rebalancing' was started with the Matignon Agreements of 1988. Horowitz (2004: 307–8) noted how:

> Many people hope that the [Koniambo] project will entail an economic readjustment or 'rebalancing' (*rééquilibrage*) as promised in the Matignon and Nouméa Accords, allowing the Northern Province to be placed on more equal footing with the Southern Province.

The development of the VKP area is an example of this policy (Néaoutyine 2006). The provincial authorities wanted to turn the growing VKP area into a real urban centre, an Oceanic city to rival Nouméa. New rental housing, shops and industrial areas mushroomed (see Figures 3.4 and 3.5). One problem has been that the settlements and socioeconomic facilities, such as shops and schools, in the area are scattered in different locations. Access to facilities such as supermarkets requires the use of a car. Public transport does not really exist; there are only some shuttle buses operated by individual businessmen that link the Kanak villages to the three towns (Voh, Koné and Pouembout). The process of socioeconomic development was initially concentrated on private and state land, but the public authorities have since made efforts to integrate Kanak clans and their customary land into the process (North Province 2013).

Figure 3.4 First stone of the new cultural centre in Koné, 2008.
Source: Photo by Matthias Kowasch.

Figure 3.5 Rental housing on customary land, Baco, 2011.
Source: Photo by Matthias Kowasch.

With the construction of the nickel smelter in North Province, and a wider development of the VKP area, the independence movement has also wanted to show that Kanak people are capable of handling a major global mining project, and that the Koniambo project can indeed lead to more economic and financial independence from France. The Kanak independence leader Jean-Marie Tjibaou explained:

> The future of self-determination and the promotion of dignity require that we are not beggars. For this, it is necessary that the country helps people organising themselves to produce wealth, making them financially independent. (Pitoiset and Wéry 2008: 161)

The businessman D. Xamène from Oundjo, who has a subcontracting company involved in earthworks and equipment transport, expressed the political dimension of the project as follows:

> One day we will have our independence, we will have our factory … It is the wish of all Kanak peoples. This is a political struggle. (Kowasch 2010: 374)

Nevertheless, the focus on the mining industry is a bold move, because the general economic development strategy depends on nickel prices in the world market.

Employment, Business Creation and Subcontracting

The study conducted in the four Kanak villages in the neighbourhood of the new nickel smelter showed that many of the residents of these communities supported the Koniambo project: 60 per cent of the 239 interviewed people were *in favour* or *rather in favour* of the project; only 13 per cent were *against* or *rather against*. Kanak people generally perceive the Koniambo project to be 'their project', one that is meant to generate employment and services, economic development and political emancipation. Their hopes largely hinge on the prospect of employment (198 responses). Broader economic development of the VKP area ranks second in the number of responses (87) (Kowasch 2010: 373). The mining sector is mainly considered as a big employer, not as an instrument for development more broadly. The interviewees showed an interest in working either for the mining operator Koniambo Nickel SAS (KNS) or for one of its subcontracting companies. Among the 239 interviewees, aged from sixteen to seventy years, 73 (31 per cent) wanted to be employed by KNS and 26 (11 per cent) were already working for the mining company or for a subcontractor (Kowasch 2010: 454). Young people under thirty years of age especially expressed a great willingness to work in the mining industry. Most of the interviewees wanted to be truck drivers or cleaners, which is consistent with the job opportunities that actually exist for many people in the neighbouring villages. Being a truck driver or cleaner is popular because parents, grandparents and friends already have experience in these professions, and these jobs also allow people to stay in their home village. The customary and family relationships that are central to one's identity are easier to maintain if one can stay in the village. Furthermore, most of the interviewees who were motivated to work in the mining sector did not have the qualifications to apply for higher-paid positions. In many cases, it was cheaper for companies to employ Asian workers than to hire local workers, partly because foreign workers are exempt from many of the strict regulations that apply to local working conditions. For example, Chinese workers who were employed by the China Machinery Industry Construction Group could work 60 hours per week, and unofficially often worked even longer hours, and still earn the New Caledonian minimum salary, which is around US$1,400 per month. As pointed out in Chapter 2 (this volume), local communities feared the massive influx of foreign workers because so many were single men. KNS did make an effort to employ local people, so in late 2012, at the peak of the construction phase, over 60 per cent of the workers on the Vavouto construction site were locals. Furthermore, 77 per cent of the local employees, excluding the expatriates, were natives of North Province, 77 per cent were Kanak, and the most common employee profile was that of the young male Kanak who was a native of North Province (see Chapter 2, this volume). This helps to explain why so many of my own interviewees were interested in working for KNS.

The education level of interviewees in the villages of Netchaot, Baco, Oundjo and Gatope who indicated that they were interested in working on the Koniambo project is shown in Figure 3.6. The majority of young Kanak people between sixteen and twenty-nine years old (55 per cent of the interviewees) had been to secondary school, but had not obtained a higher school certificate. Often, they continued with specialised training for jobs like that of technician or truck driver. Only 22 per cent had obtained a higher school certificate, and a small minority (5.5 per cent) had spent a minimum of two years on a university degree.

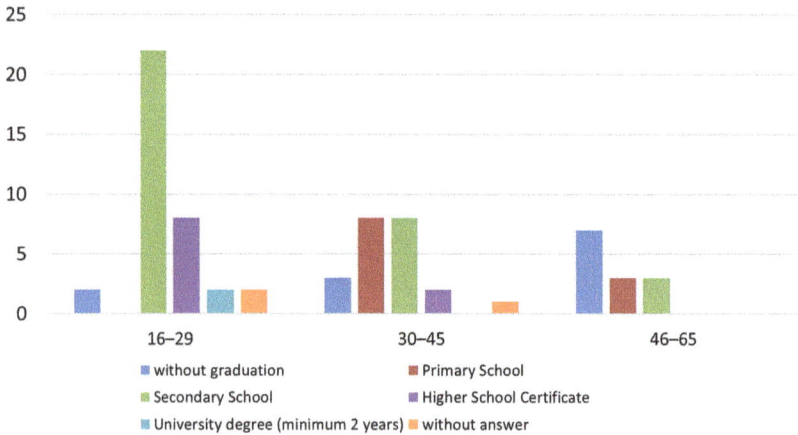

Figure 3.6 Education level of the peoples in the four villages motivated to work in the Koniambo project.
Source: Kowasch (2010: 457).

These results are consistent with those from a study conducted by the provincial Observatoire de la Santé et des Actions Sociales. That study found that just 4 per cent of young Kanaks in North Province obtained a higher school certificate, which permits one to enter university, as against 10 per cent in non-Kanak communities (North Province 2008: 26). Among interviewees aged between thirty and forty-five years, no one had completed a university degree. The majority (72 per cent) had finished their education at the end of primary school (36 per cent) or secondary school (36 per cent). In keeping with the colonial history of the country, no one between the ages of forty-six and sixty-five years in the four studied villages held either a higher school certificate or university degree. This study also showed that young Kanaks had lower annual wages than their European fellow citizens with the same skills: 'We have to note that the young Kanaks who work have an income less than half than all other communities: US$8,677 against US$19,500' (ibid.: 18). The study also took the example of businessmen and noted that non-Kanaks earned three times more than Kanaks (US$30,136 against US$10,665).

In their working document, Le Meur et al. (2012: 16–20) explain that Kanak business development started late, but accelerated with the construction of two new nickel smelters in the north and south of New Caledonia. During the first nickel boom (1967–71), the creation of small and medium enterprises did not involve the Kanak population, but was dominated by new immigrants (ibid.: 16). In 1993, Isabelle Leblic (1993: 240–51) reported a growing demand for subcontracting in the mining sector in North Province. Kanak subcontracting in the mining sector finally accelerated with the construction of the Goro and Koniambo smelters at the beginning of the new millennium.

In North Province, there were more than 550 registered companies by 2008, compared with 340 in 2005 and less than 300 in 1995 (North Province 2013: 35). Even if not all newly established companies were subcontractors in the Koniambo project, the increase in business activity was still a consequence of the development of the nickel sector. From 1998 to 2009, nearly 90 subcontracting companies were founded to service the Koniambo project, most of them with the support of KNS (Le Meur et al. 2012). Examples of subcontractors created with such support are Société des Travaux du Massif Koniambo (STMK) and Société Webuihoone Maritime SAS (SOWEMAR) (ibid.: 45).[2] The main activity of SOWEMAR was maritime transport, which included the renting of two tug boats in order to bring container ships from the lagoon entrance to the port of Vavouto (Kowasch 2012a, 2012b). It was also responsible for security around the plant site, and had a role to play in the marine emergency plans of KNS, such as intervening in the event of accidental pollution.

---

2   Shares in SOWEMAR were split between a clan company from Gatope (46 per cent), a collection of coastal clans (5 per cent), and SOFINOR, an SMSP subsidiary (49 per cent) (see Chapter 2, this volume; also Kowasch 2010).

KNS also assisted in the foundation of SAS Vavouto, which is a federation of various local companies and businessmen (see Chapter 2, this volume). It was founded in 2008 as the interface between local subcontractors and the mining company. The goal of SAS Vavouto was to seek, manage and distribute contracts by dividing them into several subcontracts for its members. Indeed, in accordance with its development strategy, North Province does not allow royalty payments to Kanak communities but instead supports their direct participation in the Koniambo project by means of business contracts and employment. Not all local subcontractors are members of SAS Vavouto but, as of August 2012,[3] the company had federated 126 mostly local (often indigenous) enterprises (personal communication, SAS Vavouto, August 2012). Voh, where the smelter was built, had the highest concentration of subcontractors—54 enterprises (see Figure 3.7).

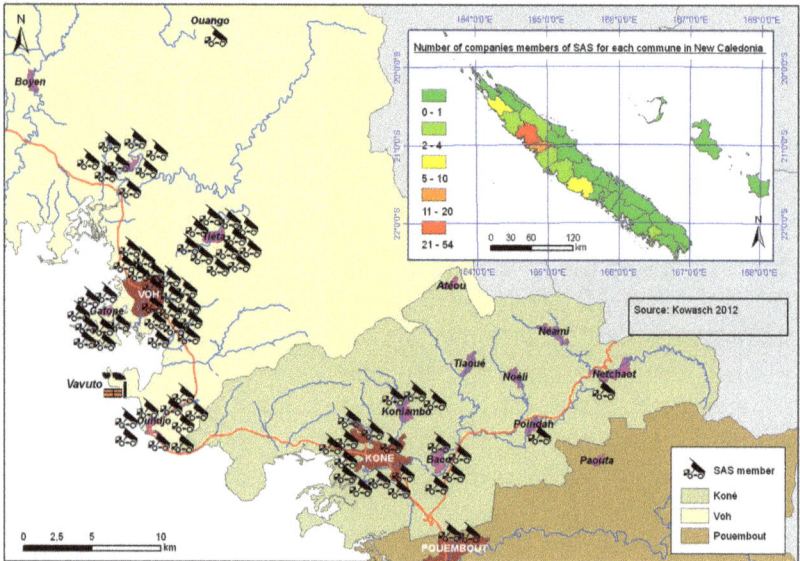

Figure 3.7 Subcontracting companies belonging to SAS Vavouto in August 2012.

Source: Author's map, based on information supplied by the Bureau des Entreprises du Nord.

3 Around 6,000 people were working on the Vavouto construction site at the peak of the construction phase.

Of the 126 enterprises belonging to SAS Vavouto at the time, nine were from Oundjo, eight from Gatope, two from Baco, and one from Netchaot. The further the village from the nickel smelter site, the fewer the number of SAS Vavouto members. People living closer to the construction site felt more involved and were better informed about job and business creation opportunities. Furthermore, KNS gave more support to business creation in the neighbourhood of the project, which meant that business creation and local employment were both higher in the neighbouring villages of Oundjo and Gatope than in Netchaot, which is more than 30 kilometres from Vavouto Peninsula.

Most Kanak subcontractors were involved in three types of business: earthworks and the transportation of material and personnel. The business types of all small enterprises located in Baco and Oundjo were identified in the revision of the regional planning and development program for the VKPP area (VKP and the commune of Poya) in 2012 (North Province 2013). The types of business operating in Baco are shown in Figure 3.8 (though not all of them were subcontractors in the Koniambo project).

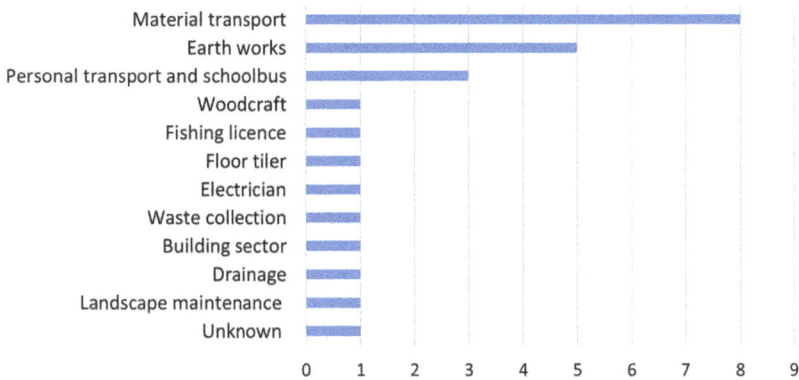

Figure 3.8 Number of enterprises per working area in Baco,
August 2012.
Sources: Kowasch (2010), North Province (2013).

Customary Land Claims, Sacred Places and Economic Projects

During French colonisation, the indigenous Kanaks were driven into reservations, just like the indigenous populations of North America. The reservations occupied about 10 per cent of Grande Terre. Land reform in New Caledonia started in 1978 with a view to buying land from settlers and returning it to the Kanak clans. Kanak clans create local associations in order to claim and receive these estates. In 2010, the results seemed to be satisfactory to the Agence de Développement Rural et d'Aménagement Foncier (ADRAF), because a balance of landownership had been achieved: customary land and European private estates both occupied 295,300 hectares (Kowasch 2012a: 206). Some 63 per cent of the surface area was classified as state land, while 1 per cent remained subject to negotiation by ADRAF. As a result, Kanak clans theoretically have the opportunity to establish economic projects on their land. ADRAF attributes estates to local associations licensed to do business. These associations are known as 'local groups with special rights', *groupements de droit particulier local* (GDPLs), rather like Aboriginal corporations in Australia. But, until now, much of the redistributed land has been left fallow because the primary motivation of the clans is usually not economic (Doumenge 2003). The allocations are generally based on clan histories and identity values. The anthropologist Michel Naepels (2006: 50) notes that the 'social identity and legitimacy of any group is an itinerary, a series of places where the group passed'. From the Kanak viewpoint, the legitimacy of land claims results from the places where the clan has lived during its history. The 'first occupants' have given a name in their language to those places. But the land can be given to another clan for farming or house building, so different legitimacies can overlap. The legitimacy of land claims becomes negotiable, characterised as a 'game of power and influence' (Le Meur 2010: 109). Moreover, the displacement of clans during colonisation now makes it difficult to say who the 'first occupant' was.

This historical context has led to land conflicts that restrain potential investment. Another handicap for investors is the fact that customary land is inalienable and non-transferable, so investors sign a 'customary deed' with the traditional landowners, which gives them some security of tenure, but which can theoretically be removed at any time. In the absence of a legal framework, the Kanak-dominated North Province supports and provides economic projects on customary land. That is why the initiative for projects on customary land generally comes from the public authorities, not from Kanak clans. Project implementation rests in the hands of state actors, so the involvement of customary representatives can be described as 'guided participation' in socioeconomic development.

As already mentioned, the provincial authorities have encouraged neighbouring Kanak communities to participate in the Koniambo project. The VKP area, including its customary land, should be turned into an urban centre, so the public authorities have made efforts to integrate Kanak clans and their customary land into the process of urban development. It is the larger tracts of customary land that are more likely to be the target for potential economic projects. In South Province, in the neighbourhood of the Goro nickel project, customary land is confined to a narrow coastal zone, and development of this land for new sociocultural facilities or industrial parks is almost precluded. In the north, the proportion of customary land is significantly higher. Around 50 per cent of the district of Koné, 28 per cent of Voh, and 12 per cent of Pouembout is classified as 'customary land' (ADRAF 2012: 5) (see Figure 3.9). Land reform has resulted in the doubling of the amount of customary land in the district of Koné over a period of 25 years, from 9,820 hectares in 1978 to 18,720 hectares in 2004 (ADRAF 2004: 6). When we compare the four villages of Gatope, Oundjo, Baco and Netchaot, the situation of Baco, whose territory is cut lengthwise by the north–south Territorial Road No. 1 and also by the road connecting the east and west coasts, is particularly interesting (North Province 2013: 43). The customary estate of Baco covers a total area of 4,264 hectares—229 hectares in the tribal reservation and 4,035 hectares transferred to customary ownership. In the village of Baco, the clans are grouped into four 'broader clans': Baco, Poavidapthia, Poaxu and Wabealo. Each of these broader clans has its own local association (GDPL) to manage its customary land. Thanks to its geographical location between the towns of Pouembout and Koné, in front of the parliament house of North Province, Baco is ideally positioned for development projects.

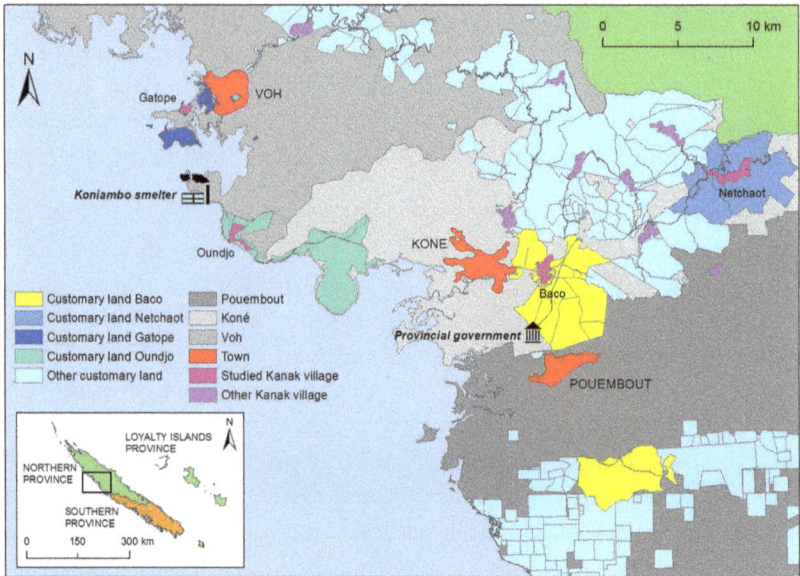

Figure 3.9 Size of customary lands of the villages of Netchaot, Baco, Oundjo and Gatope.

Source: Author's map, based on information supplied by the Direction des Infrastructures, de la Topographie et des Transports Terrestres and the Agence de Développement Rural et d'Aménagement Foncier.

One example is the development project on the customary land of the Baco GDPL (517 hectares). The provincial authorities approached the customary representatives in order to create sociocultural and economic facilities such as rental housing, a commercial centre, a bank, a cinema, an art school, a Lapita (pottery) museum, a bus terminal and an area for shops and industrial services on a total area of 48 hectares. Construction started in August 2010 and the first 15 houses were occupied in 2012. The Baco association then embarked on a second economic project: the establishment of a 17-hectare industrial park. This time, the project was supported by the council of elders representing all the clans living in Baco because the park was to be built on the village's customary reservation, next to Koné-Tiwaka, the connecting road between the east and west coast of New Caledonia. The first stone for the new industrial park, that would host only non-polluting industries and crafts, was laid in February 2009. The goal of both projects was to develop local employment and to benefit from the broader development of the VKP area. Baco has often been described as a 'flagship project' for development on customary land in New Caledonia.

Other Kanak villages do not have the same advantages. Netchaot has spacious estates, but the village is more than 30 kilometres from the construction site and around 15 kilometres from the provincial capital at Koné. Gatope's customary land is close to the nickel smelter but is limited in extent—286 hectares in the tribal reservation and only 32 hectares that has been added by redistribution—but this latter area did become the site of newly built rental housing. Oundjo has a large area of customary land that is located between the town of Koné and the nickel smelter. Originally, the mining operator SMSP wanted to build its processing plant on Pinjen Peninsula (1,610 hectares), which is owned by the clans living in Oundjo. In 1999, SMSP organised a meeting in Oundjo to officially ask for permission to do this.

The peninsula had several advantages for this purpose. It was located near the town of Koné, it had good road access, because Territorial Road No. 1 passes just behind the peninsula, and it was also accessible by boat at different points. In addition, the terrain was flat, which would make it easier to construct the smelter, and since it was located between two entrances to a single lagoon, container ships could enter the lagoon through one passage and leave through the other.

However, the clans did not agree on the proposal, which led to the so-called 'Pinjen conflict' (Horowitz 2003a, 2003b; Kowasch 2010, 2012a, 2012b). While some of the clans argued in favour of the financial benefits of land leasing, others were opposed because the peninsula contains sacred places, where ancestors reside, and was used for hunting and crab fishing. In addition, Pinjen had become a symbol of Kanak struggles for land and cultural recognition, having been transferred to the clans in the 1980s after several claims and contests, so those opposing the new development proposal did not want to 'lose' this land for a second time. In the end, the majority of the clans voted against leasing the land to SMSP, so the company bought a site further north, on Vavouto Peninsula.

Meanwhile, the conflict between the clans in Oundjo continued, but we have to note that this was not just a 'mining conflict'. Pinjen Peninsula has a long history of settlement, colonisation and expulsion (Kowasch 2012b: 205). Several families from clans now living in Oundjo, Gatope and Népou[4] passed through Pinjen at one time in their clan itinerary. The legitimacy of land claims in Pinjen is still a subject of struggle, even

---

4   A Kanak village in the commune of Poya, in the south of Pouembout.

if ADRAF attributed the peninsula to an association including all the clans from Oundjo in 1989. The suggestion that every current 'mining conflict' is not only about resources and benefits, because it has a longer history, fits well with the observations of the geographer Glenn Banks, who argued that:

> conflicts are never finally 'resolved' ... current conflicts are likely to form the background to future conflicts, even when it appears that the parties have 'settled' their differences (Banks 2008: 26).

Development projects on customary land are generally implemented outside of local people's actual living space. A majority of the Kanak people do not want the urban transformation of the VKP area to affect their own villages. The maintenance of traditions and community values is very important. As the businessman D. Xamène explained, Kanak people know that their lifestyles will change under Western influence, but this does not mean that all Kanak people will support the change (Kowasch 2010: 460). At the same time, it is important to note the lack of space available for development projects inside the Kanak villages. The tribal reservations are the bitter fruit of colonial land dispossession and displacement. From the end of the nineteenth to the mid-twentieth century, they represented a restricted space for the Kanak population. Land reform involved a strategy of demographic de-concentration, and the 'reconquest' of land in favour of Kanak clans. In the current logic of negotiated decolonisation, development projects on customary land outside the villages represent a reappropriation of this land.

Conflicts and Social Transformations

The extent of local people's approval of the Koniambo project seems to decrease with their proximity to the site of the new nickel smelter (Figure 3.10). In Oundjo, only 2 kilometres from Vavouto, 28 per cent of the interviewees were *against* or *rather against* the project. In Gatope (12 kilometres away), 24 per cent gave these responses; in Baco (around 15 kilometres away), the figure was 12 per cent; and in Netchaot (30 kilometres away) it was 2 per cent.

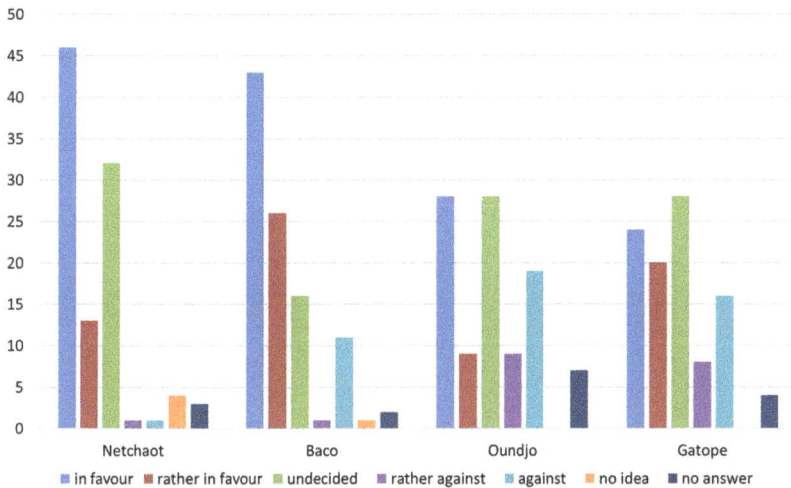

Figure 3.10 Difference in perceptions of the four neighbouring Kanak villages vis-à-vis the Koniambo project (per cent per village).
Source: Kowasch (2010: 367).

Gatope and Oundjo are the two villages that have experienced the greatest environmental impacts. These included the destruction of the mangroves where crabs were hunted. Foreign construction workers who spent their free time on the beach at Gatope or on the coast near Oundjo made disruptive noises and left waste behind. Furthermore, the Chinese and Filipinos did not speak French, so could not communicate with local people, while the latter did not seem very interested in meeting the foreign workers in any case. Residents of Gatope helped to dig the access channel from the port of Vavouto to the open sea, while fishermen from Oundjo saw the container ships arriving and leaving every day. Furthermore, the territorial road bisects the village of Oundjo, and the trucks transporting equipment to the building site damaged the road and made it dangerous for children to cross.

The participation of local Kanak people in the Koniambo project has also depended on the level of social cohesion and solidarity within the clan and the village. As we have seen, development projects on customary land represent one way to benefit from the project. The establishment of sociocultural and economic facilities on customary land is not just a matter of interest to the individual businessman, but also to the clan or even the whole village. This 'collective business participation' requires group consensus. Where there has been disagreement, the project has not moved forward. There are several examples of projects that were put on

121

hold, including a large rental housing project in the village of Koniambo and the establishment of facilities on the customary land of the Wabealo GDPL in Baco.

When they use the natural environment, or reclaimed land, to participate in the economic development of the VKP area, Kanak communities transform the value of the land itself. Sacred places lose their value when replaced by a rental housing project. Mineral extraction can destroy sacred places with strong cultural heritage value. The values of other sacred places increase, because clans want to preserve them for traditional activities or to respect the ancestors and or maintain their identity value. Traditional landowners want to benefit from mining activities when they make customary land available to foreign and local companies. To be members of the 'contact clan' for public authorities or the mining company is a kind of recognition of their status as 'landowners'. As we have already seen, customary land is characterised by the superposition of different legitimacies: the property is not unique or exclusive, but rather consists of competing titles and claims. Every development on customary land entails a remodelling of the territory, which often leads to land disputes (Kowasch 2012a). But the expansion of a group's prestige and influence can be realised only in consensual manner, not just through remodelling its territory (Herrenschmidt 2003: 322). Progress that is disconnected from the land, such as employment in the nickel sector, will not lead to a change of identity or a growing influence for the clan. It is the recognition of a clan at a certain place that can change the perception of what causes territorial remodelling. Land claims are part of this game of influence and power.

Nevertheless, the growth of mining benefits, participation and material wealth in the VKP area are not enough to establish harmony and peace. Horowitz (2004: 309–10) emphasises this aspect:

> History has shown, for New Caledonia as for elsewhere in Melanesia, that the generation of economic benefits is not a sufficient condition for harmony, either within the nation at large or within small-scale communities.

Everywhere in the area, the distribution of benefits and contracts has become a source of conflict. Since not everyone can have a contract or a job, socioeconomic disparities within the clans and communities neighbouring the project have increased as the mining revenues have filtered down. New inequalities are visible within these communities as the 'new wealthy' invest in consumer items, including cars and electronic goods. This imbalance has generated conflicts and jealousies because some Kanaks have become businessmen and have accumulated personal wealth. The problem is not always a lack of benefits to other clan members, but rather that people are not following traditional norms in terms of the distribution of such benefits. In Oundjo and Gatope, for example, social conflicts contributed to an increase in the number of purely individual businesses, and thus to 'favouring individualism' (see Chapter 2, this volume). Growing disparities within Kanak communities reflect the theory of fragmented development (Scholz 2002), in which social and regional disparities widen with the decline of social consensus and solidarity in the context of globalisation and societal fragmentation. The president of the council of elders in Oundjo spoke in July 2009 about development and business creation in his village, saying: 'Today, I say: If you want to do something, just do it! But keep it to yourself' (Kowasch 2010: 344). Greater individualism increasingly erodes traditional community life. The Pinjen conflict is still present in Oundjo, and the clans who were in favour of construction of the smelter on Pinjen Peninsula still seem to be more involved in the project. Most initiatives to create businesses are taken individually, just like the decision to apply for a job, not only in Oundjo but also in other Kanak villages. Development projects on customary land, however, are still based on collective decision making and support.

In comparison to Oundjo and Gatope, Netchaot is further from the mining site and more remote. The council of elders that meets regularly is dynamic and engaged. The customary district of Baco/Poindah, where the villages of Baco and Netchaot are located, founded a subcontracting company in order to derive a collective benefit from the Koniambo project. However, some development projects, such as the construction of rental housing for construction or mining workers, could not be implemented because Netchaot is too remote. Traditions and customary ceremonies have been preserved, with over 100 toponymic places in the village territory proving the vitality of its traditions, and the Kanak languages Paicï and Cemuhi still being spoken in the village.

The participation of Kanak people in the Koniambo project can be qualified as 'fractional' because traditional activities, including fishing, hunting and subsistence farming, are still practised. Ninety-two per cent of interviewees in the four villages who were employed had businesses were still engaged in subsistence farming of crops like yam, taro, mango, papaya and banana. Fifty-seven per cent fished in the rivers and 48 per cent went hunting. There are no rivers in the coastal villages of Gatope and Oundjo, where most of the residents instead fish at sea. According to large surveys conducted by the Institut Agronomique Neo-Calédonien (Guyard et al. 2013), 'multi-activity' is a characteristic of all residents in the studied villages, but employees and businessmen naturally have less time to work on farming, fishing and hunting. Sometimes they even have to buy yams for customary ceremonies. The disjunction between traditional activities and salaried employment reflects the transformation of Kanak communities in the neighbourhood of the mining project and can be a source of conflict, because the planting of yams has an identity value.

A similar situation in Papua New Guinea was described by Benedict Imbun (2011: 51–5), who wrote of the ambivalent attitude of communities near the Porgera gold mine confronted with salaried employment, business creation and wealth accumulation. Their ambivalence resulted from the juxtaposition of the status of traditional 'landowner' with the social class status generated by the mining industry. In this instance, they developed an anti-risk strategy by combining business creation with employment.

The disjunction between traditional activities and employment concerns not only farming, fishing, hunting and customary ceremonies; it also affects daily activities such as the care of children and older people. There are not enough nurseries and retirement homes in the VKP area, and this has constituted a new challenge for the local communities.

Conclusion

Historically, indigenous Kanak peoples were largely excluded from employment in the mining industry in New Caledonia. Mining employment has been the principal aspiration of Kanak people from the villages neighbouring the Koniambo project. The surveys showed that the majority of the Kanak communities support a project they consider to be 'their project', that will provide local employment but also bring broader development. For the political authorities of North Province, the mining industry is an instrument for economic independence and political emancipation. The public authorities are struggling to push the urbanisation of the VKP area where the four studied villages are located. Furthermore, the provincial authorities and SMSP have begun applying a 51/49 per cent shareholding split with multinational companies in other countries, an arrangement that is unique to New Caledonia and uncommon for indigenous mining companies worldwide. Their aim is to be 'masters of the exploitation of mineral resources' so as to take responsibility for wider economic development.

Meanwhile, we have to distinguish between the perceptions of mining among Kanak village people and Kanak political leaders, even if the latter also live in the villages. Sometimes, members of the provincial assembly are at the same time customary chiefs or members of the local council of elders, but in each role they have to defend other interests. Nevertheless, the public authorities have encourage the tribal populations to participate in the development process, whether through employment or through the establishment of rental housing or other sociocultural and economic facilities on customary. The construction of such facilities brings more financial benefits for the next generation because the traditional landowners must first amortise their bank loans. Collective participation in such ventures also depends on consensus, social cohesion and solidarity within the clan and the village.

But development and new socioeconomic disparities coexist. Not everybody can work in the mining sector, and not every businessman can get a contract. The struggle for contracts and employment in the mining sector is very strong, as local subcontractors have to compete with international companies and more experienced businessmen from Nouméa. Individual business creation has grown in the neighbourhood of the Koniambo project, while collective participation is often hindered by internal discord and problems with the distribution of benefits. When individual businessmen are obliged to manage 'collective enterprises', this can result in favouritism because they occupy several positions. Employment and business creation require specific professional skills and qualifications. The improvement of educational opportunities remains a challenge for North Province. Kanak people who moved to Nouméa several years ago have returned to the villages in order to benefit from the nickel boom in the north. Most of the conflicts in the villages concern the distribution of mining benefits, land claims and legitimacy. The unequal distribution of benefits can create jealousy within the communities and lead to intra- and inter-clan conflicts. Customary 'ownership' of estates leads to a flow of financial benefits, so that land claims increase when public authorities or private businessmen are planning economic projects on these estates. However, this chapter has also shown how the conflict over Pinjen Peninsula has its own history deriving from colonisation, displacement and customary struggles between families. The request by SMSP to build the nickel smelter on the peninsula simply 'awoke' an existing conflict whose resolution will take a long time.

Socioeconomic disparities within the indigenous community will increase in northern New Caledonia, and social differentiation and marginalisation will create new problems. But the development of the mining industry also enables new opportunities for marginalised peoples in remote villages. Employment, business creation and the establishment of rental housing and facilities generate revenues and wealth. At the same time, Kanak people continue to practise traditional activities such as subsistence farming, hunting and fishing, even when they have a regular income.

The environmental value of the estates where economic projects are implemented is changing. In terms of the environmental value model (Figure 3.3), the territorial perception of the Kanak population is moving away from the 'traditional' Melanesian model in which the cultural heritage value is considerably higher than the practical value. Greater involvement in the formal economy explains the loss of value of certain sacred places and their cultural heritage value. Young people who are working no longer have the time to care about such things, and are less interested in 'traditional' knowledge. The cultural heritage value of land is impaired by the loss of this historical or mythical knowledge (Kowasch 2012a: 203), as local perceptions are transformed into a transitional or ecological model.

The Caledonian journalist and filmmaker Anne Pitoiset describes the role of the Koniambo project and the resulting societal transformations in an interview for the journal *Hebdo*:

> For the pro-independence movement, the step towards independence is engaged for a long time, the plant is a financial instrument and also a symbol of their ability to fit into the economy. For the non-independence parties, the project represents a key factor for the maintenance of civil peace … But as Ian Pearce, the president of Xstrata says, the problems will really start with the beginning of production [of nickel metal]. The neighbours will see the factory fumes, they will experience increased traffic, their lives will be transformed in a way they struggle to imagine today. (Pitoiset 2013)

In early 2013, at the moment of the first pouring of nickel metal and one year before reaching full production capacity, the life of the communities was already changing. A new shopping centre had recently opened, the first traffic lights in Koné would soon be in service, more clan members had salaried positions, rental yields from houses and apartments were increasing, four-wheel drive vehicles could be seen everywhere in the Kanak villages, and customary weddings were being held on weekends or during holidays in order to avoid working hours. Traditional knowledge and language skills should be preserved, but work and a regular salary seem to have a growing importance for most Kanak people nowadays.

References

ADRAF (Agence de Développement Rural et d'Aménagement Foncier), 2004. 'Etat des Lieux des Terres de GDPL.' Koné: ADRAF.

——, 2012. 'Etude Foncière Voh-Koné-Pouembout-Poya. Partie 1: La Situation Foncière.' Nouméa: ADRAF.

Amin, A., 2002. 'Ethnicity and the Multicultural City: Living with Diversity.' *Environment and Planning A* 34: 959–980. doi.org/10.1068/a3537

Auty, R.M., 1993. *Sustaining Development in Mineral Economies: The Resource Curse Thesis.* London: Routledge.

Banks, G., 1996. 'Compensation for Mining: Benefit or Time-Bomb? The Porgera Gold Mine.' In R. Howitt with J. Connell and P. Hirsch (eds), *Resources, Nations and Indigenous Peoples: Case Studies from Australasia, Melanesia and Southeast Asia.* Melbourne: Oxford University Press.

——, 2008. 'Understanding "Resource" Conflicts in Papua New Guinea.' *Asia Pacific Viewpoint* 49: 23–34. doi.org/10.1111/j.1467-8373.2008.00358.x

Bebbington, A., L. Hinojosa, D. Humphreys-Bebbington, M.L. Burneo and X. Warnaars, 2009. 'Contention and Ambiguity: Mining and the Possibilities of Development.' *Development and Change* 39: 887–914. doi.org/10.1111/j.1467-7660.2008.00517.x

Blaikie, P., 2012. 'Should Some Political Ecology Be Useful? The Inaugural Lecture for the Cultural and Political Ecology Specialty Group, Annual Meeting of the Association of American Geographers, April 2010.' *Geoforum* 43: 231–239. doi.org/10.1016/j.geoforum.2011.08.010

Constanza, R., R. D'Arge, R. De Groot, S. Farber, M. Grasso, B. Hanonnon, K. Limburg, S. Naeem, R. O'Neill, J. Paruelo, R.G. Raskin and P. Sutton, 1997. 'The Value of the World's Ecosystem Services and Natural Capital.' *Nature* 387: 253–260. doi.org/10.1038/387253a0

David, G., J.-B. Herrenschmidt and E. Mirault, 2007. *Valeur Sociale et Économique des Récifs Coralliens du Pacifique Insulaire.* Nouméa: South Pacific Commission, Coral Reef Initiatives for the Pacific.

Demmer, C., 2007. 'Autochtonie, Nickel et Environnement en Nouvelle-Calédonie: Une Nouvelle Stratégie Kanake.' *Vacarme* 39: 43–48. doi.org/10.3917/vaca.039.0043

Doumenge, F., 2003. 'La France Confrontée au Trou Noir du Pacifique, la Face Inconnue de la Question Calédonienne.' *Conflits Actuels: Revue d'Étude Politique* 10: 101–112.

Emergences, 2009a. 'Baco—Approche des Dynamiques sur l'Espace Coutumier.' Koné: Document du dossier (Diagnostic—Actions des Terres Coutumières de Koné-Pouembout)/Volet Tribu/GDPL.

———, 2009b. 'Netchaot—Approche des Dynamiques sur l'Espace Coutumier.' Koné: Document du dossier (Diagnostic—Actions des Terres Coutumières de Koné-Pouembout)/Volet Tribu/GDPL.

Freudenburg, W. and L. Wilson, 2002. 'Mining the Data: Analysing the Economic Implications of Mining for Nonmetropolitan Regions.' *Sociological Inquiry* 73: 549–575. doi.org/10.1111/1475-682X.00034

GoNC (Government of New Caledonia), 2016. 'Tableau de l'Économie Calédonienne (TEC).' Nouméa: Institut de la Statistique et des Études Économiques.

Graff, S., 2012. 'Quand Combat et Revendications Kanak ou Politique de l'Etat Français Manient Indépendance, Décolonisation, Autodétermination et Autochtonie en Nouvelle-Calédonie.' *Journal de la Société des Océanistes* 134: 61–83. doi.org/10.4000/jso.6647

Grochain S. and D. Poithily, 2011. 'Sous-traitance Minière en Nouvelle-Calédonie, le Projet Koniambo.' Nouméa: CNRT Nickel (Document de Travail 4, Gouvernance Minière).

Guyard S., L. Apithy, S. Bouard, J.-M. Sourisseau, M. Passouant, P.-M. Bosc and J.-F. Bélières, 2013. *L'Agriculture des Tribus en Nouvelle-Calédonie : Résultats d'une Enquête de 2010 sur la Place et les Fonctions de l'Agriculture, l'Élevage, la Pêche et la Chasse pour les Groupes Domestiques Résident en Tribu*. Koné: Institut Agronomique Néo-Calédonien/Centre de Coopération Internationale pour la Recherche Agronomique pour le Développement.

Herrenschmidt, J.-B., 2003. 'Territorialité et Identités en Mélanésie.' In D. Guillaud, C. Huetz de Lemps and O. Sevin (eds), *Îles Rêvées – Territoires et Identités en Crise dans le Pacifique Insulaire*. Paris: Presses de l'Université de Paris-Sorbonne/Institut de Recherche pour le Développement.

Horowitz, L.S., 2003a. Stranger in One's Own Home: A Micropolitical Analysis of the Engagements of Kanak Villagers with a Multinational Mining Project in New Caledonia. Canberra: The Australian National University (PhD thesis).

——, 2003b. 'La Micropolitique de la Mine en Nouvelle-Calédonie: Analyse des Conflits Autour d'un Projet Minier au Sein d'une Communauté Kanak.' *Journal de la Société des Océanistes* 117: 255–271.

——, 2004. 'Toward a Viable Independence? The Koniambo Project and the Political Economy of Mining in New Caledonia.' *The Contemporary Pacific* 16: 287–319. doi.org/10.1353/cp.2004.0047

Imbun, B., 2011. *Anthropology of Mining in Papua New Guinea Greenfields*. New York: Nova Publishers.

Kowasch, M., 2010. Les Populations Kanak Face au Développement de l'Industrie du Nickel en Nouvelle-Calédonie. Université Montpellier III/Université de Heidelberg (PhD thesis).

——, 2012a. 'Le Développement de l'Industrie du Nickel et la Transformation de la Valeur Environnementale en Nouvelle-Calédonie.' *Journal of Political Ecology* 19: 202–220.

——, 2012b. 'Les Lieux Toponymiques en Nouvelle-Calédonie – un Champ de Recherche Ouvert.' *Enquêtes Rurales* 14: 187–208.

——, 2012c. 'La Zone Voh-Koné-Pouembout.' In J. Banvallot and J.-C. Gay (eds), *Atlas de la Nouvelle-Calédonie*. Montpellier: IRD Editions.

——, 2014. 'Fieldwork in a Context of Decolonization: The Example of a French Overseas Territory (New Caledonia).' Erdkunde 68: 251–264. doi.org/10.3112/erdkunde.2014.04.02

Langton, M. and O. Mazel, 2012. 'The Resource Curse Compared: Australian Aboriginal Participation in the Resource Extraction Industry and Distribution of Impacts.' In M. Langton and J. Longbottom (eds), *Community Futures, Legal Architecture: Foundations for Indigenous Peoples in the Global Mining Boom*. London: Routledge.

Leblic, I., 1993. *Les Kanak Face au Développement: La Voie Étroite*. Nouméa: Agence de Développement de la Culture Kanak/Presses Universitaires de Grenoble.

Le Meur, P.-Y., 2010. 'Réflexions sur un Oxymore: Le Débat du "Cadastre Coutumier" en Nouvelle-Calédonie.' In E. Faugère and I. Merle (eds), *La Nouvelle-Calédonie, Vers un Destin Commun?* Paris: Editions Karthala.

Le Meur, P.-Y., S. Grochain, M. Kowasch and D. Poithily, 2012. 'La Sous-Traitance Comme Interface: Rente Minière, Contrôle des Ressources et Arènes Locales en Nouvelle-Calédonie.' Nouméa: CNRT Nickel, Programme Gouvernance Minière (Document de Travail 10).

Le Meur, P.-Y. and T. Mennesson, 2011. 'Le Cadre Politico-Juridique Minier en Nouvelle-Calédonie: Mise en Perspective Historique.' Nouméa: CNRT Nickel, Programme Gouvernance Minière (Document de Travail 3).

Naepels, M., 1998. *Histoires de Terres Kanak*. Paris: Editions Belin.

——, 2006. 'Réforme Foncière et Propriété dans la Région de Houaïlou (Nouvelle-Calédonie).' *Etudes Rurales* 177: 43–54.

Néaoutyine, P., 2006. *L'Indépendance au Present: Identité Kanak et Destin Commun*. Paris: Syllepse.

North Province, 2008. 'Etre Jeune en Province Nord: Vivre la Jeunesse.' Koné: Observatoire de la Santé et des Actions Sociales de la Province Nord.

———, 2013. 'Actualisation du Schéma Directeur d'Aménagement et d'Urbanisme (SDAU) des Zones de Voh, Koné, Pouembout et Poya.' Unpublished report.

Pitoiset, A., 2013. 'Interview in "Hebdo".' *Nouvelle-Calédonie 1ère*, 20–26 April: 11–12.

Pitoiset, A. and C. Wéry, 2008. *Mystère Dang*. Nouméa: Le Rayon Vert.

Poithily D., 2010. Les Stratégies et Motivations des Entrepreneurs VKP. Nouméa: Université de la Nouvelle-Calédonie (Masters thesis).

Robbins, P., 2006. 'Research Is Theft: Rigorous Enquiry in a Post-Colonial World.' In G. Valetine and S. Aitken (eds), *Philosophies, People, Places and Practices*. Thousand Oaks (CA): Sage.

Sachs, J. and A. Warner, 1995. *'Natural Resource Abundance and Economic Growth.'* Cambridge (MA): National Bureau of Economic Research (Working Paper 5398).

Scholz, F., 2002. 'Die Theorie der Fragmentierenden Entwicklung.' *Geographische Rundschau* 10: 6–11.

Taylor, J. and B. Scambary, 2005. *Indigenous People and the Pilbara Mining Boom: A Baseline for Regional Participation*. Canberra: The Australian National University, Centre for Aboriginal Economic Policy Research (Research Monograph 25).

Tolia-Kelly D.P., 2006. 'Mobility/Stability: British Asian Cultures of "Landscape and Englishness".' *Environment and Planning A* 38: 341–358. doi.org/10.1068/a37276

Winslow, D., 1995. 'Indépendance, Savoir Aborigène et Environnement en Nouvelle-Calédonie.' *Journal of Political Ecology* 2: 1–19.

4. The Boakaine Mine in New Caledonia: A Local Development Issue?

CHRISTINE DEMMER

Introduction

Following several years of severe tension between Kanak separatists on one hand, and the French state and descendants of the colonists on the other, the Matignon-Oudinot Agreements, which were signed in June 1988, brought peace to New Caledonia. Two of the three provinces established at that time—North Province and Islands Province—have since been administered by the separatists. They set themselves the objective of developing these areas, which are populated mainly by Kanaks and had previously been considerably less affluent than South Province, which includes the capital Nouméa and is home to the bulk of the white population. However, the acceptance of these agreements by the Kanak nationalists from the Front de Libération Nationale Kanak Socialiste (FLNKS), and the associated concept of 'economic rebalancing', only made sense to them in the context of what remains their ultimate objective—the achievement of independence.

Naturally, nickel was the abundant resource that the elected representatives of North Province immediately tried to develop for this purpose. In 1990, the FLNKS established an entity called the Société de Financement et d'Investissement de la Province Nord (SOFINOR), an ad hoc finance and investment vehicle that immediately purchased the mining company Société Minière du Sud Pacifique (SMSP), which was formerly a subcontractor of New Caledonia's oldest nickel mining company, Société Le Nickel (SLN). After 20 years of management by the separatists, SMSP had become a holding company that was no longer directly involved in the mining of nickel ore, but instead financed and developed a number of industrial and metal-processing projects. Since 1995, it has been the world's primary exporter of nickel and holds 51 per cent of the shares in the companies it manages in partnership with the Swiss company Glencore (formerly Xstrata), the South Korean company POSCO, and the Chinese company Jinchuan (see Chapters 2 and 3, this volume). Through these arrangements, it provides revenues to North Province via SOFINOR, which holds 87 per cent of the company's shares, and contributes to the establishment of other companies and also to the development of the local tourist industry and economic infrastructure.

The focus of this chapter is a sequence of political events surrounding the forced closure of one nickel mining operation in the Kanak municipality of Canala in 2002. Canala is located on the east coast of North Province, adjacent to the border with South Province (see Figure 2.1, Chapter 2). In 2009, its resident population was more than 3,300, and its territory hosted four major colonial chieftaincies (*chefferies*): Bwaaxéa, Kèrèduru, Nôôme and Penyîmê Nèkètè (or Nakéty). This municipality found itself in a strong position in the new economic and political system because it contained three major nickel deposits—Nakéty, Bogota and Boakaine— that had been mined since the late nineteenth century, and accounted for 25 per cent of New Caledonian nickel reserves in 2000 (Boisard 2002).

The exploitation of these deposits was in some ways symbolic of the new direction adopted by the separatist struggle. Indeed, Canala was selected as the location of the discussions held in August 1989 that led to the purchase of SMSP by North Province. These discussions were held in the wake of the Ouvéa Cave hostage crisis of May 1988, when 19 Kanak hostage-takers died in an operation mounted to express their opposition to proposals made by Bernard Pons, then Minister for Overseas Departments and Territories, during the course of a French presidential election. The new minister, Louis Le Pensec, accompanied Prime Minister Michel Rocard in this new round of discussions.

The French politicians clearly chose Canala as the site of these discussions because it was home to some of the key actors in *les événements* of 1984. These 'events' had included the active boycott of the legislative elections held in that year, as well as blockades, land occupations and expulsions of colonial settlers. They visited the municipality, which was home to the separatist leader Eloi Machoro, the former general secretary of the separatist party Union Calédonienne. They also visited the traditional lands of Léopold Jorédié, a member of the same party, who became one of the leaders of the FLNKS after Machoro was assassinated in 1985. Jorédié had recently been appointed as mayor of the municipality of Canala, and was elected as the first president of the newly established North Province, having previously been director of the North 'Region' that preceded it between 1985 and 1988.[1]

The message conveyed to the representatives of the French state by the separatists in August 1989 was quite clear: it was essential that they be given the resources necessary to develop their provinces and municipalities. The agricultural micro-projects that had been developed in preceding periods would no longer satisfy them. The acquisition of a mining company and mining concessions was far more in line with their economic and political demands.

Having won that argument, the new directors of SMSP purchased 22 concessions over the Boakaine deposit from SLN in 1991.[2] Mining activity started in the following year, with local workers being recruited to participate in what was seen as part of the economic struggle for independence. However, ten years later, in 2002, the operation came to an abrupt end, despite the presence of extensive nickel reserves and its enormous symbolic resonance. The workforce was temporarily laid off after the suspension of the contract to supply a Japanese smelting company (PAMCO),[3] and this event triggered a general strike and the ransacking of the site.

---

1   During that period, New Caledonia was divided into four 'regions'—North, Centre, South and Loyalty Islands.

2   In the local language, Xârâcùù, the name Bwakaiinè (Boakaine) is derived from the words *bwa* ('head'), *kaii* ('light up'), and *nè* ('fire'). This name alludes to the torch-lit processions that were held on the range and previously marked the opening of the yam harvest festival. Boakaine was the site of one of the very first concessions in the Grande Terre to be mined by a colonial settler (in 1874), and he was also one of the founders of SLN (in 1880). Like Nakéty, it has been more or less intensively mined at different periods since then.

3   This was before SMSP established a new partnership with the Korean company POSCO.

Difficult working conditions, inadequate salaries and safety problems were cited by the local branch of a trade union as justifications for these actions. However, another kind of argument opened with a reading of the conflict in terms of the relationship between social classes. Furthermore, customary leaders (*coutumiers*) who embodied the Kanak political order also reminded people that SMSP had failed to keep promises made when the mine was opened. These involved the progressive handover of mining concessions or—at the very least—of operational control to the local population.

Despite the relaunch of negotiations between SMSP and local actors in recent years, the situation was not resolved in the decade following this strike. Attempts were made to establish an operating company whose shares would be held by the local clans, but the municipal authorities insisted that they should be the primary local partners of SMSP because of their responsibility for local economic development. So was this really a conflict between two legitimate political authorities—the elected municipal (and provincial) separatist leaders on one side, opposed by customary leaders from the four big chieftaincies on the other side?

In my attempt to answer this question, I adopt an approach from political anthropology that sets out to describe how colonised and formerly stateless, segmentary societies currently deal with the problem of segmentation. 'Segmentary societies' are here understood to be those in which the 'political system' consisted of groups organised by models of kinship (clans, lineages, etc.), whether or not they were headed by 'chiefs'. In New Caledonia, chieftaincies have been associated with specific territories (reserves) and organised into 'tribes' (*tribus*) and customary 'districts' since the early colonial period, with the former headed by 'small chiefs' and the latter by 'big chiefs'.[4] The representative bodies are the 'council of elders' (now known as the 'council of the clan chiefs') at the tribal level, and the (customary) 'district council' at the level of the municipality. The district council comprises the big and small chiefs within the district, plus any presidents of the tribal councils who are not chiefs in their own right, and it elects its own president. The occupants of these various positions are all

---

4   Reference to the big chieftaincies as separate 'districts' is avoided in the rest of this paper, since the term is also applied at a higher level, where the boundaries of the (customary) district are roughly aligned with those of the municipality. There is one small chieftaincy (Koh) that belongs to the neighbouring commune of Kouaoua but is part of the customary district of Canala.

known as 'customary leaders' (*coutumiers*), but that term is also applied to the descendants of the first occupants of a place who may have ceded their authority to a 'stranger chief'.[5]

This is the context in which I explore the connections between two forms of legitimate authority (see Horowitz 2009). Does a conflict between them mean that 'imported' political legitimacy is never fully accepted, or that the prospect of economic benefits at a local level enables the reactivation of endogenous political legitimacy? My approach to this particular problem—which is based more broadly on an approach to social change in the colonial and post-colonial context—is in line with that of the so-called 'Manchester School', which was originally developed during the late colonial period in Southern Africa (Gluckman 1940a, 1940b, 1942; see also Chapters 1 and 5, this volume).

Beyond the event represented by the closure of the Boakaine mine, my analysis aims to contribute to an explanation of the current status of sociopolitical relationships in New Caledonia at different levels, especially at the municipal level. Given that chieftaincies were colonial spatial and political constructs intended to reflect the traditional form of Kanak political organisation, linked to the legal framework under which native reserves were established through the second half of the nineteenth century,[6] the municipal space (and not the municipality as a formal body) should be considered as a hybrid unit of analysis compared to Western political models. I also propose to consider the place of the municipality within its provincial context and, beyond this, in the context of an 'overseas state' that is still formally part of the French state but also in the throes of an already advanced movement towards independence. However, unlike Gluckman, who approached the political situation of the Zulu people in Zululand through a detailed account of the changing relationships between 'colour groups' in the South African state, the situation I describe involves relationships between groups of Kanaks. This certainly applies at the level of the relationship between the municipal authority and the chieftaincies of Canala, from which the last colonial settlers were expelled during *les événements* of 1984, but also applies with respect to the relationship between the mining company and the 'local populations' at the provincial

---

5   In Canala, people in this latter category are known as 'great subjects' (*kwara aéé*).
6   Reserved lands were decreed as inalienable (see Merle 2000). Since 1998, they have become 'customary lands' as part of the process of negotiated decolonisation, and are still not subject to the common law.

level. Accordingly, I not only take account of the effects of historical legacies and/or the reactions to foreign domination in my analysis of this intra-Kanak political game; I also consider, in particular, the question of the historical depth of Kanak participation in the territory's political life since the 1950s, which makes them 'political actors' in a double sense (see Soriano 2013).

The current political context includes Kanak control of the provincial authorities (in two provinces), which has been seen as a springboard for the recovery of a sovereignty that has yet to be attained. The Boakaine conflict can only be fully understood against the backdrop of this very specific transitional context. From the perspective of the Kanaks, this is primarily what positions the operation of this mine—and the whole of SMSP— at the interface between an economic and political issue. And this is what also results when competition for responsibility for local development based on the mining industry can set different political levels or figures against each other, even though the Kanak political parties regard this development as a contemporary objective in their struggle for independence.

A Conflict Between Separatist Parties

In reality, an understanding of this conflict during the period before the mine was closed must go beyond the question of how Kanak actors have mobilised and prioritised the different sociopolitical repertoires available in a competition for control of the nickel economy. It depends more broadly on an understanding of Kanak involvement in the process of decolonisation, which in turn involves consideration of the relationship between traditional political legitimacy and partisan political struggles. When the mine was reopened in 1991, the future protagonists in the strike of 2002 presented themselves as separatists, not as customary leaders, and were proud to show that their party, the Union Calédonienne (UC), was strongly positioned as an actor involved in the new strategy that was adopted in the pursuit of political power. Gaetan Dowadé, who was Jorédié's 'right-hand man' and deputy mayor in Canala at that time, explained this to me as follows:

> At that time [in 1991], we took SMSP. Léo [Jorédié] said to me: 'Set up a company in Canala, get the people together, we will see how we shall set about reopening Boakaine. We will take SMSP. We will see which mine will be opened after that.' That's when I knew that Jean Marie [Tjibaou]

presented proposals to [Jacques] Lafleur … for the repurchase of SMSP. So each municipality had to set up a municipal mining company. We would set the example. The miners set up there before, and they didn't ask [permission] for anything. After the 'events', they had to go through us. (personal communication, Gaetan Dowadé, 26 July 2011)

Like other statements of the same nature, this quote from an interview leaves no doubt as to the capacity of the municipality's militants to fulfil the objectives of the FLNKS *by themselves*. They shared, and indeed were among the initiators of, the vision of development through the mining sector that was known as 'the vision for the country', and was to be realised at the level of North Province in anticipation of the recovery of national sovereignty. This was not a vision that set the indigenous and non-indigenous communities in opposition to each other, like the one that proposed to capture the mining revenues at the level of what were referred to in the 2000s as the 'mining chieftaincies' (Demmer 2007). Instead, to identify the logic of this conflict, it is necessary to look beyond a claim based purely on political identity, as represented in the argument that 'the chieftaincies should be able to assume responsibility for development'. One must also consider the reference to customary political institutions as one that was made to serve ends other than the simple defence of a certain kind of sub-national political legitimacy.[7] In the present case, the explanation may perhaps be found in the particular—and sometimes tense—relationship between the municipality of Canala and North Province after the mine had been reopened in 1991.

As already mentioned, Léopold Jorédié was not only a native of Canala, but in 1989 he was both the mayor of this municipality and the first president of North Province. However, as an elected representative of the UC, he lost both of the mandates he had acquired in 1989—first at the municipal level in 1995, and then at the provincial level in 1999. In 1998, shortly before the signing of the Nouméa Agreement, Jorédié wanted to negotiate directly with the loyalists without being bound by the 'mining prerequisite' that stipulated construction of a nickel processing plant in North Province (see Chapter 5, this volume). Supported by François Burck, another elected representative of the UC, and Raphaël Mapou, representing the Parti de Libération Kanak (PALIKA),

7   Incidentally, in the context of Kanak emancipation and involvement in the nickel economy, my remarks highlight the extent to which an interpretation of actions taken to restrict control of natural resources in terms of the affirmation of indigenous people's rights cannot fully reflect the dynamics of political relationships between mining companies and local populations in New Caledonia.

he therefore established a new coordinating body known as the Comité de Coordination des Indépendantistes, which subsequently became the Fédération des Comités de Coordination Indépendantistes (FCCI).

These developments in the political career of Jorédié—and above all his electoral disappointments—prompted several of my informants to claim that he was trying to get some of his supporters to launch a strike at Boakaine with a view to creating difficulties for PALIKA, which was then the dominant party in the province, which was in turn the main shareholder in SMSP via SOFINOR. Given that he could not aspire to take control of the company again, this was a way of dealing a severe blow to a party that had achieved a leading role in the attainment of the new political and economic objectives established by the FLNKS, for it was the UC that had been able to take credit for this achievement in the late 1980s. In the words of Jorédié's chief lieutenant, recalling the reopening of the mine in 1991:

> We became indispensable on the level of the mine. I speak of us, the UC, the strongest people in Canala. It will be said … you have to go through the FLNKS. But we were UC FLNKS.

Removed from power in both politics and business, Jorédié would then attempt to damage SMSP and the party that controlled the province, both of which were heavily reliant on the Boakaine mine to supply the nickel required by their new Korean business partners.

Viewed in this way, the conflict is very specific—rooted as it is in the micro-political history of North Province—but also highly personalised. Accordingly, it will come as no surprise to learn that no other mine owned by SMSP experienced a conflict as serious as this one. Even if the supporters of parties other than PALIKA worked in the mines or were leaders of the municipalities or chieftaincies that hosted the mine sites, there was no situation like that which emerged in Canala, where the mayor who had represented the municipality had also been president of the province during negotiations over the role of mineral resource development in the struggle for independence.

However, it is necessary to go further in the analysis of this conflict, for it did not merely involve the settlement of a personal account by a politician whose political career was in decline. Examination of the facts leaves no room for doubt that the leaders of the strike that led to the closure of the mine were in the FCCI camp. Apart from the blow dealt to their political opponents, the leaders of the strike clearly intended to position themselves as new partners in dialogue with SMSP, either in the short or medium term. It is in part the same people, the former leaders of the FCCI (which no longer exists),[8] who proposed in 2011 to establish a clan-based company in Canala to manage the mine. One informant summarised the point as follows:

> There was a political concern because Léo [Jorédié] belonged to the UC and the UC started the struggle, and he wanted to reassume control of Boakaine as a source of financial support for his party.

But beyond the question of party finance suggested by this informant, the conflict offers manifest evidence of the centrality of the issue of control over the mines for all the separatist parties, first from the time of the Matignon-Oudinot Agreements in 1988, and even more so from the time of the Bercy Agreement that contained the 'mining prerequisite' to the Nouméa Agreement of 1998.

In this regard, the logic of this strike—if not all of its immediate pretexts— is liable to recur in the party political relationships found in all the mining municipalities. What is at stake today for each of the separatist parties is the need to demonstrate their capacity to respond to the current demands of the struggle: being the actors who know how to mine nickel means being those who are closest to the logic of decolonisation as this has been defined for almost 25 years. This contrasts with the earlier period of the 'regions', when micro-development projects to 'build Kanaky' flourished, and many Kanak political actors were mainly concerned to promote the interests of their own party at a purely local level (Demmer 2016). The exercise of party political control over nickel mining has proven to be a means of winning points in the struggle for power at all levels of political organisation.

In the majority of cases, it would also be a mistake to simply interpret the attempts of Kanak political parties to access mineral revenues in terms of a quest for limited economic benefits. Although it is not possible to rule out such motives completely, it remains crucial to keep in mind the centrality

---

8   Support for the FCCI waned after 2004, although Jorédié (who died in September 2013) organised a list of candidates for the North Province elections in 2009. Today, his right-hand man in Canala, Gaetan Dowadé, is an active member of Parti Travailliste.

of mining issues for the separatists, in the context of decolonisation, in order to understand the ideological logic that has governed these 'rent-seeking' practices.

There is another kind of economic logic that could help to explain the appearance of ultra-local development claims originating from political entities as varied as clans, chieftaincies, municipalities and newly defined 'customary areas'.[9] There are no tax revenues from the operation of nickel mines that are allocated to local public expenditures. The creation (in 2009) of a 'nickel fund' that is supposed to be financed from corporate income taxes and to pay for various forms of social insurance and environmental remediation in the mining sector is the only existing tax mechanism with direct local impact.[10] These claims have their roots in the deeper problem of redistributing the wealth arising from the exploitation of the country's primary natural resource.

This leads me to highlight an additional aspect of the political party contest over Boakaine. The elected representatives of Canala see themselves as being positioned at the end or bottom of the SOFINOR aid chain. As one of them put it:

> SOFINOR! It took money from Boakaine and invested it in other structures. You should see the organisation chart for SOFINOR. There are plenty of subsidiaries but none in Canala, not for development or anything else!

Although I was unable to verify this particular allegation, the issue is one of concern to both the elected representatives and ordinary people of Canala. Moreover, the concern grew with the change in the province's ruling majority, since Canala is one of the historical fiefdoms of the first political party that attracted popular Kanak support. The chieftaincies of this district were evangelised by Père Luneau between 1920 and 1940, and it was he who established the Union des Indigènes Calédoniens Amis de la Liberté dans l'Ordre, which later became the UC. Many of the municipality's inhabitants have remained loyal to this party, and some

---

9   Each of the eight 'customary areas' includes a number of customary districts and chieftaincies. They were established under the Matignon-Oudinot Agreements, and are defined by a mixture of linguistic and political criteria.

10   New Caledonia has struggled to abandon a fiscal regime that grants special favours to the mining industry. Registered mining companies are entitled to almost complete exemption from taxes for 15 years after the point at which their actual production reaches a threshold of 80 per cent of a nominal target, with a 50 per cent exemption for the following five years.

believe that this is the reason why it has suffered discrimination at the hands of the PALIKA representatives at the provincial level, who think of Canala as the municipality of Léopold Jorédié. Some informants thought that the municipality had problems in securing provincial support for its development priorities because it was particularly 'well served' in the earlier period when Jorédié was the president, so this either represented the 'fair' backswing of a partisan pendulum or a simple rebalancing between the municipalities in North Province over time.

My analysis of the conflict could conclude here if the question did not remain as to why the claim for control of Boakaine was initiated in 2002 by FCCI members and supporters, but increasingly passed to the municipality's chieftaincies. It is not only the big chieftaincies but the small chieftaincies of Mèèwèè (Méhoué), Bwakaiinè (Boakaine), Nôôwé (Nonhoué) and Mèrènèmè (Mérénémé), which have the strongest territorial links to the nickel deposits, that claim the mine for themselves. This insistence on the chieftaincies as legitimate beneficiaries is part of the contemporary debate about reopening the mine that has placed the municipality and the customary authorities in opposing positions.

How the Militants Became Customary Leaders

Following the purchase of the Boakaine mining concession in the wake of 'the events', Raphaël Pidjot, then a director of SMSP, and André Dang Van Nha (commonly known as André Dang), the company's chief executive officer, launched a series of local discussions about the participation of the people of Canala in its operation. It was explained to me that: 'The customary people of the municipality were always involved in the discussions ... It was said that the mine was a political and customary asset'. This discourse was easily sustained because the people who had been at the barricades in 1984 and 1988 included elected representatives of the municipality who were also customary leaders. However, what was mainly sustained in this process was the emergent belief of the separatists in a 'Kanak way' of organising production and establishing companies.

It was on this basis that the Compagnie Minière Kanak de Canala (COMIKA), the local partner of SMSP, was established in 1992 with capital amounting to one million francs. All of the *tribus* (small chieftaincies) within the municipality held shares in this company through different types of corporate entity—whether limited liability companies (*sociétés à responsabilité limitée*), 'economic interest groups' (*groupements d'intérêt économique*), or the hybrid entities known as 'local groups with special rights', *groupements de droit particulier local* (GDPLs).[11] Together they held 20 per cent of the shares, while the remaining 80 per cent was held by SMSP.

The president of COMIKA was also the deputy mayor of Canala, so the municipal authority had close links with the *tribus*. His list of preferred candidates for the municipal election had been compiled through the selection of a candidate from among the leaders of each *tribu*. The economic ideal of the time was respected: the company involved segmentary groups and was identified as a communal structure.

COMIKA's dividends were supposed to be paid into an investment fund dedicated to municipal development, and distribution of this money was entrusted to the councils of elders. According to one informant: 'They wanted a bank like SOFINOR but just for Canala.' Economic rebalancing at the level of North Province presupposed development at the level of the mining municipalities, and it was they who were initially seen as having a legitimate interest in SOFINOR, the main shareholder of SMSP. In Canala, this could further consolidate the position of the UC and the municipal authority led by Jorédié.

According to my informants, COMIKA was dissolved in 1995. Officially, this was because SMSP pushed for its liquidation, having waited in vain for Canala's inhabitants to increase their level of investment. Its failure was said to have been hastened by management problems. This event also coincided with the victory of the Union Nationale pour l'Indépendance (UNI), an FLNKS coalition that included PALIKA but not the UC, in the municipal elections of that year. This was the point at which Léopold Jorédié lost his own seat.

Following the liquidation of COMIKA, SMSP established Boakaine Mining with a view to enlisting the mineworkers as shareholders in the mining operation. André Dang was the company's first president, and there

---

11   A GDPL is a legally recognised form of organisation, with a legal personality, that is meant to promote economic activities on customary land. It brings together individuals who are connected through customary bonds (within a family, clan or *tribu*). Thus the GDPL is primarily constituted of persons with customary civil status and is managed under customary law. It may also be used in the context of land claims.

were five locally based directors—Patrice Mwâsadi, Henri Suènô, Pwêêdi Chanel, Christian Onyiari and André Dowadé—who were 'nominated by the people of Canala'. Some of these men joined the FCCI when it was established in 1998, while others remained loyal to the UC. Boakaine Mining was liquidated in 2002 following the strike. Some people did not regard it as a municipal entity but simply as an SMSP subsidiary.

In 1998, Patrice Mwâsadi, who was the president of the council of elders of the Mèèwèè chieftaincy, and close to Jorédié, became the second president of Boakaine Mining. He is one of the *kwara bwatù* ('subjects who lower themselves'), which refers in this context to the members of a host clan who are supporters of a 'stranger chief'. However, he claims that he was accused of lacking customary legitimacy by others groups asserting competing claims over the deposit. On this latter score, he pointed out that the big chieftaincy of Bwaaxéa had not had its own big chief for some time, and he was not only representing his own (small) chieftaincy of Mèèwèè in his capacity as president of its council of elders, but also represented all of the municipality's *tribus* because Mèèwèè is the seat of the big chieftaincy of Bwaaxéa, and Bwaaxéa is itself the most senior of the four big chieftaincies in Canala.[12] Furthermore, since the chieftaincy of Nôôwé had only been separated from that of Mèèwèè in 1940, and that of Bwakaiinè had only been separated from that of Nôôwé during the 1990s, he could also present himself as being, in a sense, the 'customary leader of Boakaine'. Mwâsadi explained his position as follows:

> There was a political [and] customary line: Mèèwèè and Gélima (Xürüchaa) were the two recognised tribal entities. In relation to the Boakaine mine: it was Mèèwèè. There was a huge consensus on the nomination [of the board of Boakaine Mining]: Suènô is from Gélima, Pwêêdi Chanel is the chief of Mîâ/Xwinnè, and Dowadé is the president of the council of Mèrènèmè/Kakö. Onyiari was the jack-of-all-trades of Mèèwèè. Myself, I came back from Australia, I spoke English. Within the organisation of the UC in Canala, I was below Léo. With Dowadé, I assisted Léo in the North Province. And on the level of Mèèwèè, my clan was the most important clan of the chieftaincy. We are the *kwara bwatù*.

---

12 Most people agree with this last point. The Kèrèduru chieftaincy is said to be directly subordinate to Bwaaxéa, while Penyîmê Nèkètè is said to be subordinate to Nôôme, and Nôôme is ranked lower than Bwaaxéa.

The other vehicle for dialogue between the mine's operators and the local population was the so-called 'recruitment committee', also known as the 'customary committee', which was established at the same time as COMIKA. This committee was composed of six persons who all had traditional links with the Boakaine deposit—two each from the *tribus* of Mèèwèè, Nôôwé/Bwakaiinè, and Kakö/Mèrènèmè (also part of the big chieftaincy of Bwaaxéa). They constituted the link between the mining operation and three of the four big chieftaincies of Canala.[13] The committee members were paid about XPF18,000 each month for their attendance at two or three meetings with the mine's management to discuss local people's complaints (mainly on questions of pollution), and to find out what jobs needed to be filled and how current workers were performing. The chiefs and presidents of the council of elders of each *tribu* contacted by the committee then selected the local people who would fill any vacancies in the workforce. Each of the committee members paid XPF1,000 from their monthly stipend into a common fund intended to contribute to the purchase of a minibus for transporting the workers to and from their workplace. This payment was clearly requested by the committee members in exchange for maintaining social peace.

Prior to 1999, this committee had also ensured that the trade unions did not gain a foothold in the mine. As its first president explained: 'We did not want this … because we did not want to defend the lazy people. We fought to have this instrument, it was necessary for the people to work.'[14] It was in the context of the breakdown of the dialogue between the mining company and the 'local circle' of former UC members around Léopold Jorédié that the trade unions gained access to the mine in 1999. Given that the leadership of the municipal authority had by then passed to the UNI, and Boakaine Mining was under the control of individuals who had remained loyal to the UC, the FCCI members attempted to regain control in a different way. It was they who encouraged two trade unions, first the Union des Syndicates des Ouvriers et Employés de Nouvelle-Calédonie and then the Confédération Syndicale des Travailleurs de Nouvelle-Calédonie, to establish their presence at the mine site. It was the second of these unions that initiated the strike in 2001.

13 The big chieftaincy of Penyîmê Nèkètè (or Nakéty) was left out of this process because it has its own mining operations, and its inhabitants would prefer to work on these.
14 Much of the work involved filling the ore carriers for PAMCO (the Japanese foundry) as quickly as possible, and even working on Sundays due to the 'urgency of the economic rebalancing'.

The discourse concerning the lack of recognition of the chieftaincies emerged at the same time. The strike led to the recollection of commitments supposedly made by André Dang to the customary leaders in the chieftaincy of Mèèwèè in 1991 that entailed greater participation of the chieftaincies in the actual operation of the mine. Some informants referred to a promise that operational responsibilities would be transferred to COMIKA or its successor, Boakaine Mining, after ten years—or even earlier. Other informants recalled a proposal to increase COMIKA's shares in the operating company to 51 per cent. Dang was also said to have promised that COMIKA would acquire ownership of the mine's vehicles—mainly trucks—and thus become a primary subcontractor. More peripherally, in the area of corporate social responsibility, he was supposed to have promised help for the renovation of a church and a bridge and the remediation of a former spa. Accordingly, the municipal authority had established an association through which revenues from the mine would be used to fund these projects, but this was wound up after the municipal elections of 1995.

Proposals to increase the level of customary shareholding in the mine were seemingly abandoned in 1995, since they were not supported by the majority of newly elected representatives who then assumed control of the municipal authority. They were raised again in 1998, when Patrice Mwâsadi became the president of Boakaine Mining, and then resurfaced in 2002 as a justification for the previous year's strike. While these demands made reference to local people's connection with the land, they were not based on a discourse of indigeneity that places colonised peoples in opposition to their colonisers, but rather on a 'localist' discourse that was less concerned with resistance to a form of external domination than with the reaffirmation of a political and genealogical history, even though some of Canala's clans could not claim that their ancestors had crossed the mineral tract of Boakaine at one time or another.

What was being demanded here was the recognition of a 'local' political authority based on the claim of some clans to precedence in the occupation of the land. This authority claims respect for those who were living in the environment of the mine, more than for those who accommodated the operators in a space to which they had distinctive rights based on their clan histories. From the way that the notion of 'respect' was invoked in my interviews, it became clear that this was a standard form of Kanak political logic—the logic of welcoming and hosting strangers according to specific rules and duties, and that this was the mindset within which COMIKA was established in 1992, albeit at the level of the *municipality* rather than the chieftaincy.

To understand the previous claim to ownership of the mining company, it is important to remember who was making it. The trade union leaders involved in the strike negotiated with the local leaders of the FCCI because the customary leaders who were most involved in the claim were also close to this party. The FCCI backed its claim by reference to the recognition of customary land rights in the Nouméa Agreement, and the former UC militants who moved to the FCCI thereafter presented themselves as customary leaders:

> Before the split [between the UC and the FCCI] we worked to have the mine reopened. We, the former members of the UC municipal authority … SMSP wanted to continue with us [FCCI] but we were outside the municipal authority. It [SMSP] preferred to continue to negotiate with us. It was they [the municipal authority] who blocked it. They wanted to erect a kind of barrier between us and Dang. They said: 'What happens at the mine is the municipal authority's problem! It is not the [problem of] the customary leaders.' Because we became customary leaders when we left the mine. They [the municipal authority] tried to separate us. But we were still working with the people there [in SMSP].

What this FCCI supporter is saying is that the status of 'customary leader' presented his group with an opportunity to reassert their rights after they had lost control of both the mine and the municipality (not to mention the province), so the former UC militants 'became customary leaders'. However, this was still a contentious designation, since the post-colonial segmentary logic that prevails in the customary domain meant that there were many different customary positions within the chieftaincies, partly originating in the structure of the colonial administration, which allowed for the assertion of competing claims. Therefore, conflicts between the directors of Boakaine Mining resulted in the emergence of a distinction between 'legitimate' and 'illegitimate' customary leaders, while those who claimed legitimacy in the context of their initial support for the FCCI closed ranks around the strikers and then wanted to be placed at the top of a clan-based shareholding company to recover control of the mine site.

Analysis of the first phase of the Boakaine conflict, up to the point of mine closure, provides a clear example of the instrumentalisation of the customary model of political organisation in a situation involving partisan political competition. Furthermore, rather than argue that a Western form of political legitimacy is unattainable in a segmentary form of political organisation, it would seem more reasonable to say that individuals accept the state and its local authorities as fully integrated institutional frameworks while not discounting other political models and institutions that may be mobilised in certain situations.

The virtual disappearance of the FCCI from the New Caledonian political landscape in the mid-2000s has altered the terms of more recent debates between the municipality, the 'customary leaders' and SMSP about the project of reopening the Boakaine. For the political parties, mining Boakaine is a matter of development from the perspective of independence, but it has also become an issue between the political figures authorised to represent development at that level. Former FCCI leaders initially argued for control of the mine to be located at the customary district level because its boundaries matched those of the municipality. In the new phase of debate about reopening the mine, the conflict is more definitively located in the debate supported by the indigenous movement, and reflects the recent evolution of the wider political debate in which the question of independence is superimposed on that of institutional pluralism (Demmer 2016).

In 2011, Martial Tyuienon, then president of the customary district of Canala, and acting as the official spokesperson for claims being made against SMSP, was still insisting that no one could claim ownership of the nickel ore. Only in 2014 did he begin to claim material or intellectual property rights in the mineral resource, echoing the claims made by various indigenous movements all over the world. In 2011 he said:

> You can say you are the owner of the mining land. That's true. But not in relation to the nickel; you cannot know. We know who knows how to make yarn, grow taro, build houses, hunt animals, fish ... but did anyone burn stone to make iron?

During my fieldwork in 2014 he stated:

> With help from family magic, even if they had no use for the nickel, my ancestors knew where to find it. They had the gift of seeing. They maybe knew about the nickel and used it for other purposes: yam and taro ...[15]

---

15   Martial Tyuienon died in 2016.

The Contest over Reopening the Mine

Initially, in 2010, the council of the customary district of Canala, supported by Jorédié's former vice-mayor (and most loyal ally), Gaetan Dowadé, was attempting to create a clan-based joint venture in the form of a *société par actions simplifiée* (SAS) with a view to reopening the Boakaine mine. This local development strategy would have bypassed both the municipality and the tribal councils. The clans in question were represented by 'clan chiefs' who are legally recognised in legislation that became effective in January 2007. Based on the sums paid by each clan at the outset (between XPF50,000 and XPF500,000), it was envisaged that the board of the SAS would distribute some of the dividends based on an evaluation of the clans' needs, while reinvesting the rest in collective projects within their *tribus* (or small chieftaincies). In addition, these 'customary' shareholders, through the district council and its president, would assume the role of the former 'recruitment committee' by selecting people to work at the mine. Decontamination and afforestation measures would be implemented by the district's 'mining committee', which would charge an 8 per cent levy on contracts concluded between the SAS and SMSP. The *tribus* close to the mine would receive preferential treatment in these subcontracting arrangements.

While some of the former FCCI leaders may already have planned to establish a private company that would operate like a 'municipal bank', redistributing dividends among the clans rather than the *tribus*, the 'masterminds' of this more recent project appeared to want to turn the customary district council into an administrative body that would be a real alternative to the municipal authority.[16] Apart from the 'mining committee', the district would have a 'training and education' committee, an 'environment committee', a 'social committee' and yet another committee devoted to 'customary affairs'. As far as the proponents of the 'clan-based company' were concerned, the Boakaine mine would act as a financial reservoir for the development of the district rather than the municipality of Canala.[17] However, the policy model here was still

---

16   This idea gained currency with the publication (in 2014) of the *Charte du Peuple Kanak* (Kanak People's Charter), which inspired by of the discourse of indigenous people's rights.

17   Customary leaders in Canala sometimes say that they are ready to share power with the municipal and provincial authorities, but sometimes say that they want to take it away. In this respect, they have had some difficulty coming to terms with the definition of legal pluralism contained in the Kanak People's Charter.

based on a 'segmented' conception of the general interest, in which the priority was not to serve the individual citizens who would be represented by the municipal authority, but to ensure the well-being of the clans through the elders who held power in the traditional political hierarchy (Demmer 2009).

This would conflict with the UC-run municipality's conception of development. As previously noted, when the UNI secured its majority in the municipal elections of 1995, its assumption of responsibility for municipal development was combined with a challenge to the status of the recruitment committee 'because that is not the law and the mine is the domain of the state'. Instead, the party proposed to deal with mining issues through the municipal mining committee, albeit with participation by the president of the customary district council and the president of the larger 'customary area' to which the district belonged. When the UC regained control of the municipality in 2001, the mayor, Gilbert Tyuienon, complained about the lack of benefits derived from local mining operations. The following comment was made at the mayor's office:

> In Nakéty, the country is given 19 billions of nickel per year. It should be the most beautiful and tranquil *tribu* … but it is often in the news for bad reasons … The people of Canala are tired of being forced to be nothing more than metal workers who borrow 17 or 20 million and pay XPF350,000 in instalments every month!

Such sentiments were linked to an expectation that the New Caledonian congress (the country's legislature) would enter into serious discussions about increasing the level of taxes imposed on the industry and redistributing the benefits to the municipalities according to the principle of equalisation (or derivation).

In the first half of the current decade, the municipality's own plans to reopen the Boakaine mine were made contingent on the assumption of majority (51 per cent) ownership of the operation by a joint venture (*société par actions simplifiée*) known as Société Anonyme d'Économie Mixte Locale de Canala (SAEML Canala). This had been established in 2010 to promote agricultural development in the area, and would henceforth be able to use mineral revenues to fund such activities. The customary leaders would be represented in this entity through four GDPLs (*groupements de droit particulier local*)—one for each of the municipality's four big chieftaincies. Their involvement was seen as both a matter of principle, allowing for a distinctively Kanak form of participation, and also as a way of allaying the latent tensions between the elected representatives and the customary leaders.

However, the proposed GDPLs were subject to criticism by the district council, as the advocate of the clans recognised by the 2007 law, on the grounds that they were 'superficial', 'disconnected from real history' and would not reflect the reality of the segmentary political system. Conversely, some of the people who opposed the new joint venture were equally sceptical of the legitimacy of the 'new' clans. As one of them put it:

> The clan chiefs. That was done from the top down. It should be the other way around. That comes from the [Customary] Senate.[18] The Senate is the form the French gave to this chamber. They thought of the French system. Suddenly, they subjugate from top to bottom. But it's not like that. The people made the clan chiefs without a single thought! None at all!

Accordingly, representatives of the municipal authority and the customary district each can regard themselves as the most legitimate political actors, and believe that they should therefore control the benefits generated by the operation of local mines. During this most recent period, each side has asserted the longevity of its own institutions. The elected representatives might talk about the 'immortality' of the municipality, or their own capacity to get things done, while the customary leaders might respond by stating a preference for consensus over the principle of universal suffrage, or complaining that elected leaders are too far removed from their 'roots'. Both sides also complain that the other cannot be trusted in the management of money. However, the customary leaders will sometimes admit that 'the municipality is the body that is closest to the population', while the elected representatives will say that they do not want to exclude the customary leaders who are part of the political landscape in all Kanak municipalities.

In this particular case, the elected representatives also wanted to keep their distance from the group of actors who were identified with the FCCI but had aligned themselves with the customary sphere. The municipality could therefore say that it had nothing against the customary leaders themselves but distrusted this particular group during the course of the

18　The *Sénat Coutumier* is an assembly of 16 senators, with two representing each of the eight customary areas.

negotiations over the reopening of the mine. One Kanak businessman whom I interviewed did not even attempt to unravel these problems of identity and political legitimacy, but simply said that he understood the desire of the people living near a mine—whoever they were—to control the resource in the absence of a taxation system from which they derived any benefit (see Grochain and Poithily 2011).

A resolution was thus adopted in 2012, at a meeting held in in Mèèwèè, the 'headquarters' of the grand chieftaincy of Bwaaxéa, which stipulated that the customary leaders, the municipal authority and SMSP should all be involved in reopening the mine. For its part, SMSP understood the local people's desire to get control over the resource, especially in the absence of localised taxation, and wanted to encourage the creation of subcontracting companies, as well as local employment, training and shareholding. Although SMSP was unwilling to transfer the mining titles demanded by some of the customary leaders in Canala, it was prepared to negotiate a solution with local actors—whether customary leaders or elected representatives—who had not been party to the conflict of 2002.

The elected representatives themselves decided to take a back seat in the negotiations in 2012. At the end of the day, no agreement was reached: the municipality has failed to hold discussions with the Boakaine *tribus*, while the district council still disagrees with the proposal to establish SAEML Canala, and SMSP is unwilling to transfer or farm out the mining titles to either the municipality or the district. The mine is still closed, and it will not reopen without a clear local agreement between the municipality and customary authorities and a negotiated arrangement between SMSP and the UC-led municipality. The first of these conditions seems more likely to be met, since the municipality is at odds with the UNI-run North Province over the redistribution of mining revenues and the support provided to Kanak private mining companies.

Conclusion

The conflict that led to the closure of Boakaine is a perfect entry point for understanding contemporary social relations in the Kanak world. In the context of the progressive decolonisation of New Caledonia, the conflict demonstrates the potency of this political issue for the pro-independence parties: they all see it as one that involves their status as leaders in the implementation of a project to prepare the economic grounds for independence. What this conflict also reveals is that the mines are tools of regional development for all political actors, both Kanaks and non-Kanaks. Local development depends largely on SMSP, a company owned and controlled by North Province, and tensions emerge from the (feeling of a) lack of direct benefits at the municipal level, as has been explained in the case of Canala. Whereas the FLNKS thinks of resource management at the scale of the whole territory, the local redistribution of mining revenues at the local level remains a contentious issue—all the more so with the complex coexistence of elected and customary political entities.

The direct and indirect employment generated in and around a mine is not enough to meet local people's expectations on this score. In the era of neoliberal democracy, actors from the mining areas invite themselves to participate in the debate on this subject. Two closely related (though different) registers are mobilised in Canala. On one hand, people refer to a 'localist' legitimacy that reasserts the rights of 'the people from the place', meaning both the municipal residents and the subjects of the chieftaincies, or even the members of clans whose ancestors once traversed the mineral deposits. Due to the complex layering in the politics of recognition of Kanak customary entities since the colonial era, the definition of 'the place' can extend from the clan to the tribe, the district or even the customary area encompassing them. On the other hand, people adopt the most recent language of indigenous mobilisation, defending in a generic and well-structured manner the rights of the first occupants of a settler colony. This ethnocultural discourse asserts more clearly the ideological cleavage between elected representatives and customary authorities—the latter now seen as more legitimate bodies for the management of the mines and indigenous populations. The proximity of the referendum on self-determination can explain the flowering of this type of discourse as an alternative path towards decolonisation: the achievement of internal sovereignty through the legal recognition of cultural difference.

A change is worth noting here. In 2002, the need to consider customary areas, districts, chieftaincies and clans was expressed by Kanak politicians when they had lost their audience at a national or provincial level and tried to re-establish themselves at the local level. What is new here is that this strategy is not necessarily deployed to facilitate a return to electoral politics but that it still serves the political purposes of the customary actors themselves. Customary authority is invoked in its own right as an alternative form of power entitled to legal recognition. This evolution opens up avenues for the establishment of differentiated political rights, which would reconfigure once again the relations between Kanaks and non-Kanaks, as well as among the Kanaks themselves.

References

Boisard, M., 2002. *Cartographie des Zones Dégradées dans la Région de Canala, Nouvelle Calédonie.* Nouméa: Université Bordeaux 3, Institut EGID.

Demmer, C., 2007. 'Autochtonie, Nickel et Environnement en Nouvelle-Calédonie: Une Nouvelle Stratégie Kanake.' *Vacarme* 39: 43–48. doi.org/10.3917/vaca.039.0043

——, 2009. 'Secrets et Organisation Politique Kanake (Nouvelle-Calédonie): Pour Sortir des Catégories Privé/Public.' *l'Homme* (190): 79–104.

——, 2016. *Socialisme Kanak. Une Expérience Politique à Canala (Nouvelle-Calédonie).* Paris: Karthala.

Gluckman, M., 1940a. 'Analysis of a Social Situation in Modern Zululand—A: The Social Organization of Modern Zululand.' *Bantu Studies* 14(1): 1–29. doi.org/10.1080/02561751.1940.9676107

——, 1940b. 'Analysis of a Social Situation in Modern Zululand—B: Social Change in Zululand.' *Bantu Studies* 14(1): 147–174. doi.org/10.1080/02561751.1940.9676112

——, 1942. 'Some Processes of Social Change Illustrated with Zululand Data.' *African Studies* 1: 243–260. doi.org/10.1080/00020184208706592

Grochain S. and D. Poithily, 2011. 'Sous-Traitance Minière en Nouvelle-Calédonie: Le Projet Koniambo.' Nouméa: CNRT Nickel, Programme Gouvernance Minière (Document de Travail 4).

Horowitz, L.S., 2009. 'Environmental Violence and Crises of Legitimacy in New Caledonia.' *Political Geography* 28: 248–258. doi.org/10.1016/j.polgeo.2009.07.001

Merle, I., 2000. 'De l'Idée de Cantonnement à la Constitution des Réserves: La Définition de la Propriété Indigène.' In A. Bensa and I. Leblic (eds), *En Pays Kanak: Ethnologie, Linguistique, Archéologie, Histoire de la Nouvelle-Calédonie*. Paris: Éditions de la Maison des Sciences de l'Homme (Ethnologie de la France 14). doi.org/10.4000/books.editionsmsh.2788

Soriano, E., 2013. *La Fin des Indigènes: Le Colonial à l'Épreuve du Politique en Nouvelle-Calédonie*. Paris: Karthala.

5. Conflict and Agreement: The Politics of Nickel in Thio, New Caledonia

PIERRE-YVES LE MEUR[1]

Introduction

In late July 1996, the entrances to Plateau and Camps des Sapins, the two main mines in Thio, were blockaded by the commune's Kanak inhabitants. Blockades were also placed on the ore transfer belt in nearby Thio Mission and on a Japanese ore carrier, the *Tango Gracia*, which was in the process of being loaded by its Filipino crew. The stand-off between the Thio *coutumiers* ('customary representatives') and the company Société Le Nickel (SLN) lasted two weeks and culminated in a wide-ranging written agreement which, despite being local in scope, went beyond the context of work and touched on environmental and social issues, the question of the control of resources, and the redistribution of mining revenues (see Filer 1997; Bainton 2009).

1   This study was carried out as part of a research program financed by the National Centre for Technological Research, under the title 'Nickel Politics between Local Governance and Corporate Governance: The Comparison of Mining and Industrial Development in New Caledonia'. I should like to sincerely thank Jean-Michel Sourisseau for his critical and constructive reading of an initial version of this text, and Susan Cox for translating it into English.

Thio, a commune located on the east coast of South Province, was the country's 'nickel capital' for a long time, and SLN, which was established in 1880 and based in Thio during the interwar period, enjoyed a dominant position in the production and processing of nickel in New Caledonia.[2] The town of Thio was also one of the focal sites of *les événements* ('the events'), a period of violent clashes between 1984 and 1988 that combined elements of civil war between the Kanak separatists and New Caledonian loyalists with the Kanak decolonisation struggle, so the location and protagonists in this particular tale are not without significance. Equally significant is the unfolding of this social movement—the blockade itself, the willingness to extend or escalate the conflict beyond the boundaries of the mine site and then the eventual agreement. Finally, the discursive structure of this conflict as a *conflit des coutumiers* also merits some explanation.

I shall start from the hypothesis that this conflict and the resulting agreement mark a turning point in local mining governance. This chapter will describe the configuration of actors, issues and discourses involved in the conflict, and will situate it in the previous history of local conflicts and in relation to the changes in mining affairs and the political situation in New Caledonia in the 1980s and 1990s. I shall then analyse the reconfigurations of the local arena prompted by the conflict negotiations from three perspectives. The first is that of landownership, specifically the contest between the political (or historical) and legal recognition of land rights. The second is that of politics—the redefinition of affiliations and locality, the reorganisation of chieftainship and the positioning of the town hall. The third is that of business, involving questions of subcontracting, local employment and business development.[3]

---

2   This monopoly was toppled in the 2000s, particularly in 2010, with the establishment of two new mining and industrial projects in the north and south of the island (see Chapters 2, 3 and 6, this volume).

3   The analysis is based on retrospective interviews carried out in Thio between 2008 and 2014, and on various documentary sources: newspaper articles published in Les Nouvelles Calédoniennes between 26 July and 7 August 1996; minutes of the consultation meeting of 9 August 1996; the text of the protocol to the agreement of 6 August 1996; the letter sent on 18 October 1996 by the presidents of the two districts of Thio (Philippe Nekaré and Charles Moindou) to the government delegate, Dominique Bur, and the chief executive of SLN, Jean-Jacques Mouradian, through Jean-Michel Arlié, Deputy Commissioner of South Province, requesting compensation for the land washed away or polluted through the activities of SLN in light of the earlier agreement; the letter sent by the sous-préfecture (subdivision) to the deputy commissioner on 30 October 1996; a letter sent by the deputy commissioner to the chief executive of the Agency for Rural Development and Regional Planning; and other correspondence and minutes of follow-up meetings relating to the August agreement.

The Conflict and the Agreement

The Conflict-Based Approach

The heuristic productivity of the conflict-based approach has long been acknowledged, in particular since the work of Max Gluckman, Victor Turner and other members of the 'Manchester School' (see Turner 1957). No society is exempt from conflict, so it constitutes a general empirical phenomenon. Conflicts constitute moments in time that crystallise and highlight opposing positions; they do not emerge haphazardly. They constitute spaces for the expression of arguments or registers of justification (revealing judgments, knowledge and information) that tend to 'harden' and gain in visibility over the course of a dispute. Finally, conflicts are not simply 'revealing' in the photographic sense of representing a pre-existing sociopolitical topography; they are factors that reconfigure the rules of a game—mining governance in the case in point—and therefore constitute factors of institutional change. They are part of broader sequences of events that include their modes of resolution (in this case a written agreement), as well as measures of avoidance or patterns of re-emergence. Moreover, and in relation to the context of New Caledonia, disputes like that under consideration here are part of a long-term process, that of the consubstantial 'meta-conflict' of settler colonialism. We shall see clearly how this historical legacy is inscribed in the manner in which the question of landed property is negotiated. In this sense, conflicts fulfil a function, even if it is not that ascribed to them in advance. The point of view adopted here is processual and not functionalist, and leaves open the specific empirical question as to whether this was a conflict 'about' the control of resources, or whether control of resources became the pretext for a conflict about other things.

We can take inspiration here from the suggestions made by Glenn Banks in his analysis of mining conflicts in Papua New Guinea (Banks 2008). Banks stresses the overlaps between questions surrounding resources and affiliations and the continuity of a sociopolitical order. Secondly, he highlights their processual dimension: what is involved in most cases is the restoration of a balance rather than the regulation of a situation in the strict sense—in other words, a question of 'peace' rather than 'justice', to use the terms employed by Sten Hagberg in a completely different context (Hagberg 1998). Along the same lines, the compensation eventually paid is more 'relational' than 'transactional' in nature and, more often than not, the requested and/or self-appointed mediators are interested and involved in the conflict arena, as opposed to being neutral and external, as would be desirable in a 'United Nations' style of mediation. Banks proposes, thirdly, that mining conflicts are both strongly embedded at a local level—in terms of actor configurations and normative repertoires—and marked by, or the products of, the exogenous rupture constituted by the precipitous development of the mining project, as tragically shown in the case of the Panguna copper mine in Bougainville (Regan 1998, 2014; see also Allen 2013 and Chapter 12, this volume). I argue that this rupture is expressed in a tension between inflows (of actors, material and immaterial resources) and a 'normative gap' that derives from failure to establish the 'rules of the game' in the early stages of mining projects. This tension, which should be viewed relative to the historicity of mining activity in a given location, is a powerful driver of conflict. It could also be argued that mining conflicts question the boundaries of the company, as an entity located on the interface between corporate governance and local arenas, and the enclave logic that inherently underlies a mining operation. The company is formed through the definition of a physical, institutional and ideological action perimeter—the 'inside' and 'outside' (or externalities) of the company—and this perimeter is contested by certain actors.[4]

The Context

The context in which this localised conflict arose was that of the sequence of *les événements* and the resulting political agreements—the Matignon-Oudinot Agreements in 1988, the Nouméa Agreement in 1998—the latter marking the advent of an original negotiated process of decolonisation. The context is also that of the increasing significance of mining as an issue, which the pro-independence parties placed at the heart of the political debate in the 1990s, resulting in the negotiation of a *préalable minier* ('mining prerequisite') in the Bercy Agreement of February 1998.[5]

---

4   On the different forms of industrial pollution (e.g. 'spillage' versus 'overflow'), see Letté (2009). On the notion of the mining enclave, see Le Meur et al. (2013) and Allen (2017).

5   The signature of the Bercy Agreement as a 'mining prerequisite' to the Nouméa Agreement of 1998, which prompted the initiation of a process of negotiated decolonisation in New Caledonia, should not be seen retrospectively as the start of a harmonious and linear process. Its roots go back further, to the nationalisation of the Société Minière du Sud Pacifique in 1990. Debates among separatists on the opportunity to start thinking in capitalist terms were tempestuous, and the story of the appropriation of the issue of mining by the pro-independence groups was yet to unfold. I thank Jean-Michel Sourisseau for drawing my attention to this point (see also David et al. 2016: 33–39).

This was followed by the arrival of the mining multinationals in the 2000s and the rise of environmental and indigenous rights discourses. In this regard, the conflict of 1996 may be seen as 'pre-indigenous rights' and, as we shall see, represents a turning point in relation to a number of issues—mining, land, indigeneity, customary authority, and so forth (see Chapters 4 and 6, this volume).

The Story

The conflict of 1996 started with a lockout that would last for 13 days. The lockout was not the only form of expression used; it was also accompanied by written documents. On 25 July, a package of demands (*cahier de revendications*), signed by representatives of the two customary districts of Thio (Thio and Borendy),[6] was submitted to SLN, to the High Commissioner of the Republic (the representative of the French central state in New Caledonia), to South Province and to the parties represented in the Congress of New Caledonia 'because, according to the customary authorities, the demands concern all of the partners' (Huillet 1996a). The package of demands was divided into eight sections relating to: (1) recruitment; (2) the granting of mining licences by SLN to the customary authorities; (3) subcontracting, with a demand that Nouméan companies establish offices in Thio; (4) assignment to the commune of the Baie de la Mission housing development scheme (the issue had been blocked for three years); (5) the restitution of land and compensation for arable areas; (6) the establishment of an anti-pollution team at the mine; (7) financing for rehabilitation of environmental damage caused by the operation; and (8) the creation of infrastructure that would 'open up' the commune.

A meeting was held in Thio on 29 July 1996 with a view to negotiating these demands. The leaders of the movement rejected an offer by SLN to transfer the meeting to its headquarters in Doniambo, near the Nouméa processing plant. They demanded that the meeting be held in the open air, on 'their land', on the access road to the Plateau mine.

---

6   The organisation of indirect colonial government through the administrative chieftaincy was gradually established with the creation of the 'tribe' in 1867, the chieftaincy ('*petit chef*', that is, 'little chief' or tribal chief) in 1868, the institution of the *régime de l'indigénat* ('indigenous legal code') in 1887, and the district headed up by the '*grand chef*' ('big chief') in 1897. The transfer of the Kanak population to reservations was carried out progressively during the same period, and was systematised with the *grand cantonnement* at the turn of the twentieth century under Governor Feillet (see Dauphiné 1987; Merle 1998, 2004).

On 6 August 1996, an agreement was signed at the La Foa *sous-préfecture* ('subdivision') office between SLN, the commune of Thio, and the two councils of the customary districts of Thio. The agreement basically constituted a response to the grievances outlined in the package of demands presented in July, and included:

- provisions for the resettlement of people located in the flood zone, based on the argument that the mining slag heaps had caused the silting up of the Thio River;
- provision by SLN of insurance policies to local farmers to cover the eventuality of damage to their crops; and
- the possibility of quadripartite mediation between the state, the commune, SLN and the two customary districts in cases of dispute.

Another meeting was held at the town hall on 9 August, following hot on the heels of the signing of the agreement. The agenda for this meeting also referred to the package of demands and the terms of the agreement. However, rather than focusing on questions directly related to the mine—recruitment, mining licences, subcontracting—it highlighted, in particular, the issues of land use planning, residential development and land redistribution. A monitoring committee was established and was active for around a decade, particularly in relation to the recruitment of the inhabitants of Thio. However, the mining licence took a long time to materialise. The new Kanak mining company, Société Minière des Kanak de Thio (SOMIKAT), was established in 2007 and commenced production in 2010 (Seyrane Belliot and Léopold Gnahou, interviews, 29 March 2012).

The blockades were not located in the operational area of the mine, and there was no occupation of mining sites or offices as is often the case in labour conflicts. In this instance, the theatre of the conflict—which was not a traditional labour conflict—shifted towards the interfaces and thresholds between the town and the mine (the entrance to the Plateau and Camp des Sapins mines) and between the town and the outside world (the wharf). The locations of the negotiations were also carefully selected: under no circumstances would they be held on SLN ground, whether in Thio or at the headquarters in Nouméa. The protesters insisted that they be held on the territory of the native inhabitants, that is, in front of the mine, at the town hall, which had been under the control of the pro-independence Front de Libération Nationale Kanak et Socialiste (FLNKS) since 1986, or in the subdivision, in the 'house of the state'.

Some of the actors involved in the conflict, and some people who were still children at the time, recall the atmosphere in those two weeks as being rather festive and non-violent. Seyrane Belliot, the manager of SOMIKAT (until 2014), remembers playing with her friends amongst the adults, in a joyful hubbub involving political debates and barbecues in front of the entrance to the mine (interview, 19 November 2012).[7] At the same time, women from the tribes at the lower end of Thio worked in shifts to prepare meals for the blockaded Filipino sailors on their ore carrier, who were beginning to run out of supplies.[8]

The choice of the locations and the declarations of the local leaders express the same desire for access and opening up: it was about getting away from the usual form of confrontation with SLN, which always tried to resolve problems through the adoption of a private, bilateral logic of compensation.[9] While condemning the lack of environmental awareness on the part of SLN, and the devastating effects of this negligence on the landscape of the commune, the customary leaders insisted that this state of affairs was equally the concern of 'all parties', that is, all political-administrative levels of organisation. This strategy of 'publicising' the conflict leads us to the identification of the actors involved, and hence also to pinpointing important absences.

Actors and Absences

First, it is clear that the customary representatives were the driving force behind the process. The individuals in question were the two major chiefs, Damas Toura (Thio) and Philippe Nekaré (Borendy), who both enjoyed significant historical and 'customary' legitimacy beyond their positions as administrative chiefs, who were current or former employees of SLN, and who belonged to the loyalist political party, the Rassemblement pour la Calédonie dans la République, like the chief of Kouaré, Moïse Mapéri,

---

7 Seyrane's mother, Ithupane Tiéoué, a PALIKA militant and, at the time, secretary to the mayor of Thio, completely concurred with this account (interview, 26 November 2010).

8 The tension was clearly palpable at times—this was also an argument used in the negotiations—but it is astonishing to note the extent to which the accounts provided in the anti-independence local daily newspaper *Les Nouvelles Calédoniennes* are partial in this regard, and insist solely and vehemently on the 'tension that emanates from the strikers. A machete, a club or a rock in their hands, they guard the road blockade and rise to meet the least approach from outside' (Huillet 1996a). The 'primitivist' tone of this account (e.g. the 'club') is noteworthy.

9 This had been the typical company response to all claims for damages, be it that involving the Thio Mission during the interwar period (Winslow 1993), or later involving European livestock breeders (R. Lacrose, interviews, 28 August and 11 September 2008).

who was also active in the process. However, the leading personality in the conflict was Charles Moindou, the president of the district council of Thio, and also an active member of the Union Progressiste Mélanésienne (UPM), which was part of the pro-independence FLNKS alliance established in 1984. He and his two brothers, Joseph and Albert,[10] formed a group with considerable accrued political and economic capital as members of the UPM, even though some people contested the legitimacy of their positions as customary leaders.

SLN was involved through its office in Thio, the governance of which was—and still is—based on a pairing formed by the head of the office, generally a metropolitan 'expatriate', and the human resources manager, who was normally integrated into the local social fabric, or at least of New Caledonian origin, though not actually a member of the mine-affected community.

The central or 'metropolitan' state was represented by the deputy commissioner (*commissaire délégué*) of South Province and, less directly, by three government departments—the Agence de Développement Rural et d'Aménagement Foncier, the Direction des Infrastructures, de la Topographie et des Transports Terrestres and the Direction du Développement Rural (the last two being part of New Caledonia's territorial administration).[11]

Two other groups of actors may be characterised in more sociological terms. First, the town hall played the role of the 'political-state intermediary' due to its pivotal position as the bottom rung of the state apparatus and its elected leadership, especially as it had been taken over by the pro-independence FLNKS alliance after *les événements*.[12] In addition to the town hall, which would remain a rather secondary force in the process, two men were chosen as official spokespersons by the customary leaders—Amédée Tiéoué and

---

10   Joseph was an elected representative of the FLNKS in South Province from 1989 to 1995. Albert would become the mayor of Thio in 2001.

11   New Caledonian usage tends to equate the 'state' with the government of France, neglecting the fact that it contains different levels of political organisation, especially in the New Caledonian context, where the three provinces (South, North, Islands) established by the Matignon-Oudinot Agreements of 1988 enjoy wide-ranging powers, in particular with respect to economic development and the environment. The Nouméa Agreement of 1998 installed a collegial government at the level of the territory of New Caledonia, and the territorial congress, elected through indirect suffrage by the provincial assemblies, has its own legislative powers.

12   The current mayor, Louis Mapéri, was a member of the Parti de Libération Kanak (PALIKA), who had been elected to replace the former mayor, Roger Galliot, after a year-long period of municipal subjection to central state control.

Joseph Moindou (the brother of Charles). Both of them belonged to economically influential families, and in 1982 had been co-founders of UNICONCEQ, one of the first Kanak enterprises in Thio (see Pitoiset 2002; Le Meur et al. 2012). So their legitimacy as 'political-economic' intermediaries was based on their entrepreneurial expertise and (above all) their interpersonal political skills, rather than their customary authority, which was either weak or disputed. Nor did their position as 'mediators' in such a conflict entail a position of neutrality (see Banks 2008).

The second group consisted of the 'youth'. This was a somewhat fluid category that involved a generational positioning of juniors against elders and a degree of political stigmatisation of 'young people' as being essentially 'unruly'. They emerged as a group in the conflict of 1996 through an association for the unemployed (*association des chômeurs*), which strongly supported the demand for local recruitment but found itself gradually ousted from the negotiations by the elders representing the customary authorities.

Other groups were either completely absent from the conflict, as in the case of the trade unions, or largely so, as in the case of the political parties. The absence of the trade unions confirms that this was not a labour-related conflict, even though some of the actors were union members, and despite the demands made in relation to recruitment and subcontracting.[13] As for the marginal role of the political parties, it may be noted that support from the separatists who dominated the local authority was counterbalanced by the discursive definition of the conflict as a 'conflict of the customary people'. Furthermore, despite the fact that the conflict took place in the wake of the process of provincialisation established by the Matignon-Oudinot Agreements, the authorities of South Province were not involved in the process.[14]

---

13 In the same year, the dominant trade union of the time, SOENC, formed an alliance with the customary people of the Xârâcùù area in a dispute with the Société Minière du Sud Pacifique and a group of small mining companies (*petits mineurs*). It should also be noted that SLN made tactical use of 'trade union sources' when it suggested that the Thio blockade put the jobs of the workers at the Doniambo nickel processing plant temporarily at risk (Hervieu 1996).

14 The same absence would be observed in the conflict between the local population and multinational mining company in the neighbouring commune of Yaté a few years later (see Horowitz 2012). North Province took a backseat position during the negotiation of the agreement between SLN and the river clans in the mining area of Tiébaghi, but later intervened in the establishment of the public–private partnership created to manage the local subcontracting arrangement. This point has been confirmed by the former head of the SLN office in Tiébaghi, who negotiated that agreement, as well as the 2008 agreement between SLN, the commune of Thio and South Province (René Féré, interview, 20 February 2012).

The absence of the provincial authorities can be interpreted in several different ways that are not mutually exclusive. First, the provincial administration was still recently formed and underdeveloped. Second, responsibility for mining was only really devolved to the provinces with the passage of the Organic Law of 1999, and even then was subject to de facto control by the territorial Direction de l'Industrie, des Mines et de l'Énergie de la Nouvelle-Calédonie (see GoNC 2009). Third, the majority of provincial politicians were opposed to independence, so had no sympathy with the concerns of a pro-independence commune. And, finally, the local protagonists themselves wanted to localise the conflict, which gave the central state the possibility of assuming a custodial role in the process.

Localisation of the Conflict and Redefinition of Social and Political Affiliations

It should be emphasised here that the tension at the root of the conflict was the local version of a more general dispute, which was expressed in the political sequence of events extending from *les événements* to the Matignon-Oudinot Agreements and (ultimately) to the Nouméa Agreement.

The historical perspective points to a series of events characterised by the deepening, extension and coalescence of local disputes, which, for many people, centred on the environmental damage caused by SLN's poorly monitored mining activities—uncontrolled dumping, river silting, floods, the displacement and resettlement of certain tribes (Saint-Philippo I and Ouroué, for example)—and regulated by bilateral agreements between SLN and the affected persons. It should also be noted that some of the more serious conflicts made a deep impression, based among other things on the violence and repression involved, and had a significant impact. These included the N'Goye conflict in 1978, conflict around the Koindé sawmills (in the neighbouring commune of La Foa) in 1981, and that in Kouaré (on the other side of the central mountain range, but still in Thio) in 1982–83, which had both environmental and economic dimensions, since it involved forest resources and employment opportunities. And shortly before the conflict described here, on 3 February 1996, there was the list of grievances compiled by the customary area council of Xârâcùù, which contained the same types of demands but on a broader scale than those made in the present case. The Xârâcùù customary council demanded compensation for Kanaks from the tribes living in the coastal zone from Mara Cape (in Canala) to the left bank of the Kakoué River (in Thio), the transfer of the mining licences to the customary authorities, and the transfer of land downstream of the mines to the customary landowning clans (Huillet 1996a).

At the same time, it is important to note the national context, which was characterised in the period 1995–96 by the negotiation of the follow-up to the Matignon-Oudinot Agreements. The FLNKS placed the issue of mining at the centre of the negotiations, demanding a 'strong gesture' on the part of the state, and a 'mining precondition' for continuation of the debate. It was this process that culminated in the Bercy Agreement signed on 1 February 1998 (before the Nouméa Agreement of 5 May 1998), which involved an exchange of mining concessions between SLN and the Société Minière du Sud Pacifique and the promise to build a smelter in North Province (see David et al. 2016; also Chapter 2, this volume).

Interviews conducted with two of Thio's political leaders produced agreement on two points.[15] The first, paradoxically, was that mining itself was not an issue in *les événements* in Thio, despite the burning down of the SLN offices in Thio Mission and the occupation of the company's residential quarters. At the same time, however, the leaders insisted that this battle was their battle, this was the story of Thio, and it was not influenced by national arguments or slogans. This desire to separate and localise the struggle should be related to the emphasis placed on the role of the 'customary people' as protagonists. As Charles Moindou put it in a newspaper interview:

> It has nothing to do [with the mining precondition]. Our movement is independent of politics. The customary people were the ones who took the decisions … From '84 to '88, SLN negotiated with the customary people. Why did it stop doing this? (Huillet 1996a)

This very common discourse concerning localisation and separation may be interpreted as denoting a concern for the recognition of a Kanak capacity for action, or agency, which is very often denied to them in (white) New Caledonian statements, especially when these refer to the 'insurrection' of 1984–85, when the town was occupied by separatists. According

15 The leaders in question were François Burck, a native of Thio and president of the Caledonian Union until 1996 (interviews, 17 June and 25 August 2011) and Louis Mapéri, the PALIKA member who was mayor of Thio from 1986 to 2001 (interviews, 22 June and 24 August 2011).

to widespread discourses among non-Kanak people, the relationships between the communities in Thio were supposedly good, and the factors or actors that came to disrupt them originated from elsewhere, partly in the person of separatist leaders like Eloi Machoro, who came from neighbouring Nakéty, but also because the separatists had the backing of the French socialist government from 1981 to 1986 (Roger Galliot, interview, 21 June 2011).[16] SLN actually did look for Kanak mediators to help restart its production lines and restore its relationship with the local population at the time of the 'insurrection'. One of them was Camillo Ipéré, who has been working for SLN since 1970, and invested his customary and unionist skills in the task of restoring a minimal level of confidence (interviews, 15 July and 25 September 2008), but was hardly a neutral mediator, given his role as a UPM militant and active participant in the pro-independence struggle.

The other side of this localisation is constituted by the rhetoric of custom. The discursive framing of the conflict as a 'conflict of the customary people' enables the elimination of two forms of internal division. The first of these is that of party affiliation: despite the fact that they enjoyed strong customary legitimacy, the two 'big' chiefs and some of the 'small' chiefs in the area belonged to the 'loyalist' Rassemblement pour la Calédonie dans la République. The second harks back to the pre-colonial or early colonial period, especially the war of 1878, when the clans of the central mountain range (mainly Owi and Koua) took the side of the insurgents in opposition to those of the lower valley of Thio (e.g. Toura, a chiefly clan of that area) (Dousset-Leenhardt 1976; Le Meur 2013). This touches on a paradox of the administrative chieftaincy, for when reference was made to the 'customary people' in 1996, what was in fact intended was the administrative chiefs who were signatories to the agreement signed with SLN, and the partly alien (colonial) origins of their own status enabled the relegation of the old lines of division to the background (Louis Mapéri, interview, 24 August 2011). However, land claims and allocations had revived these divisions in the context of the land reform launched in 1978.

This discursive framing presented a form of indigeneity that is in some ways 'pre-indigenous' from the point of view of the United Nations Declaration on the Rights of Indigenous Peoples (see Demmer 2007;

---

16 This theme of 'betrayal' by the metropolitan state is a recurrent feature of the New Caledonian interpretation of the events, as revealed in interviews I conducted with M. Fels (a former SLN employee), G. Santacroce (an elderly farmer), and C. Bull in 2010 and 2011.

Levacher 2016; also Chapters 6 and 9, this volume). At the same time, it enabled the affirmation of a desire to regain control over natural resources and 'traditional' hierarchical principles, which sometimes ran 'counter' to the demands of the youth and pro-independence political parties who were at the forefront during *les événements*. As a result, the social question was in some sense put on hold, even though it might have been expected to occupy a strategic position in a conflict between local Kanak people and the major industrial enterprise and source of employment in the area. An alternative interpretation, which would be consistent with the theory of the reaffirmation of traditional hierarchical principles, is that the social question was not really put on hold, but was reformulated within a 'Kanak cultural framework' that was mindful of the hierarchies based on antecedence and seniority.

This takes us to the strictly socioeconomic dimensions of the negotiations. With respect to subcontracting, it was possible to observe a shift between the original grievances and the concluding agreement in terms of a greater insistence on local employment. However, there was never any question of supporting the creation of a local subcontracting company, and this reluctance is difficult to explain, given the prevailing dependence on companies with headquarters elsewhere. The Nouméan companies were simply asked to establish offices in Thio. Local employment in the strict sense of 'recruitment' constituted the first point of the agreement:

> SLN undertakes to recruit in the next four years at least 16 young people (males and females), with or without qualifications, from the tribes of Thio. SLN undertakes to provide theoretical and practical on-the-job training in Thio for those without qualifications. Moreover, SLN undertakes to recruit three dockers within a period of six months, the first from September 1996.

The question of local employment was specifically raised by the association of the unemployed and, despite the fact that members of the association sometimes felt that they were being excluded from the negotiating table by their elders, results were achieved on this issue and were sustained through the establishment of a local recruitment monitoring committee.[17] The intergenerational dimension of conflicts that are not necessarily organised around this particular topic has been observed in several other cases in Thio, including those associated with *les événements*.

---

17   There were some differences of opinion in this point (Ithupane Tiéoué, interview, 26 November 2010).

Apart from its intergenerational component, the question of local employment is by definition a 'local' matter. The history of mining in New Caledonia involved a 'nickel boom' in the period 1967–72, followed by a period of decline in the later 1970s (Freyss 1995). Many Kanaks moved to Thio at the time to work in the mines or for the subcontractors, and a good number were employed by companies of French origin. Some of them settled there permanently, married, and eventually became involved in local associations or militant movements. Prior to that, a large number of Kanaks from the Loyalty Islands, in particular Lifou and Ouvéa, had come to work for SLN. These workers inhabited an encampment beside the Plateau mine up to the 1970s, and later lived in the SLN housing estates near the wharf at Pawani. The Association des Kanak de la Grande Terre et des Îles emerged from these migratory movements. This 'ethnonational' association, initially led by pro-independence militants, later evolved into a neighbourhood association, the Association du Botaméré, which took its name from the sugarloaf-shaped mountain that looms over the Thio Mission estate. Both the association and its militant leaders, like Pierre Ayawa,[18] remained very much in the background during the conflict of 1996. The end of the nickel boom had given rise to certain tensions between the Kanaks who were and were not natives of Thio.

At the same time, the promotion of employment and local entrepreneurship followed a logic of intertribal equity at the level of the commune of Thio. The objective was to balance the distribution of the mining revenues between the different tribes, even if the tribal entity—an administrative unit of colonial origin—is sometimes circumvented by a logic based on networks and alliances. This was the case in the choice of the new manager of SOMIKAT to replace Amédée Tiéoué, who resigned several years after the conflict, and was stigmatised as a 'foreigner' despite the fact that his grandfather had been a resident of Thio, and despite his role as both entrepreneur and official spokesman during the conflict itself.[19] The new manager came from Nakéty in the neighbouring commune of Canala (see Chapter 4, this volume; also Figure 2.1), but his alliances meant that he could represent the tribe of Ouindo, which is based in Thio.

---

18 A native of Ponérihouen, Pierre Ayawa came to work for SLN in Thio in 1959 and has remained there since (interview, 7 October 2010).

19 He was known as 'Lifou' after the birthplace of his paternal great-great grandfather, though his clan originally came from the area of Poindimié on the north-east coast.

Hence, in the long term and, more specifically, through the conflict and agreement of 1996, it is possible to observe a process of reconfiguration of local citizenship, as has been the case in other mining situations. However, this was unlike the case of Lihir in Papua New Guinea (Bainton 2009; also Chapter 11, this volume), where the struggle over access to mining revenues triggered the construction of a local apparatus based on kinship, and hence with strong ethnic connotations. In Thio, it was more the issue of residency and communal affiliation that came to the fore, possibly because the racial ideological space was already occupied by the 'meta-conflict' of settler colonialism, and since a considerable number of non-Kanak Caledonians had left the area in the period 1984–88, the residential administrative unit of the tribe then achieved greater prominence in public debate. In line with what has already been said about the role of the administrative chieftaincy, this also enabled the circumvention of certain potential dividing lines within the local community.

The Localised Globalisation of the Conflict

The local appropriation of the conflict went hand in hand with a desire to 'globalise' it from the perspective of the issues and actors. Following Latour (2005: 173ff), this process can be understood in terms of a tension between three complementary aims: (1) to 'localise the global' by situating the latter as a form of the local characterised by a greater density of connections with other sites, and not by a position of superiority; (2) to 'redistribute the local' by showing how it is generated through a series of mediation and translation processes generating the intervention of exterior elements or agents; and (3) to 'connect the [local and global] sites' by means of measurement, classification, categorisation and standardisation procedures.

The objective here was, first and foremost, to get beyond the confrontation with SLN, which meant getting out of the mining enclave: 'SLN, the state, the provinces, and the territory each have their part to play'. The argument played out on several levels. First, it was a way of denouncing the past and present collusions between SLN and the territorial administration. It should also be seen as a way of restating the point that the scope of the conflict was not limited to the labour question within the boundaries controlled by the company. As an interviewed customary representative said:

> Thio has been exploited for 100 years with consequences for the environment, the displacement of communities, the problems of pollution and supply of drinking water (Huillet 1996a).

Finally, the objective was tactical: it was also a matter of seeking alliances. Hence, the environmental argument, which had been deployed for a long time by the Kanaks of Thio, as in the case of the N'Goye conflict of 1978, was taken up again by the head of the subdivision, Jean-Jacques Arlié, who raised the question of compensation when he asserted that 'the customary people are right'. While SLN supported the establishment of what it called an 'economic interest group' (*groupement d'intérêt économique*) with the task of combating pollution, the company refused to accept responsibility for a century of negative impacts on the environment, and preferred to pass the buck to the state.[20] While pollution of the environment could be seen as a mode of appropriation in its own right (Le Meur 2010), SLN also deployed the environmental argument in a technical register in the land negotiations: the lands being claimed by local people were not actually required for mining, but they would act as a buffer against the damage associated with the activity.

The politics of land was central. As Charles Moindou was quoted as saying: 'This country must be restored to the landowners' (Huillet 1996a). In this regard, the localised conflict of 1996 is certainly part of the long trajectory of the 'meta-conflict' of settler colonialism, updated by demands for independence and land reform. The agreement of 1996 has several clauses dealing directly with land, concerning the Baie de la Mission residential development and claims over Koua, N'Goye and other land areas. These cases were to be dealt with by land allocations in the years that followed. However, the demand made by Charles Moindou referred to a different 'layer' of land—that is, not the legal one dealt with through statutory claims, acquisitions and allocations,[21] but the deeper, historical and identity-based layer that calls for a response in terms of the recognition of Kanak antecedence, indigeneity and equal dignity (see Honneth 2000). In this regard, the conflict of 1996 constituted a turning point as it was situated in the logic of the land reform that was in full swing by the 1990s, while entirely prefiguring this demand for recognition without

---

20 Jean-Michel Valois, the director of the company's mining operations, was quoted as saying that '[t]his problem is enormous but it is up to the state to take responsibility for it' (Huillet 1996b).
21 The status of reserves, reserve extensions, clan-based transfers and the entities known as *groupements de droit particulier local* ('local groups with special rights') would be transformed by the provisions on customary lands in the Organic Law of 1999.

necessarily involving any claims for the allocation of specific areas of land.[22] This demand would arise in several subsequent conflicts and agreements that were not confined to the mining sector, but lay at the root of other large-scale economic projects (Le Meur and Mennesson 2012; Le Meur et al. 2013).

Although the conflict of 1996 manifested a tendency to publicise local disputes more widely, as well as the increasingly complex politics of the mining arena, it was still characterised by a localised framework that had not yet been influenced by the global discourses that did not come to prominence in New Caledonia until the next decade. As previously indicated, there was no explicit mention of 'indigenous rights' in the sense defined by the United Nations, but this did not hinder the emergence of a discourse on identity, antecedence and local citizenship. Likewise, environmental non-governmental organisations had no part to play in the conflict, but this did not mean that environmental issues were absent from the conflict. There were no lawyers on the horizon, and that may be related to the presence of a centralised power represented by the *sous-préfecture* (subdivision), expressing a political-administrative logic rather than one based on legalistic forms of conflict management.[23] SLN had not yet woken up to the idea of corporate social responsibility, and sustainable development had not yet emerged as the conceptual mediator of divergent interests (Sourisseau et al. 2015). The very brief preamble to the 1996 agreement confirms these absences, but in some respects, they are also hollow presences:

> In consideration of the claims expressed by the councils of the two districts of Thio of 25 July 1996, in consideration of the responses provided by SLN, not wishing to further penalise the economic and social climate of the commune, mindful of the need to provide the population of this commune with a long-term development project through the management of natural resources, the following has been agreed.

---

22 The land reform process, which was launched in 1978 and continued in various institutional forms up to the 2000s, was based on the acquisition of areas of private land by a public agency and their redistribution in the legal form of customary land to Kanak collectives (tribes, clans or groups of clans) based on criteria that combined the logics of proof (historical group legitimacies) and negotiation (current land requirements) (Le Meur 2011).

23 This may be contrasted with the negotiation of the agreement negotiated for the Grand Sud (Goro) project, which was signed by the customary authorities, the indigenous association called Rhêêbù Nùù, and the Brazilian mining company Vale in 2008 (Horowitz 2012).

A Turning Point in Local Mining Governance?

Twenty years have passed since the agreement of 1996, a period that gives us the distance required, first, to ascertain whether it has been complied with and had any effect and, second, to evaluate the extent to which it really was a turning point, or 'bifurcation' (Mahoney 2000), in the history of mining governance in Thio.

With regard to the first question, despite the non-binding character of the agreement, it cannot be denied that its various points have been translated into concrete action, albeit at varying speeds. The measure that took longest to implement was the granting of the mining licence to SOMIKAT. More than ten years passed before the local Kanak mining company was formally established in 2007. Priority was given to the question of local employment, and the committee established to monitor the implementation of the agreement functioned for around ten years.

The question of subcontracting leads to the second question. In this regard, the undertakings in the 1996 agreement were ultimately exceeded. Several local subcontractors were established and, in some cases, replaced well-established Nouméan companies. One such example is the local security company (APST) that was established in 2007 by Alcide Tiéoué, the brother of Amédée Tiéoué, after the latter had fallen out with the management of the Nouméan company that had previously dominated the local market for security services.[24] Several of the local subcontracting companies established after 1996 opted for a logic of intertribal equity, but were also partly owned or operated by Wallisian, Indonesian and 'European' elements of the population. One of them was even called the Société de Développement des Ethnies Réunies—the Company for the Development of United Ethnic Groups. Others, like those owned by members of the Moindou family, adopted a more individual and/or familial strategy, thereby risking the veiled criticism of some relatives, but not losing sight of this issue in their recruitment of personnel. In all cases, irrespective of the actors involved, the local dynamics exceeded the point reached by local people's views and demands in 1996.

---

24   Amédée had been that company's branch manager in Thio, which was one of the entrepreneurial roles he had played during the conflict.

During the APST episode, SLN sided with Alcide Tiéoué and supported his enterprise, very probably out of concern to keep the social peace. So did things change completely? As Ithupane Tiéoué, former secretary to the mayor of Thio, reminds us: 'With SLN, you always have to fight to get something' (interview, 26 November 2010). The case of the tribe of Ouroué, whose history of environmental damage and displacement is closely associated with mining, further illuminates this point. Having obtained an agreement similar to that of 1996, also following a conflict, their leaders had to 'increase the pressure again' for the agreement to be implemented, even to the extent that new negotiations had to be held and a new text put on the table in 2007. Not insignificant in this regard is the fact that Narcisse Mapéri, the 'small chief' of this tribe, wears a range of different hats: as a prominent member of the Union Syndicale des Travailleurs Kanak et des Exploités (USTKE), political party member (Parti Travailliste), and as the first president of the Comité Autochtone de Gestion des Ressources Naturelles (CAUGERN).[25]

The transformation of mining governance, inspired in part by the conflict and agreement of 1996, is obvious, and is located in the reconfiguration of the relationships between SLN, the commune and the customary authorities. SLN gradually withdrew from functions that did not have a direct economic significance and that it had long performed by acting as a substitute for the state (Le Meur 2009). Local networks and infrastructure were 'communalised', and the company's housing schemes were privatised. The tripartite agreement between SLN, the commune and South Province, which was signed in 2008, further consolidated this trend.[26] The fact that this agreement marked the entry of the province into the political process, and the relative marginalisation of the customary representatives, may be seen as a result of the electoral victory of the anti-separatist Avenir Ensemble ('Future Together') political party, which formed as a result of a split in the loyalist block, and which was representative at the time of a more 'autonomist' vision of New Caledonia's destiny.[27]

---

25 CAUGERN was established in Thio in 2005. If it has a lasting political influence in the commune, that influence has been largely 'subterranean' (Narcisse Mapéri, interview, 2 October 2013).

26 This was the precursor of other agreements of the same nature with North Province that were signed in Poya and Koumac in August 2011.

27 This agreement was also far less localised than that of 1996 (or that of 2004 in Tiébaghi), and also reflected a distinctly 'urban' perspective (René Féré, interview, 20 February 2012).

The mining governance regime that was developed in Thio after 'the events' is placed under the constantly evoked and constantly rejected horizon of the 'post-nickel era'—primarily by SLN playing on the possible end of mining to calm down social movements, but also by other actors, especially the pro-independence mayors who have constantly striven to outline a post-mine future for Thio. This mining governance regime is characterised both by the mutual repositioning of SLN and the town hall, against a background of frequently difficult relationships,[28] and by transformations in the customary world. The latter changes have arisen both in respect of land, with the growing importance of the demand for recognition without a direct link to claims for land redistribution (as seen earlier), and in respect of the chieftaincy, with the emergence of a new generation of younger chiefs who may have had careers as entrepreneurs.[29]

The conflict of 1996 was structured by the tension between localisation (of situations, conflicts and arrangements) and globalisation (of the reference frameworks, arenas and discourses). Its emblematic nature as a turning point in mining governance should be appreciated on two different levels that reflect this tension. We have seen how, even if it is clear from a local perspective that this conflict constituted a turning point in terms of the scope of the mechanisms it triggered, the assessment of it should still be qualified. Viewed from a wider national perspective, and in the historical context featuring the Matignon-Oudinot and Nouméa agreements, the conflict of 1996 may be viewed retrospectively as the harbinger of a transformation of the mining policy domains, both from the discursive point of view, as in the qualification of the 'pre–indigenous conflict' or the 'pre–sustainable development' preamble to the agreement itself, but also in terms of the actor configuration. The way in which land issues and customary issues were combined in the framework of a renewed politics of recognition was also new.

---

28 The difficulties were documented in interviews with the current (post-2014) mayor, Jean-Patrick Toura, and his predecessors, Thierry Song (2005–14), Albert Moindou (2001–05), Louis Mapéri (1985–2001) and Roger Galliot (1971–85).

29 'The time when the Kanaks watched the trucks rolling past has gone,' as M. Boéhé, the mayor of Houaïlou, another east coast mining commune, said in relation to the youth movements in his commune which opposed the pollution caused by SLN (Huillet 1996b).

Final Remarks in the Light of Recent Events

To advance the analysis further, it is necessary to develop the comparative work initiated in the context of the research program on mining governance (e.g. Le Meur et al. 2012; Grochain 2013), and to shift the focus to other regional situations that are not exclusively associated with mining. For example, in New Zealand, a form of Maori entrepreneurship has emerged in a context characterised by the recognition of biculturalism, political-tribal reorganisation and neoliberal ideology (Rata 2000). The case of Papua New Guinea is also interesting in this respect, with the role (and shortcomings) of development forums, in which the local impacts and benefits of mining projects must be negotiated in advance (Filer 2008), and with the development of local entrepreneurship in the mining sector (Bainton and Macintyre 2013), which in many ways resembles the dynamics observed in New Caledonia.

Recent local developments in New Caledonia are also revealing, especially the chain of confrontations triggered in Thio by heavy rains and environmental damage in 2013. The ensuing mobilisation—with a mine blockade reminiscent of the 1996 conflict—encountered other sequences of events: one was the local implementation of mine site regulations stipulated by the 2009 mining code; another was the exploration campaign that several mining companies wished to launch on the 'forgotten coast'— south of Thio and north of Yaté—where mining operations had ceased 30 years previously in another climate of conflict. These interlinked events confirm the turning point hypothesis in terms of a more complex mining arena with proliferating and recombining discursive repertoires (Le Meur 2015a, 2015b). The 2013 conflict, with its focus on mining impacts, actually pushed companies into acting as honest corporate (and local) citizens. It represented at once a continuation of the 1996 dynamics— with local people trying to regain control over their resources, territories and lives—and a major shift from away from the *conflit des coutumiers*, as the 2013 mobilisation adopted an explicitly transethnic stance, including Thio residents of European and Polynesian descent who played no part in the 1996 conflict. This inclusive strategy has not prevented communal leaders—some of them CAUGERN members[30]—from making use of

---

30 This point is based on interviews with two CAUGERN members—the previously mentioned Narcisse Mapéri, USTKE member and former chief of the Ouroué tribe (2 October 2013), and Jean-Guy M'bouéri, former customary senator and leader of the association created in the wake of the 2013 conflict (4 October 2013 and 17 September 2014).

the indigenous toolbox, for instance by invoking the principle of 'free, prior and informed consent' (Szablowski 2010). At the same time, the two-year mining ban on the 'forgotten coast' declared by Thio and Yaté customary authorities and indigenous organisations in 2014, after a series of meetings among local people and with mining companies, has followed a different pathway rooted in the discourses of clanship and indigeneity. However, these collective actions all display a strong sense of local identity and express the multilayered nature of citizenship and sovereignty (Joyce 2013). By the same token, they also question the boundaries of the firm and the intricacies of corporate and local governance.

References

Allen, M.G., 2013. 'Melanesia's Violent Environments: Towards a Political Ecology of Conflict in the Western Pacific.' *Geoforum* 44: 152–161. doi.org/10.1016/j.geoforum.2012.09.015

——, 2017. 'Islands, Extraction and Violence: Mining and the Politics of Scale in Island Melanesia.' *Political Geography* 57: 81–90. doi.org/10.1016/j.polgeo.2016.12.004

Bainton, N.A., 2009. 'Keeping the Network out of View: Mining, Distinctions and Exclusion in Melanesia.' *Oceania* 79: 18–33. doi.org/10.1002/j.1834-4461.2009.tb00048.x

Bainton, N.A. and M. Macintyre, 2013. '"My Land, My Work": Business Development and Large-Scale Mining in Papua New Guinea.' In F. McCormack and K. Barclay (eds), *Engaging with Capitalism: Cases from Oceania*. Bingley (UK): Emerald Group Publishing (Research in Economic Anthropology 33). doi.org/10.1108/s0190-1281(2013)0000033008

Banks, G., 2008. 'Understanding "Resource" Conflicts in Papua New Guinea.' *Asia-Pacific Viewpoint* 49: 23–34. doi.org/10.1111/j.1467-8373.2008.00358.x

Dauphiné, J., 1987. *Chronologie Foncière et Agricole de la Nouvelle-Calédonie, 1853–1903*. Paris: L'Harmattan.

David, C., Sourisseau, J.-M., Gorohouna, S., Le Meur, P.-Y., 2016. 'De Matignon à la Consultation sur l'Indépendance. Une Trajectoire Politique et Institutionnelle Originale.' In S. Bouard, J.-M. Sourisseau, V. Geronimi, S. Blaise and L. Ro'i (eds), *La Nouvelle-Calédonie Face à Son Destin: Quel Bilan à la Veille de la Consultation sur la Pleine Souveraineté.* Paris: Karthala.

Demmer, C., 2007. 'Autochtonie, Nickel et Environnement: Une Nouvelle Stratégie Kanake.' *Vacarme* 39: 43–48. doi.org/10.3917/vaca.039.0043

Dousset-Leenhardt, R., 1976. *Terre Natale, Terre d'Exil.* Paris: Editions Maisonneuve & Larose.

Filer, C., 1997. 'Compensation, Rent and Power in Papua New Guinea.' In S. Toft (ed.), *Compensation for Resource Development in Papua New Guinea.* Boroko (PNG): Law Reform Commission (Monograph 6). Canberra: The Australian National University, National Centre for Development Studies (Pacific Policy Paper 24).

——, 2008. 'Development Forum in Papua New Guinea: Upsides and Downsides.' *Journal of Energy & Natural Resources Law* 26: 120–150. doi.org/10.1080/02646811.2008.11435180

Freyss, J., 1995. *Economie Assistée et Changement Social en Nouvelle-Calédonie.* Paris: Presses Universitaires de France.

GoNC (Government of New Caledonia), 2009. *Le Schéma de Mise en Valeur des Richesses Minières de la Nouvelle-Calédonie.* Nouméa: Direction de l'Industrie, des Mines et de l'Énergie.

Grochain, S., 2013. *Les Dynamiques Sociétales du Projet Koniambo.* Nouméa: Editions IAC.

Hagberg, S., 1998. *Between Peace and Justice: Dispute Settlement between Karaboro Agriculturalists and Fulbe Agro-Pastoralists in Burkina Faso.* Stockholm: Almqvist & Wiksell (Uppsala Studies in Cultural Anthropology 25).

Hervieu, P., 1996. 'À Thio, les Tribus Pourraient Lever les Barrages Aujourd'hui.' *Les Nouvelles Calédoniennes*, 30 July.

Honneth, A., 2000 [1992]. *La Lutte pour la Reconnaissance.* Paris: Editions du Cerf.

Horowitz, L., 2012. 'Translation Alignment: Actor-Network Theory, Resistance, and the Power Dynamics of Alliance in New Caledonia.' *Antipode* 44: 806–827. doi.org/10.1111/j.1467-8330.2011.00926.x

Huillet, F., 1996a. 'Thio: Les Contumiers Bloquent la SLN.' *Les Nouvelles Calédoniennes*, 26 July.

——, 1996b. 'Tensions sur les Mines.' *Les Nouvelles Calédoniennes*, 27 July.

Joyce, R., 2013. *Competing Sovereignties*. London: Routledge.

Latour, B., 2005. *Reassembling the Social: An Introduction to Actor-Network Theory*. Oxford: Oxford University Press.

Le Meur, P.-Y., 2009. 'Opérateurs Miniers, Gouvernementalité et Politique des Ressources à Thio, Nouvelle-Calédonie.' Paper presented to the Pacific Science Intercongress, Papeete, 3–6 March.

——, 2010. 'La Terre en Nouvelle-Calédonie: Pollution, Appartenance et Propriété Intellectuelle.' *Multitudes* 41: 91–98. doi.org/10.3917/mult.041.0091

——, 2011. 'Politique et Savoirs Fonciers en Nouvelle-Calédonie: Retour sur une Expérience d'Anthropologie Appliquée.' *Journal de la Société des Océanistes* 132: 93–108.

——, 2013. 'Locality, Mobility and Governmentality in Colonial/Postcolonial New Caledonia: The Case of the Kouaré Tribe (xûâ Xârâgwii), Thio (Cöö).' *Oceania* 83: 130–146. doi.org/10.1002/ocea.5009

——, 2015a. 'Making Peace with the Mining Past? The Politics of Value and Citizenship in Thio, New Caledonia.' Paper presented to the 10th European Society for Oceanists Conference, Brussels, 24–27 June.

——, 2015b. 'Anthropology and the Mining Arena in New Caledonia: Issues and Positionalities.' *Anthropological Forum* 25: 405–427. doi.org/10.1080/00664677.2015.1073141

Le Meur, P.-Y., S. Grochain, M. Kowasch and D. Poithily, 2012. 'La Sous-Traitance Comme Interface: Rente Minière, Contrôle des Ressources et Arènes Locales en Nouvelle-Calédonie.' Nouméa: CNRT Nickel, Programme Gouvernance Minière (Document de Travail 10).

Le Meur, P.-Y., L. Horowitz and T. Mennesson, 2013. '"Horizontal" and "Vertical" Diffusion: The Cumulative Influence of Impact and Benefit Agreements (IBAs) on Mining Policy-Production in New Caledonia.' *Resources Policy* 38: 648–656.

Le Meur, P.-Y. and T. Mennesson, 2012. 'Accords Locaux, Logique Coutumière et Production des Politiques de Développement en Nouvelle-Calédonie.' *Revue Juridique, Politique et Économique de Nouvelle-Calédonie* 19: 44–51.

Letté, M., 2009. 'Débordements Industriels dans la Cité et Histoire de Leurs Conflits aux XIXe et XXe Siècles.' *Documents pour l'Histoire des Techniques* 17: 163–173.

Levacher, C., 2016. De la Terre à la Mine? Les Chemins de l'Autochtonie en Nouvelle-Calédonie. Paris: École des Hautes Études en Sciences Sociales (PhD thesis).

Mahoney, J., 2000. 'Path Dependence in Historical Sociology.' *Theory and Society* 29: 507–548. doi.org/10.1023/A:1007113830879

Merle, I., 1998. 'La Construction d'un Droit Foncier Colonial: De la Propriété Collective à la Constitution des Réserves en Nouvelle-Calédonie.' *Enquête* 7: 97–126.

——, 2004. 'De la "Légalisation" de la Violence en Contexte Colonial: Le Régime de l'Indigénat en Question.' *Politix* 17(66): 137–162. doi.org/10.3406/polix.2004.1019

Pitoiset, A., 2002. *L'Actionnariat Populaire en Province Nord de la Nouvelle-Calédonie: Société de Profit dans une Société de Partage.* Nouméa: Université de Nouvelle-Calédonie.

Rata, E., 2000. *A Political Economy of Neotribal Capitalism.* Lanham (MD): Lexington Books.

Regan, A.J., 1998. 'Causes and Course of the Bougainville Conflict.' *Journal of Pacific History* 33: 269–285. doi.org/10.1080/00223349808572878

——, 2014. 'Bougainville: Large-Scale Mining and Risks of Conflict Recurrence.' *Security Challenges* 10(2): 71–96.

Sourisseau, J.-M., S. Bouard, C. Gaillard, P.-Y. Le Meur, T. Mennesson and G. Pestaña, 2015. 'Entre Neutralisation et Requalification: Les Limites de la Diffusion du Référentiel du Développement Durable en Nouvelle-Calédonie.' In S. Blaise, C. David and V. David (eds), *Le Développement Durable en Océanie: Vers une Éthique Nouvelle?* Marseille: Presses Universitaires de Provence.

Szablowski, D., 2010. 'Operationalizing Free, Prior and Informed Consent in the Extractive Industry Sector? Examining the Challenges of a Negotiated Model of Justice.' *Canadian Journal of Development Studies* 30: 111–130.

Turner, V., 1957. *Schism and Continuity in an African Society.* Manchester: Manchester University Press.

Winslow, D., 1993. 'Mining and the Environment in New Caledonia: The Case of Thio.' In M. Howard (ed.), *Asia's Environmental Crisis.* Boulder (CO): Westview Press.

6. Contesting the Goro Nickel Mining Project, New Caledonia: Indigenous Rights, Sustainable Development and the Land Issue

CLAIRE LEVACHER

Introduction

Since the middle of the 1990s, indigenous groups have become increasingly active in the international arena with a view to publicising the social, environmental, health and political impacts of mining projects on their lands, territories and resources. The extent of this activity contributed, in particular, to the formulation of Articles 26–32 of the United Nations Declaration on the Rights of Indigenous Peoples, which was adopted in 2007.[1] These articles establish the right to redress, compensation, protection of the environment and, beyond this, the right to self-determination

---

1   The articles in question concern the rights of indigenous peoples to lands, even in the absence of the allocation of formal property rights by the state; the recognition of this by the international community of states and the protection of the territories involved; the right to redress, restitution and compensation; the conservation of the environment and the production capacity of the lands; the mitigation of the adverse effects of production on the lands; and the absence of military activity. Article 32 states, finally, that: 'Indigenous peoples have the right to determine and develop priorities and strategies for the development or use of their lands or territories and other resources.' The states

in terms of development options. In fact, the mobilisation of environmental issues in the context of global mining conflicts frequently constitutes an argument for the rejection of industrial mining projects. And yet, the association of indigenous struggles with environmental battles is not systematic, particularly in countries in which the mining sector can constitute a development tool for the indigenous population (Ali 2009). This is precisely what I propose to investigate here through the examination of a conflict involving the company INCO (subsequently Vale) in relation to the Goro nickel mining project in the South Province of New Caledonia, which concluded in 2008 with the signing of a 'sustainable development pact' between the local Kanak population and the company—the Pacte pour un Développement Durable du Grand Sud.

Based on the results of a qualitative study carried out as part of a doctoral thesis from 2011 among the inhabitants of the municipality in question, institutional representatives, employees of the Goro Nickel Project, and other political actors, I aim to revisit the phases of a conflict with the mining company that has already been examined from both environmental (Horowitz 2009, 2010, 2012) and political (Demmer 2007, 2012; Djama 2009; Le Meur 2010) perspectives. The various twists and turns in this mobilisation process, between 2001 and 2008, illustrate the progressive formalisation of an argument relating to the rights of indigenous peoples, which links the question of development with management of the natural environment.

The different actors who emerged during this period expressed different representations of links between people and land. The questions posed here concern the extent to which these representations enable us to gain a better understanding of the 2008 agreement, and what they tell us about the different actors' visions of development. I begin this analysis with an account of the contemporary history of New Caledonia that will facilitate an understanding of the emergence of a discourse of indigenous rights in the country. I shall then revisit the different phases of the conflict with a view to comparing the economic and environmental aspects of the project, hence enabling a reinterpretation of the 2008 agreement. Thus I hope to highlight the issues on which it was silent, and investigate the role of the local public authorities and their own concept of development.

---

are obliged to consult them in relation to the use or exploitation of minerals, water and other resources and must establish redress mechanisms to mitigate harmful effects on the environmental, economic, social, cultural or spiritual levels.

New Caledonia and the Indigenous Question

New Caledonia is currently involved in a process of 'negotiated decolonisation' triggered by the Matignon-Oudinot Agreements of 1988 and confirmed by the Nouméa Agreement of 1998. These agreements contained provisions for the organisation of a referendum between 2014 and 2019 that would define the country's future and organise the transfer of powers from the French state to New Caledonia. The country has since been divided into three provinces (North, South and Loyalty Islands) with a view to promoting its economic, social and political 'rebalancing'. The Nouméa Agreement provides for the legal and political recognition of the historical legitimacy of the Kanak people's ties to the land. It recognises the existence of a Kanak people distinct from French people, and designates them as the basis for the construction of a 'common destiny' with the territory's other ethnic groups. The recognition resulting from these agreements constituted a new stage in the history of the indigenous question in New Caledonia.

During the colonial period, the indigenous people were subjects of the French empire. Their categorisation by the French colonial administration was based on two factors: the recognition of a separate (customary) legal system described as a *droit civil particulier* ('special civil law') (Saada 2003: 12–15), and the need of the European colonists to more clearly mark the legal boundaries between French citizens and the subjects of the empire. These boundaries were defined by the decree of 22 January 1868, which established the collective ownership of Melanesian territory by instituting the 'reserve' and the 'tribe' (*tribu*) as both land-based and administrative units.[2] The land tenure regime resulted from the seizure of Kanak lands by the colonial settlers and caused the progressive regrouping and displacement of relatively dispersed settlements and political structures.

Prior to the colonial period, the customary Kanak chieftaincy was a political space delimited by the topographical landmarks of a 'country' (Bensa 1992), and the social identities of Kanak clans were based on group itineraries that provided the basis or justification for the appropriation of an area of land (Naepels 2006). The chieftaincies were broadly (re)structured by the colonial administration, which established a form of indirect rule in several steps over a period of almost 40 years (1860–1900) by creating a system of

---

2   The term *tribu* originates from the segregation of the Kanak populations of New Caledonia in reserves, which was first carried out in 1876 (see Merle 1999).

administration in which the *grand chefs* ('big chiefs') had authority over a district in which they nominated the *petit chefs* ('little chiefs') who were in charge of each *tribu* (Bensa 2000).

The 1970s witnessed the emergence of independence claims based on the recognition of the historical legitimacy of the Kanak people and their culture, and a demand for the right of self-determination. In many ways, the discourse on the link with the land developed by the Kanak nationalist movement during the 1970s adopted an indigenous rhetoric in the strict sense. It aimed to demonstrate the attachment of the people to their lands and the antecedence of this attachment.[3] What was involved here was the constitution of the political entity of the 'Kanak people' and the legitimisation of the right to accommodate the 'others'—the non-indigenous people who arrived with colonisation. The Matignon-Oudinot and Nouméa agreements confirmed and substantiated this recognition through the establishment of new institutions. In the political domain, the chieftaincies now have their own representative bodies—the Sénat Coutumier (formerly the Conseil Coutumier) and the eight customary area councils. Although it does not have any executive powers, the Customary Senate has a recognised right to be consulted on any proposals for laws or policies that 'concern the Kanak identity' (under Articles 1.2.5 and 2.1.4a of the Nouméa Agreement) or Kanak land (under Article 1.4). With regard to the latter, however, the recognition is mainly reflected in the establishment of the Agence de Développement Rural et d'Aménagement Foncier (ADRAF), which is responsible for implementing the rehabilitation of degraded land by means of a process of land reform—a process that is still underway today. The 'customary lands' involved in this process include the original reserves and the areas restored to customary ownership since 1946. They are governed by customary law, subject to the customary authorities, and are inalienable.

With the emergence of the United Nations discourse on the rights of indigenous peoples, the period launched by the Matignon-Oudinot Agreements saw the development of a new interpretation of the indigenous question in New Caledonia. By focusing reflection on the maintenance and defence of Kanak specificity in the context of the future state—whether independent or not—the change in the local political idiom

---

3   In French scholarship on Ancient Greece, it was the acknowledgment of a shared origin and antecedence that provided the basis for the establishment of specific rights, especially in Athens (Loraux 1996; Detienne 2003). Kanak demands were based on the same principle.

was clearly rooted in the 'politics of recognition'. This movement opened with the establishment of the Association pour la Commémoration de l'Année des Peuples Indigènes en Kanaky in 1993, and its successor, the Conseil National du Peuple Autochtone, against the background of the 'international decade of indigenous peoples' (Monnerie 2005; Demmer 2007). It testified to the multiplicity of possible interpretations of the future of New Caledonia, which ranged from the exercise of sovereignty within the French state to full and complete independence (Graff 2012).

The United Nations discourse on the rights of indigenous peoples was mobilised with particular reference to the management of natural and mineral resources. In effect, the Nouméa Agreement imposed the development of a *schéma minier* ('mining plan') through Article 39 of the Organic Law, which stipulated that the ways in which natural resources are exploited had to be redefined. The mining sector represents the country's second most important economic resource after transfers from the French Government, and was the subject of tough negotiations in the lead-up to the signature of the Nouméa Agreement. While the establishment of a nickel smelter in North Province was central to this process (see Chapters 2 and 3, this volume), the mining plan made no mention of the construction of another nickel processing plant on the Goro site in South Province. In 2002, a new association, Rhéébu Nùù ('Eye of the Country'), was established on the initiative of the chieftaincies of Goro, Unia and Touaourou in the municipality of Yaté, where the Goro project is located. The aim of this body was to support indigenous rights as defined in the United Nations framework and to protect the environment. Following several years of conflict with the developers— first INCO and then Vale[4]—the Pact for the Sustainable Development of the Great South was signed in September 2008.

This agreement rests on three pillars. The first explicitly provides for reparation through afforestation, the training of environmental technicians and the setting up of plant nurseries. The second establishes a consultative customary environmental committee to act as an intermediary between the company and the populations in the area around the project, and to ensure the participation of the customary bodies in environmental monitoring

---

4   The Canadian transnational corporation INCO acquired the mining titles to the Goro massif in 1992. This company was acquired by Vale, a Brazilian transnational corporation, in 2006.

activities. The third involves the establishment of a corporate foundation with a mission to involve the local populations 'in socioeconomically and culturally sustainable development'.

The conflict with the company appeared to be rooted in the process of decentralisation or 'provincialisation' that resulted from the Matignon-Oudinot Agreements and the unequal nature of power relations within and between the provinces. While the dominant political forces in North Province are pro-independence, South Province is dominated by loyalists who oppose the country's independence from France. This situation has enabled the fragmentation of pro-independence political positions between the provinces. Provincialisation promoted the emergence of claims for additional collective rights by the southern Kanaks in the management of issues that have a direct impact on them, especially the plant in Mont-Dore and the mine in Yaté. As a result, they got involved in an ongoing quest for the rebalancing of power relations with the mining company and the province, and the production of strategies and tools to achieve it.

On one hand, the conflict with the company was rooted in a context characterised by the weakness of the regulatory framework: it contributed to, and took shape within, a new process of 'bottom-up' public policy production (Le Meur et al. 2013). Hence, there was a growing awareness of a situation of dependence at the local level, to which the claim for collective rights to the management of resources constituted a response.

On the other hand, the conflict also represented a struggle for recognition.[5] It was the customary authorities that initiated this quest for new rights, aimed at defining the territories and modalities of the exercise of their own power. Opposition to the mining project involved different levels of power and representations of the political arenas that emerged rapidly at the interface of the development and environmental issues raised by construction of the plant.

---

5   According to Taylor (1994), the concept of recognition is critical to an understanding of social conflict in the context of modern forms of nationalism, minority claims, and the quest for new rights. Struggles for recognition borrow from numerous registers that play a role in the reconstruction of arguments based on existing hybrid concepts of community and identity (Jacob and Le Meur 2010).

Development and the Indigenous Shaping of Land Legitimacy

The desires and fears of the population with regard to this mining project emerged in the wake of the social impact study commissioned by INCO and carried out by the Géosystèmes, Environnements et Cartomatique Océaniens research unit of the Université Française du Pacifique. They can be summarised as having three focal points: jobs, environmental management and Kanak cultural specificity (Jost 1998). This and subsequent studies revealed elements comparable to the concerns of the people affected by the Koniambo nickel project in North Province (see Chapters 2 and 3, this volume). It was predicted that the Koniambo project would have 'positive economic impacts through employment and training, that these economic effects [would] be shared equally between the people living in the project area and, finally, that the negative consequences on the level of the ecological environment [would] be minimised' (Grochain 2007: 216).

The question of employment was initially considered at the level of the *tribus* in the area bordering the project. The *tribus* of Yaté were interested in the development of an economic system that would be able to combine customary principles with those of the market economy by using the institutions of the 'special civil law', especially the associations known as 'local groups with special rights', *groupements de droit particulier local* (GDPLs).[6] However, the multiple economic initiatives that emerged at this time were caught up in a context characterised by profound divisions, and involved a competition for economic activities and spatial legitimacy between the two municipalities affected by the mining project—Yaté and Mont-Dore.

From the time of construction of the pilot plant in 1999, the establishment of the mining complex rekindled divisions between the two municipalities that were closely linked with an older contest over the spatial boundary between them (Frouin 2010). The *tribus* in each of the municipalities—three from Yaté and two from Mont-Dore—initially aimed to establish their own companies, and in 1999 they tried to create a form of economic

---

6   The GDPL is a legally recognised structure, with a legal personality, originally created to promote economic activities on customary lands, but also deployed in the context of land claims. It brings together individuals who are connected through customary ties—within a family, clan or *tribu*—and is managed under customary law.

organisation based on a land claim made by the chieftaincies of the two municipalities. In 2001, the same five *tribus* made another attempt to establish a single GDPL in order to organise the subcontracts in all areas of economic activity relating to the proposed smelter—including transport and logistics, earth-moving during the construction phase and subsequent extraction of the ores (Anon. 2001). However, a GDPL representing the *tribu* of Goro quickly broke ranks with this structure.

The withdrawal of the Goro *tribu* had the effect of refocusing the earlier GDPL within the municipality of Yaté alone. The municipal authority of Yaté established a monitoring committee in 2001, which was meant to cover all of the construction activities associated with the project.[7] This was a forum through which the elected and customary representatives could regulate local individual initiatives by ensuring that they became integrated into the collective management of the mineral wealth. In 2002, this process resulted in the establishment of a model for the participation of the population of the Grand Sud ('Great South'), which would later be replicated in North Province (see Chapter 2, this volume). This involved the establishment (in 2003) of a joint venture (*société par actions simplifiée*) called SAS Goro Mines, whose main function was to allocate project-related subcontracts to local businesses.

This model (Figure 6.1) is based on the institutions associated with the 'special civil [Kanak] law', namely the customary authorities and the GDPLs, combined within a type of non-trading company known as a *société civile de participation* (SCP). In this instance, the Xéé Nùù SCP included all of the investors originating from the municipality, which meant that each *tribu* established had its own GDPL for the purpose of collective shareholding, alongside those 'small shareholders' who wished to make individual investments. The Xéé Nùù SCP itself held the majority of shares in SAS Goro Mines, and also had shares in another company called Sud Restauration. In terms of distributing the revenues generated by the mine, SAS Goro Mines was conceived as a way of combining the companies already positioned to secure subcontracts for the pilot plant, and then to accommodate other local companies that were yet to be established. The revenues that accrued to the SAS would be shared through the SCP

---

7   This body included representatives of the mining company (Goro Nickel), the Betchel HatchTechnip consortium, which was in charge of the building works; the Chamber of Commerce; government officials, including personnel from ADRAF; local experts; the customary authorities of the Djubéa-Kapume area; and the elected representatives of the municipality.

partnership between the different collective and individual shareholders. To avoid conflicts regarding the allocation of contracts between the entrepreneurs, and to ensure equal representation of all the *tribus* and clans of the Yaté municipality, the SCP also had the task of nominating managers for the companies participating in the SAS. The province was also involved in supporting the establishment of local companies, particularly in the catering sector, through a development fund called PROMOSUD, which was a separate element in the model of participation.

SHAREHOLDING

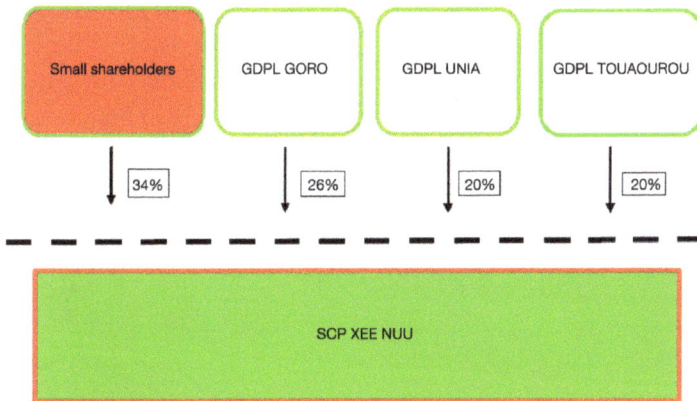

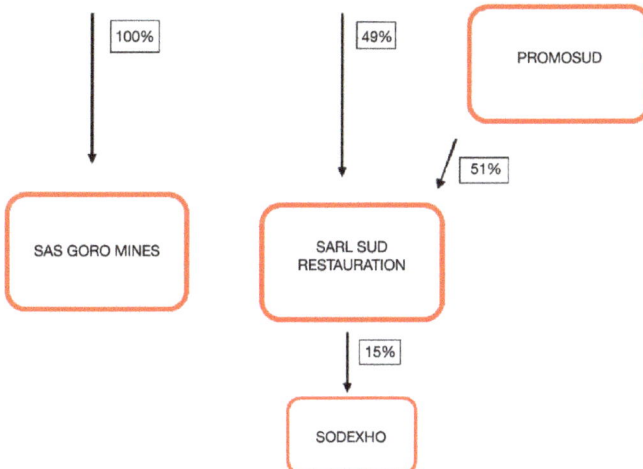

Figure 6.1 The model for participation of the population of the Grand Sud in development of the mining project.
Source: Author's diagram.

From a perspective akin to that described by Bainton (2009) in relation to the Lihir mine in Papua New Guinea, the aim of these arrangements was to provide an economic basis for the customary legitimacy of the municipality's *tribus*. The adoption of this approach made it possible to circumvent the fact that the mine is not situated on land recognised as customary by the state. Within the total area of the municipality of Yaté, customary land is limited to a small coastal strip. Accordingly, the *tribu* of Goro, which is located on the seashore, at the foot of the plateau where the mining takes place, was recognised by the municipality's other *tribus* as having a legitimate land-based claim to the area in which the mine is located. This is reflected in the model (Figure 6.1) by the greater number of shares in the SCP partnership held by the Goro GDPL. This extension of spatial claims without actual land acquisition, to achieve the recognition of customary legitimacy in the context of mining projects, was already observed in Thio (see Chapter 5, this volume).[8] In the case of Yaté, the recognition of this group's rights to the mining area enabled the negotiation of a share in the economic activities of the mining company, even while the municipality's business base remained weak in terms of volume, skills and experience. This process may be viewed as a form of compensation that acts as a driving force in the development of the customary groups and, by extension, the municipality (see Bainton and Macintyre 2013: 142). However, in practice, it encouraged a tribal appropriation of relationships with the mining company, as evident in the fact that the son of the great chief of the Goro *tribu* was appointed as managing director of SAS Goro Mines.

The establishment of the SAS did not instantly result in its official recognition by the mining company as a partner and exclusive point of contact for the management and allocation of subcontracts, nor did this recognition feature in the 2008 agreement with the customary leaders of the region. The recognition in question was only achieved in 2011, with the signing of a framework agreement between the SAS and the mining company, but one year later, there was a change in the management of the SAS, and the new management got involved in a land claim on the Goro plateau. This claim demonstrated the difficulty posed by the problem of land tenure. Making the SAS into the sole preferred partner

---

8   It should be noted that a land claim with land acquisition had been attempted in 1994, when two clans from Goro requested that South Province grant them two plots of land located in the municipality of Mont-Dore, in areas bordering or intersecting the project zone. An arbitration process resulted in the reduction of the 5,000 hectares requested to 400 (Jost 1998: 490).

of the mining company entailed a recognition of the Goro company's managing director as the 'landowner' of the site. It thus promoted competition for appropriation of the land (Kowasch 2012: 212–13), as well as the relationship with the mining company that was based upon it. Throughout this first phase in local dealings with the mining company, it appeared that the *tribus* were the instigators of economic initiatives, and that they were organised around the authority of the chieftaincies in the context of claims that extended beyond the reallocation of land rights to the appropriation of relationships with the mining corporation. The emergence of a discourse about natural resources, shaped by the international discourse of indigenous rights, confirmed this tendency to extend highly localised claims by reformulating the relationship with the land and natural environment.

International Indigenous Rights and Spatial Legitimacy

Through these initial attempts at economic organisation on the level of the chieftaincies, it would appear that the desire to participate in mining development in the region constituted a factor that determined the call for land-based legitimacy in dealings with the mining company. The transformation of the monitoring committee of the municipal authority of Yaté into an indigenous committee (Rhéébu Nùù) formalised the idea of a right to development based on a public discourse of legitimate control over geographical and political territories. Two shifts arose from this process: the first was political, and resulted from the recognition of the link with the land, while the second concerned the reformulation of Kanak perceptions of the natural environment. This dynamic was based to a large extent on claims that were both highly localised (on the level of the chieftaincies) and more extensive (at municipal and provincial levels). The appearance of elements originating from international discourse of indigenous rights also endorsed the spatial and discursive extent of the claims being made on the mining company.

When the monitoring committee of Yaté municipality became the Rhéébu Nùù Committee in 2002, the change of name was indicative of a change in structure. The collective body thereafter claimed to represent the chieftaincies of the south in a process that was supposed to overcome the aforementioned economic and political divisions, but which also invoked the international rights of indigenous peoples that were still in the process of being formalised in the United Nations arena. The move by the chieftaincies to take control of the development process came to be part of a political process that made new connections between natural and political territories. This process went beyond the political recognition provided by the agreements, which was based on the idea of a link with the land, since it involved the extension of the customary authorities' prerogatives to matters beyond those usually reserved to them. For one of the leaders of Rhéébu Nùù from the Unia *tribu*, the process of colonisation created administrative territories that did not relate to the institutions of pre-colonial society, in which territories were based not only on areas bounded by natural features and clan itineraries, but also on the political links between chieftaincies. In his view, this ancient political system established the legitimacy of the customary authorities on the land where the mining plant is located and in their relationship with the mining company.

This spatial right of the chieftaincies was incorporated into the framework of the recognition of the link with the land by the Nouméa Agreement, but the ways in which it was exercised were defined with reference to the international rights of indigenous peoples. The 'Declaration on the Right over the Space and Natural and Cultural Heritage of Kanaky', commonly known as the 'Kanaky Declaration', was produced on 22 August 2002. It refers to the concrete implementation of 'the Kanak conception of the relationship with the space and the natural and mining heritage', especially through the principle of free, prior and informed consent. This principle, as defined in Article 19 of the United Nations Declaration on the Rights of Indigenous Peoples, involves consultation of indigenous peoples by states prior to the approval of any measure that could have a direct impact on their rights, especially when it involves activities carried out on traditional indigenous territories.

The Kanaky Declaration specified the conditions for the implementation of the principle in New Caledonia, including consultation of the clans directly concerned, the chieftaincy, the neighbouring chieftaincies, the customary council and the Customary Senate. In many ways, the declaration may be considered as the founding charter for indigenous claims regarding the management of natural resources in New Caledonia. It linked the local customary authorities with the Customary Senate, and rendered visible the demand for participation in development and control of the territories that concern these authorities. The Kanaky Declaration states that:

> In future, all major economic projects involving the exploitation of renewable and non-renewable natural resources shall involve the establishment of ways that enable the Kanak customary authorities to be *actors* … The indigenous Kanak people shall be *actors* in the definition of development plans [emphasis added].

The second aspect of the chieftaincies' spatial management claim concerned the reformulation of Kanak perceptions of the natural environment. It aimed at the integration of such perceptions into the international legal and policy framework relating to indigenous peoples. The newly conceived Kanak role in the protection of nature was based, in particular, on legal texts such as those produced by the United Nations Conference on Environment and Development in 1992, by the regional conferences held in Apia (1976) and Nouméa (1986), and by Article 8(j) of the Convention on Biological Diversity, which explicitly recognises the contribution made by indigenous communities to the conservation of the biodiversity of the regions they inhabit, thanks to their traditional way of life (Djama 2009: 55). This supported a discourse that not only made Kanak populations responsible for the protection of their own natural environments, but also cast them as a driving force for the establishment of environmental awareness at the national level:

> What I would also like to clarify is that … the intensification of the exploitation of resources … raises the question of … the role of the indigenous peoples with respect to the conservation of their environment and, hence, the country's environment … In a way, there is a need for the entire country to be at peace with nature … There is a need today to get this process underway through the rehabilitation of this concept of the natural heritage, the responsibility for which goes back to the indigenous peoples. (Mapou 2002: 23)

The evolution of Rhéébu Nùù's discourse in support of environmental protection constitutes a twist in the relationship with the mining company, which involved a (re)construction of identity, while also extending the spatial extent of the claims being made. Through the environmental argument, local and regional identities, which were recognised as being based on the relationship with the land, became part of a kind of ethnogenesis (Morin 2006)—a global and legal identity within the United Nations arena. The major advantage of this (re)constructed identity was that it appealed to the creative and imaginative force of the law in a context characterised by a legal void with respect to mining and environmental issues.

The provincial plan for the exploitation of mineral wealth was only completed in 2008, and the mining and environmental regulations in 2009. The designation of the 'Lagoons of New Caledonia' as a UNESCO world heritage site was clearly enshrined in this process, and would mark the settlement of several years of technical struggles against the mining company for compliance with international environmental standards and greater respect for the environment. Hence, the environmental aspect of the Rhéébu Nùù Committee's indigenous struggle took shape in a strategic alliance formed with environmentalists—especially local environmental organisations—with a particular focus on protection of the lagoon.[9] This alliance enabled the bundling of the expertise of the environmental organisations with the local people's concerns about the risks posed by pollution and dumping of heavy metal wastes from the processing plant. The focus on protection of the lagoon thus involved a shift in the nature of indigenous interests from the mining site towards the sea. The extension of the geography of the claims therefore contributed to the entry of new actors and arguments into the ranks of the opposition to the project.

The Heritage in Question: Mining Rent at the Service of the Environment

The 'heritage' question provides a key to understanding the extent of the coexistence of environmental and development concerns expressed by Rhéébu Nùù through its interest in land issues. It links the environment with all the forms of value that can be attributed to it—economic, political or cultural. The signatories to the Kanaky Declaration defined their conception of the Kanak natural heritage as one that had to serve the causes of development and spatial management through the collection of a rent calculated on the basis of the value of the mined ore—or what

---

9   This strategic alliance raised questions regarding the environmental studies carried out with respect to the mining project—whether impact studies commissioned by the developer or the numerous counter-studies commissioned by its opponents.

would normally count as a kind of 'royalty'.[10] Hence, this patrimonial heritage assumes different dimensions that contribute to the economic, environmental, social and cultural shaping of the spaces and territories around the project. The discourse of the Rhéébu Nùù Committee established a link between the protection of the environment, the negotiation of economic rights to participate in the process of economic development, and the rights of the indigenous people through the definition of compensation for destruction of the environment that they considered as their property. This understanding of compensation was reflected in two specific demands: one for the creation of a municipal mining tax, and the other for the formation of a heritage fund for environmental rehabilitation.

Having been accused of wanting to negotiate the receipt of 'royalties', the indigenous leaders developed a more general point of view on the mineral tax regime at the national level, despite the fact that neither the plan for the use of mineral wealth nor the mining regulations, both of which were envisaged under the terms of the Nouméa Agreement, had actually been completed. The tax exemption measures of 2001 revealed the small share of benefits that the municipalities would be able to accrue from mining operations. The National Law on the Taxation of Nickel Mining for Major Metal-Processing Projects, which was passed in July that year, included corporate exemptions from business licensing taxes, property taxes, registration fees and general service taxes for a period of 15 years from the beginning of commercial operations. The mining tax payable to the municipality of Mont-Dore involved surcharges on these exempted taxes, and the reason why these surcharges only applied to Mont-Dore is that mining taxes are currently only levied on the processing of ore and not on its extraction. The demands of the Rhéébu Nùù leaders for a 'mining tax for the benefit of the municipalities and populations affected by this activity' (Anon. 2003) were made in this context of uncertainty about the definition and distribution of the economic benefits of mining both at the local level and for New Caledonia as a whole, where there were many possibilities offered by the reconstitution of the tax regime. As others saw it, a share of mining taxes constituted an alternative to the proposed acquisition of a stake in the mining company, in particular through a proposition for the buyback of part of the capital. The recognition of

---

10   While royalties are commonly defined as a kind of tax collected by the state or 'crown' in its capacity as the owner of subsurface mineral resources, customary landowners in Papua New Guinea have long been recognised as having an entitlement to a share of this type of mineral revenue.

the link with the land depended on the greater economic involvement of the local population, whether by means of equity or taxation, since these appeared to be the only possible levers for obtaining real partnership and an effective commitment on the part of the mining company to the development of the region and the country.[11]

The Rhéébu Nùù Committee's detractors accused it of negotiating a tax regime that would solely benefit the chieftaincies, without regard for the discourse of environmental protection. In 2006, Philippe Gomes, the new president of South Province, firmly opposed the payment of any kind of royalties to the owners of the land on which one or more of the mines were operating, basing his argument on the view that 'the Caledonian industrial projects lead the local populations to be actors involved in development and not rentiers' (Pitoiset and Prandi 2006). While some people now acknowledge that they had envisaged such an outcome, they say it would have been abandoned quickly in favour of the current model of participation (Figure 6.1), which was conceived as a means of distributing the benefits equitably while favouring individual and clan-based economic initiatives and financial incentives. The municipal mining tax system also made it possible to ward off or avoid financial and land claims through the integration of the concerns of the chieftaincies and the municipality for local development on the same level. In this regard, another indigenous representative body, the Comité Autochtone de Gestion des Ressources Naturelles (CAUGERN), which was formed in 2005, has initiated a process of joint reflection by the chieftaincies that wanted to be involved in development and the municipalities that wanted to benefit from additional resources, since both believed themselves to be entitled to enter into a relationship with the mining company and to benefit from the associated economic spin-offs (Demmer 2012: 36; also Chapter 4, this volume).[12]

From 2002, the demands of the Rhéébu Nùù Committee also involved the establishment of a 'heritage fund' that was, in the words of the Kanaky Declaration, to be 'determined on the basis of the intrinsic value of the processed primary resource'. So the declaration established the fact that providing employment was not enough, and that each cubic metre

11   A majority shareholding in the mining project was to be the route taken in the case of the Koniambo project in North Province (see Chapter 2, this volume).
12   This fusion of interests was achieved in Yaté through the municipal elections of 2008, which resulted in the election of a mayor from a list of Rhéébu Nùù candidates.

of nickel extracted has an economic value that should be used as a basis for the establishment of a financial entity devoted to the promotion, restoration and protection of the Kanak natural and cultural heritage, in the name of the rights of the clans and chieftaincies to their 'natural space', and under their own political control. Rhéébu Nùù defined this fund in 2003 as 'the cornerstone of the indigenous process', and stated that it should also be based on what the Kanaky Declaration called the 'value of the replacement of the 1000 hectares of forest that had already been destroyed'. In 2004, the president of Rhéébu Nùù again raised the question of an output-based mining fee, to which the mining company responded by pointing out the lack of New Caledonian legislation in this area. In 2005, the question was back in the spotlight and expressed as follows:

> It is essential that part of the mining wealth be used to compensate for the consequences of decades of exploitation. A special tax should be created to finance the environmental rehabilitation of the area damaged by the mine. (Frédière 2005)

The demands of the Rhéébu Nùù Committee and CAUGERN were completely in line with the thinking of the Comité Stratégique Industriel, which was established in 2010. This is a bipartisan body composed of government and mining company representatives, under the leadership of Anne Duthilleul, representing the French Government, which has been tasked with production of a mining and industrial strategy for New Caledonia. In 2005, a report by the Senate of the French Government on tax exemptions for the nickel processing plants expressed concern that a kind of 'freedom from tax' was being ordained, and preferred the adoption of a 'mining fee' of the kind proposed by Anne Duthilleul: 'it is essential that the local authorities be given the resources to fulfil their general interest functions; the interventions of the state cannot entirely replace the just and fair taxation of the mining sector' (quoted in Le Meur and Mennesson 2011: 9).

This suggestion was not completely rejected by the loyalist parties in New Caledonia, since one of them had drafted a proposal in 2004 for the creation of a heritage fund for the benefit of the populations affected by the Goro project. This proposal was not implemented, but there is now a de facto rehabilitation fund (the 'Nickel Fund') that redistributes the land taxes paid by the mining companies to the municipalities that have old mine sites in need of rehabilitation.

As we can see here, the indigenous collective that initiated the battle with the mining project expanded the content and scope of its demands beyond the issue of environmental protection. Through defence of the natural environment, not only was the question of environmental conservation tackled but also the means of achieving it and, beyond this, the means of attaining sustainable development based on the collection of a mining rent.

Conclusion: The Pact, Mining Rent and Sustainable Development

I set out to demonstrate that different types of arguments were mobilised by the strategic groups that have constituted and transformed Rhéébu Nùù as an indigenous collective, while also subject to change in their own internal constitution. As a result, the rights of indigenous people have become rights that link environmental protection and rehabilitation with economic development. These rights have served in the negotiation of more specific rights to the management and administration of the project-affected areas. The shift from a localised economic perspective towards a broader environmental one, in terms of both demands and actors, does not mean that the two approaches are mutually exclusive. This spatial and discursive expansion of the indigenous struggle had complementary impacts from the point of view of the legal provisions that emerged from it, and it also reflects different conceptions of sustainable development.[13]

The 'environmental turning point' in the battle against the industrial project has in a way contributed to shaping new questions for public policy by underlining the environmental effects of mining activities. The Environmental Regulations of South Province refer to the principle of consent, the establishment of an environmental compensation fund and the formalisation of environmental impact studies. The designation of the Lagoons of New Caledonia as a UNESCO world heritage site in 2008, and the establishment of New Caledonia's environmental monitoring agency, the Observatoire pour l'Environnement, in South Province in

---

13 In fact, the 'sustainable development' frame of reference only emerged on the interface between the mining sector and its local management in New Caledonia in the early 2000s. Until then, the development models were dominated by the transition from an 'assisted economy' (Freyss 1995) to economic independence based on mining rent, economic 'rebalancing' and state transfers (Sourisseau et al. 2015).

2009, are events that clearly link environmental protection to mining development. They also raise new ways of thinking about sustainable development in New Caledonia. The strategy for the integration of sustainable development into provincial policy only took shape in 2007, by linking economic and environmental issues around ecotourism projects that were conceived as a form of compensation for the negative impacts of mining activities (Sourisseau et al. 2013). In this regard, South Province's absence from the Pact for the Sustainable Development of the Great South could be indicative of a sustainable development conception that is more focused on 'rurality' and on spatial conservation than on the regulation of the mining sector (Le Meur et al. 2013).

The pact itself made sustainable development into a point connecting the diverging interests of indigenous peoples and mining companies. While the former wanted to protect their natural, social and economic environment, the latter wanted to reduce the risk of facing more stringent legislation and greater civil society mobilisation (see O'Faircheallaigh 2011). As a point of overlap between international 'soft laws'—like the 'good practice guides' developed by the International Council on Mining and Metals or the Declaration on the Rights of Indigenous Peoples— the pact tends to maintain the enclave logic inherent in projects of this type with a certain definition of sustainable development being applied to specific areas. For the customary people of the south, and for the Rhéébu Nùù collective in particular, the local implementation of sustainable development principles appeared to take root in forms of negotiation over the distribution of mineral resource rent. The local model of participation (Figure 6.1) highlighted two things: first, that landownership was not a sufficient basis for establishing a relationship with the industrial concern; and second, that the internal redistribution mechanisms would constitute a driver of development on customary lands, and especially of private economic initiatives.

The pact was also conceived as a form of compensation in the sense that it put an end to an abusive relationship. In this case, the recognition of customary rights did not unfold through the payment of rent to customary landowners in exchange for land acquisition by the state, as has been observed in Papua New Guinea (Strathern 1993; Filer 1997), but through the establishment of new mechanisms and new customary structures that enabled the negotiation and regulation of the development process (see Banks 2008). Beyond the strict economic mechanisms to which it could be reduced, analysis of the compensation arrangements in terms of relationships reveals the strength of the institutions that further optimised local control over natural resources. Thus a form of rent collection and relationship with the industrial concern depends on each of these levels of organisation and negotiation, be it the chieftaincy, the municipality, or the Djubéa-Kapume area as a whole. However, the extension of the discourse of Rhéébu Nùù to the mining tax regime aimed to leave behind this localising logic in favour of considering the mode of industrial development and the meaning of sustainable development. Like the reflection on sustainable development, the 2002 Kanaky Declaration testifies to this desire to consider in their totality the areas being developed or earmarked for development. What is involved, therefore, is the collective consideration of environmental conservation, compensation and the benefits to be gained from mining, at both the local level and that of New Caledonia as a whole.

References

Ali, S.H., 2009. *Mining, the Environment, and Indigenous Development Conflicts*. Tucson: University of Arizona Press.

Anon., 2001. '19 Milliards: Le Cher Projet des Coutumiers du Sud.' *Les Nouvelles Calédoniennes*, 24 July.

——, 2003. 'Gisement de Prony: Les Réactions.' *Les Nouvelles Calédoniennes*, 26 December.

Bainton, N.A., 2009. 'Keeping the Network Out of View: Mining, Distinctions and Exclusion in Melanesia.' *Oceania* 79: 18–33. doi.org/10.1002/j.1834-4461.2009.tb00048.x

Bainton, N.A. and M. Macintyre, 2013. '"My Land, My Work": Business Development and Large-Scale Mining in Papua New Guinea.' In F. McCormack and K. Barclay (eds), *Engaging with Capitalism: Cases from Oceania*. Bingley (UK): Emerald Group Publishing (Research in Economic Anthropology 33). doi.org/10.1108/S0190-1281(2013)0000033008

Banks, G., 2008. 'Understanding "Resource" Conflicts in Papua New Guinea.' *Asia-Pacific Viewpoint* 49: 23–34. doi.org/10.1111/j.1467-8373.2008.00358.x

Bensa, A., 1992. 'Terre Kanak: Enjeu Politique d'Hier et d'Aujourd'hui: Esquisse d'un Modèle Comparatif.' *Etudes Rurales* 127–128: 107–131. doi.org/10.3406/rural.1992.3383

———, 2000. 'Le Chef Kanak: Les Modèles et l'Histoire.' In A. Bensa and I. Leblic (eds), *En Pays Kanak: Ethnologie, Linguistique, Archéologie, Histoire de la Nouvelle-Calédonie.* Paris: Éditions de la Maison des Sciences de l'Homme (Ethnologie de la France 14).

Demmer, C., 2007. 'Autochtonie, Nickel et Environnement en Nouvelle-Calédonie: Une Nouvelle Stratégie Kanake.' *Vacarme* 39: 43–48. doi.org/10.3917/vaca.039.0043

———, 2012. 'La "Chefferie Minière" en Question: Etude de Cas à Canala.' Nouméa: CNRT Nickel CNRT Nickel, Programme Gouvernance Minière (Document de Travail 13).

Detienne, M., 2003. *Comment être Autochtone: Du pur Athénien au Français Raciné.* Paris: Seuil.

Djama, M., 2009. 'Politiques de l'Autochtonie en Nouvelle-Calédonie.' In M. Salaün, M. Thibault and N. Gagné (eds), *Autochtonies: Vues de France et du Québec.* Québec: Presses de l'Université Laval.

Filer, C., 1997. 'Compensation, Rent and Power in Papua New Guinea.' In S. Toft (ed.), *Compensation for Resource Development in Papua New Guinea.* Boroko (PNG): Law Reform Commission (Monograph 6). Canberra: The Australian National University, National Centre for Development Studies (Pacific Policy Paper 24).

Frédière, P., 2005. 'Raphaël Mapou Réussit sa Sortie du Palais.' *Les Nouvelles Calédoniennes*, 9 February.

Freyss, J., 1995. *Economie Assisteé et Changement Social en Nouvelle-Calédonie.* Paris: Presses Universitaires de France.

Frouin, A.-L., 2010. Revendications de l'Autochtonie et Contrôle des Ressources Naturelles au Sud de la Nouvelle-Calédonie: L'Exemple de la Mobilisation Kanak dans la Commune de Yaté. Université Aix-Marseille (Masters thesis).

Graff, S., 2012. 'Quand Combat et Revendications Kanak ou Politique de l'Etat Français Manient Indépendance, Décolonisation, Autodétermination et Autochtonie en Nouvelle-Calédonie.' *Journal de la Société des Océanistes* 134: 61–83. doi.org/10.4000/jso.6647

Grochain, S., 2007. Les Kanak et le Travail en Province Nord de la Nouvelle-Calédonie. Paris: Ecole des Hautes Etudes en Sciences Sociales (PhD thesis).

Horowitz, L.S., 2009. 'Environmental Violence and Crises of Legitimacy in New Caledonia.' *Political Geography* 28: 248–258. doi.org/10.1016/j.polgeo.2009.07.001

——, 2010. '"Twenty Years Is Yesterday": Science, Multinational Mining and the Political Ecology of Trust in New Caledonia.' *Geoforum* 41: 617–626. doi.org/10.1016/j.geoforum.2010.02.003

——, 2012. 'Translation Alignment: Actor-Network Theory, Resistance, and the Power Dynamics of Alliance in New Caledonia.' *Antipode* 44: 806–827. doi.org/10.1111/j.1467-8330.2011.00926.x

Jacob, J.-P. and P.-Y. Le Meur, 2010. 'Citoyenneté Locale, Foncier, Appartenance et Reconnaissance dans les Sociétés du Sud.' In J.-P. Jacob and P.-Y. Le Meur (eds), *Politique de la Terre et de l'Appartenance: Droits Fonciers et Citoyenneté Locale dans les Sociétés du Sud*. Paris: Karthala.

Jost, C., 1998. 'Perceptions Sociales et Dimensions Culturelles d'un Nouveau Projet Minier en Nouvelle-Calédonie.' In D. Guillaud, M. Seysset and A. Walter (eds), *Le Voyage Inachevé ... A Joël Bonnemaison*. Paris: ORSTOM.

Kowasch, M., 2012. 'Le Développement de l'Industrie du Nickel et la Transformation de la Valeur Environnementale en Nouvelle-Calédonie.' *Journal of Political Ecology* 19: 202–220.

Le Meur, P.-Y., 2010. 'La Terre en Nouvelle-Calédonie: Pollution, Appartenance et Propriété Intellectuelle.' *Multitudes* 41: 91–98. doi.org/10.3917/mult.041.0091

Le Meur, P.-Y., L.S. Horowitz and T. Mennesson, 2013. '"Horizontal" and "Vertical" Diffusion: The Cumulative Influence of Impact and Benefit Agreements (IBAs) on Mining Policy-Production in New Caledonia.' *Resources Policy* 38: 648–656. doi.org/10.1016/j.resourpol.2013.02.004

Le Meur, P.-Y. and T. Mennesson, 2011. 'Le Cadre Politico-Juridique Minier en Nouvelle-Calédonie: Mise en Perspective Historique.' Nouméa: CNRT Nickel, Programme Gouvernance Minière (Document de Travail 3).

Loraux, N., 1996. *Nés de la Terre*. Paris: Seuil.

Mapou, R., 2002. 'Information sur le Massif de Prony.' *Journée Mondiale des Peuples Indigènes*, 9 August.

Merle, I., 1999. 'La Construction d'un Droit Foncier Colonial: De la Propriété Collective à la Constitution des Réserves en Nouvelle-Calédonie.' *Enquête* 7: 97–126. doi.org/10.4000/enquete.1571

Monnerie, D., 2005. *La Parole de Notre Maison: Discours et Cérémonies Kanak Aujourd'hui (Nouvelle-Calédonie)*. Paris: Éditions de la Maison des Sciences de l'Homme.

Morin, F., 2006. 'L'Autochtonie, Forme d'Ethnicité ou Exemple d'Ethnogenèse?' *Parcours Anthropologiques* 6: 54–64.

Naepels, M., 2006. 'Réforme Foncière et Propriété dans la Région de Houaïlou (Nouvelle-Calédonie).' *Etudes Rurales* 177: 43–54.

O'Faircheallaigh, C., 2011. 'Use and Management of Revenues from Indigenous-Mining Company Agreements: Theoretical Perspectives.' Melbourne: University of Melbourne, Agreements Treaties and Negotiated Settlements Project (Working Paper 1).

Pitoiset, A. and M. Prandi, 2006. 'Le Projet Néo-Calédonien de Goro Stoppé.' *Les Échos*, 4 April.

Saada, E., 2003. 'Citoyens et Sujets de l'Empire Français: Les Usages du Droit en Situation Coloniale.' *Genèses* 53: 4–24. doi.org/10.3917/gen.053.0004

Sourisseau, J.-M., S. Bouard, C. Gaillard, P.-Y. Le Meur, T. Mennesson and G. Pestaña, 2015. 'Entre Neutralisation et Requalification: Les Limites de la Diffusion du Référentiel du Développement Durable en Nouvelle-Calédonie.' In S. Blaise, C. David and V. David (eds), *Le Développement Durable en Océanie: Vers une Éthique Nouvelle?* Marseille: Presses Universitaires de Provence.

Sourisseau, J.-M., S. Bouard and G. Pestaña, 2013. *Ruralité et Développement Durable en Nouvelle-Calédonie: Synthèse des Ateliers de Poindimié et Perspectives.* Province Nord: IAC Éditions.

Strathern, A., 1993. 'Compensation: What Does It Mean?' *TaimLain* 1: 57–82.

Taylor C., 1994. *Multiculturalism: Examining the Politics of Recognition.* Princeton (NJ): Princeton University Press.

7. Dissecting Corporate Community Development in the Large-Scale Melanesian Mining Sector

GLENN BANKS, DORA KUIR-AYIUS, DAVID KOMBAKO
AND BILL F. SAGIR

The chapter reports on an NZAID-funded project into corporate community development (CCD) initiatives at four Papua New Guinea (PNG) mine sites.[1] The project was undertaken by a joint team of researchers from Massey University and the University of PNG. The study was framed to examine the motivations and attitudes of mining corporations to community development, to document the activities they carried out under the banner of CCD, to see what lessons had been learned in terms of both successes and failures, and to explore the nature of interactions between local, national and international stakeholders. Before reporting on these findings and implications, we set the broader context within which CCD initiatives occur.

---

1   NZAID has been reincorporated into the Ministry of Foreign Affairs and Trade as the New Zealand Aid Program. For reasons that should become obvious, we use the term 'corporate community development' as an alternative to the term 'corporate social responsibility', which is often preferred by industry (see Banks et al. 2016).

What Drives Corporate Community Development?

A starting point is to consider the nature of the relationship between a multinational miner and the local community that is affected by it. Large-scale mining and its developmental effects are the subjects of both contention and ambiguity (Bebbington et al. 2008). The contradictions that mining poses, we believe, can be put down, in part at least, to an under-theorised approach to the nature of the relationship between these mining operations and affected communities. While there are a variety of ways of characterising this relationship (through the lens of power relations and agreements, human rights, or environmental impacts, among others) it is the economic flows from the mine that can be seen as central to shaping and driving this relationship through time. In the PNG context, these flows are shaped by the nature of the operation itself, as well as various sets of government regulations and pre-mining negotiations between the central stakeholders (including the community).

It is the compensation, wages, business contracts, royalties and equity dividends that are not only the inducements that entice communities to agree to mining developments on their land in the first place, but also drive the variety of processes that occur within communities around large-scale mines. Inward migration, one of the most destructive aspects of mining developments for local communities is, for instance, largely a consequence of people moving to the area seeking to access a share of some of the economic opportunities that the mine creates (see Chapter 11, this volume). Likewise, shifts to a cash economy, enhanced entrepreneurialism and individual ambition are all responses, through local cultural lenses, to the economic flows that spill out from the mining company. The social ills we associate with large-scale mining—gambling, prostitution, alcohol and violence—are not unconnected to these processes (Johnson 2011) (see Figure 7.1).

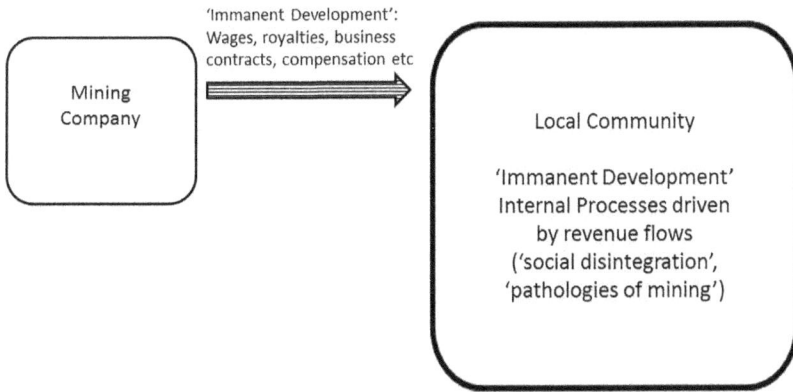

Figure 7.1 Mining and immanent development.
Source: Authors' diagram.

These internal processes can be regarded as what Cowen and Shenton (1996) describe as forms of 'immanent' development: inherent within community change, often unruly, typically energetic, always aspirational and frequently conflictual. Cowen and Shenton discuss these processes in the context of nineteenth-century European industrialisation and capitalism, describing how these broader forces recast the nature of European societies in ways that were often unpredictable and, in many cases, threatening to the established order. In the context of mining in Melanesia, terms such as 'social disintegration' (Filer 1990) or the 'pathologies' of mining (Golub 2006) are often used to characterise the rapid transformations that arise from the social effects of these mining operations on local communities. And despite many of these effects being seen as problematic by locals and observers, there is also a sense in which they are the sum of processes that people actively engage in to bring change to their lives and societies. There is an aspirational or progressive nature to much of what people are trying to do. We should also note here that these are precisely the sorts of social changes and processes that are regarded as raising social risks and threats to the mining operations themselves. The economic flows, then, drive processes that in turn are often regarded as a menace to the mining projects themselves, which are the sources of the flows. Significantly, though, when we talk of corporate social responsibility (CSR), it is clear that these immanent processes are not typically perceived as the sort of development space that resource companies can or should intervene in, or for which they should take any 'social responsibility'. This is despite Colin Filer's (1990) argument that the extent, structure and nature of the economic flows from the mine are a primary influence on the speed and form of these 'immanent' processes.

One cautionary note in relation to the language used here is the degree of convergence and overlap between the oppositional categories just outlined: 'the mining company' and 'the local community' are not entirely distinct entities. At Porgera, a 2.5 per cent share of the equity in the joint venture is held by a local entity on behalf of the mining lease landowners, and 'local' employees make up more than two-thirds of the operational workforce of the company. Equity stakes have varied at other sites, but in all cases local employees make up the dominant component of the workforce. In other words, the 'local' is an element of the capital (albeit minor) and the labour (a much more significant share) of the company and, indeed, there is also overlap to an extent in the visions for local development that many in the community and company hold.

At least since the closure of Bougainville in 1989, and certainly in the context of discourses around compliance to CSR standards and codes of the past decade, mining companies have moved to engage more directly in community development efforts (see Imbun 1994, 2006). In part we can see that these CCD programs seek to reduce the risk posed to the operation by the immanent forms of development already discussed (see Figure 7.2).

These corporate initiatives fit neatly into what Cowen and Shenton (1996) characterise as 'intentional' forms of development interventions. They are usually driven by forces external to the communities,[2] and are deliberate, strategic interventions into community processes. Cowen and Shenton's original concern was with the early form of religious and secular interventions during the industrial revolution that targeted those who had been marginalised by the 'immanent' development processes that capitalism and industrialism had sparked. They then applied this understanding to the contemporary aid and development landscape. What the authors highlighted was that these intentional forms of development are economically, socially and culturally *conservative*, and seek to essentially counter the unruly forms of change that immanent development creates. In what follows, we identify a range of corporative initiatives that fit this

---

2   Lihir is something of an exception to this rule (see Bainton 2010).

notion of conservative forms of intentional development in corporate programs that promote governance, law and order, education, health and cultural heritage.

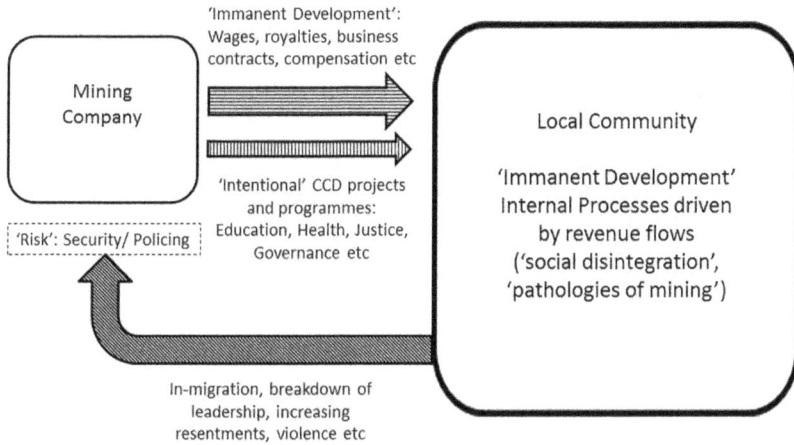

Figure 7.2 Mining, immanent and intentional forms of development.
Source: Authors' diagram.

There are two further points to be made in relation to the tensions between the immanent and the intentional. First, there is a constant tension between many of the 'intentional' projects and the 'immanent' processes they seek to counter. Corporate resource commitments to CCD are typically nowhere near as significant as what the community receives from the economic flows, and hence it is easy to identify cases where the intentional corporate development projects and schemes are undermined by, or remain marginal to, the broader immanent processes.

Second, drawing on the work of Ferguson (1990) and Li (2007) in relation to donor-funded development projects, we can also discern depoliticising tendencies at work in the intentional development work of CCD. The 'immanent' processes of community change are intensely political at local and higher levels, in part due to contests over the distribution of the revenue flows from the mine (see Burton 2014; Golub 2014). In contrast, the corporations tend to avoid becoming implicated in these political processes and, reflecting this, their intentional development programs have depoliticising tendencies: they seek to remain apolitical and focused on technical or material interventions, or what we can characterise as strengthening of infrastructure and institutions. Seeking an apolitical presence—depoliticising development—is itself, however, an intensely conservative political movement.

The Project

With this conceptual frame by way of backdrop, we now turn to report on a research project funded by NZAID, the New Zealand Government's aid and development agency, between 2009 and 2010. The project had two primary aims:

- to gain an understanding of the ways in which CCD programs reconciled tensions between community expectations and aspirations and corporate demands and constraints;
- to provide an initial investigation of the effectiveness of CCD programs at four mine sites in terms of broader measures of development effectiveness.

In broad terms, we were keen to gain and promote a greater understanding of CCD programs and use this to facilitate more integration between state, donor and corporate development initiatives. All four of us had previously worked in community development and/or the extractive sector in PNG, with both Sagir and Banks having long-standing research interests in the area, and the current project built on this existing body of research and knowledge. We visited four mine sites—Ramu and Porgera in June 2009, Lihir in February 2010, and Ok Tedi in October 2010—for periods of between three and seven days (see Figure 7.3). The sites were chosen in part to reflect the different geographical settings—from insular Lihir to the huge area impacted by the Ok Tedi mine—and in part to reflect the length of time they had been operating —Ok Tedi since 1984, Porgera since 1991, Lihir since 1997, and Ramu, which was still halfway through construction during our visit.

At each site we interviewed management, community affairs staff and other company personnel involved in local community development, local government officials, and local representatives and organisations. We also collected and collated secondary data on community development at each site, and spoke to representatives of other national stakeholders, such as the PNG Chamber of Mines and Petroleum.

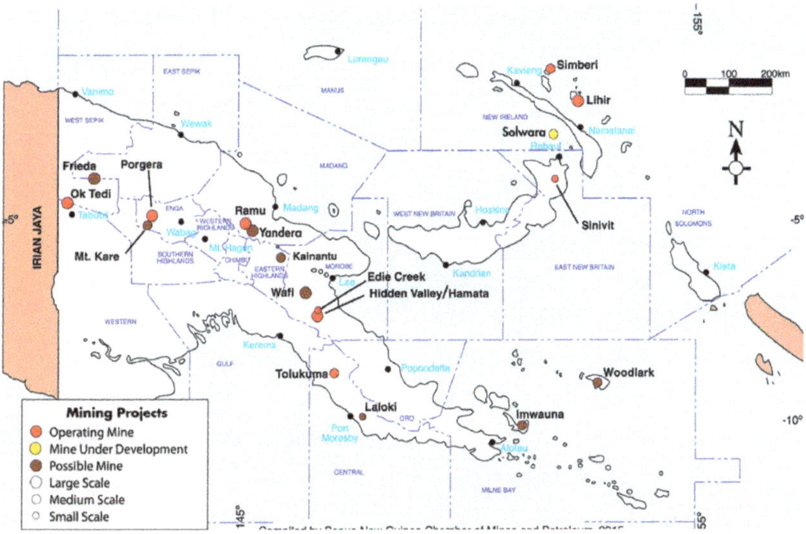

Figure 7.3 Papua New Guinea mining operations.

Source: PNG Chamber of Mines and Petroleum (pngchamberminpet.com.pg/mining-in-png/).

Research Findings

There is not scope here to provide even a brief overview of the CCD programs at each of the operations visited (see Banks et al. 2013 for more details). Instead, what follows is organised around five key themes that emerged from the research.

Scale and Range of Programs

First, it was clear that the scale and range of current CCD activities at each site made them very significant local drivers of development. Indeed, several were larger than the programs of most donor agencies operating in PNG. On Lihir, for example, the Lihir Sustainable Development Plan receives K32 million annually[3] from Newcrest Mining to support Lihirian development activities that range from health and education initiatives to cultural heritage and housing and infrastructure work. This payment is not the only support that Newcrest provides, but it is the most substantial.

---

3   The amount is adjusted each year to reflect movements in the consumer price index. The revised integrated benefits package was originally worth K20 million per year over five years.

In contrast, the Ok Tedi and Fly River Development Program Business Plan (OTFRDP 2010) outlined potential project funding sources worth up to K400 million.

The range of development activities undertaken by the companies was also notable, and included:

- law and order initiatives, particularly at Porgera (see BGC 2014);
- support for women's activities at Porgera, Ok Tedi and Lihir (see Chapter 10, this volume);
- various forms of support for local and provincial health services, especially at Lihir and Ok Tedi (see NFHSDP 2013);
- support for cultural heritage programs at Lihir (see Bainton et al. 2011);
- livelihood programs at Ramu, and to a lesser extent Ok Tedi and Porgera; and
- various forms of support for business development at all sites (see Johnson 2012; Bainton and Macintyre 2013).

While the list here is necessarily selective, all of these activities clearly align with the idea of this CCD support being targeted at conservative, apolitical initiatives: supporting law and order, women (and by implication traditional family structures), health, cultural heritage and small-scale (orderly) livelihood and business development. These forms of CCD thus fit neatly within Cowen and Shenton's model of 'intentional development'.

Motivations

There was a wide variety of motivations behind these various CCD initiatives. Indeed, different initiatives by the same mining company were typically carried out for a mix of different reasons, and even single programs or activities rarely had just one motivation. The motivations identified spanned a spectrum of 'voluntariness' that compounds the difficulty of discussing such activities as 'voluntary CSR' initiatives. We are able to identify and briefly sketch seven potential motivations for different types of activity.

The least voluntary of these community development activities are what can be labelled *negotiated obligations*, where commitments are encoded in agreements: 'we do these because we have to—it is in the agreement'. At the next level are *risk management activities*: 'we do these things to keep our operations safe'. The Restoring Justice Initiative developed by Barrick Gold in association with local district administration, provincial and national governments, and community stakeholders is a clear example of this form of community development (BGC 2014).

Less directly tied to operational safety are initiatives that can be seen to be linked to the idea of the *social licence to operate* (Owen and Kemp 2013): 'if we don't do these things the community will become frustrated and can pose a threat to the operation'. At times, this motivation can also stretch to individuals or groups within the community: the need to appease specific groups and individuals who can affect the operation arises regularly, and companies (sometimes unwillingly) feel obligated to meet these demands. It is often these focused initiatives targeted at certain groups that draw the ire of other locals who do not receive the same treatment.

These initiatives are closely aligned with, but distinctly different from, what can be more directly seen as CSR-driven initiatives. To varying degrees the multinational miners bring weighty global charters and lofty community-focused corporate rhetoric to their local programs. The Barrick response to the Human Rights Watch investigations into rape and sexual harassment by mine security personnel (HRW 2011), while rightly focused on seeking ways to reduce violence against women at Porgera, can be seen as an example, with the various initiatives seemingly tied more to global rhetoric and practices than local conditions and realities (see Burton 2014). Indeed, it appears to us that the differences between the 'social licence to operate' and CSR initiatives boil down to the degree to which they reflect local conditions as opposed to global corporate fixations.

In some cases, and in line with much of the critical literature on CSR and mining, the *public relations* value of some community development programs was certainly identified by both corporate and external interviewees as being of importance. The glossy photo for corporate sustainability reports and websites was a factor that provided a motivation for the corporate support for some programs, in part at least because it could be used to try to negate negative criticism of their presence by other (usually international) groups such as non-governmental organisations. Rarely, though, was it the sole or even primary motivation for activities or programs.

Also discernible was a strong element of altruism behind some corporate initiatives. Most of the 'community development' sections of the companies were staffed and managed by people with experience in, and commitment to, community development. These people often want to *do the right thing* by the community and empathised with local people's situation: at least one operation used the slogan 'leaving behind a better future' for its CCD strategy. There is, then, a 'will to improve' (Li 2007), even if this does not always translate into sustainable forms of development. Critics—both local and external—appear to overlook the extent to which staff in these sections are genuinely trying to bring about positive change for the affected communities, working within contexts that are not always conducive to this outcome.

Finally, in terms of motivation, business development assistance is an example of CCD initiatives that can have a relatively direct economic payoff for the companies involved. These activities can certainly lead to reduced contracting and operating costs, given that local contractors are likely to be significantly cheaper than national or international companies. Hence there can be an *economic* motivation behind some initiatives, although, like public relations, this was not the sole or primary motivation for any of the activities we identified.

What we can see, then, in terms of these 'intentional' development activities is a variety of motivations, some of which are clearly 'defensive' and some deliberately depoliticising and socially conservative (see Table 7.1). As previously noted, within any single program or activity there is often a degree of overlap between these categories—they are certainly not mutually exclusive. It is also worth pointing out two further points related to motivations.

First, in some senses, motivations matter less than outcomes, and there clearly are some CCD projects that, even if designed with strong elements of self-interest in mind, have delivered benefits to elements of some communities. There is also the view, among some of the corporate actors, that self-interested and self-managed corporate programs are more likely to be completed, and hence to deliver 'outcomes', than those that rely on government or local parties for their implementation. Sometimes such outcomes are additional to what was planned, and sometimes they are completely unintended but, regardless, these opportunities and improvements in livelihoods for individuals, and sometimes whole communities, have been significant at different points in time.

Second, and by way of contrast, there is good evidence from the world of development that undertaking activities with a mix of motives (e.g. self-interest *and* poverty reduction) is likely to be less effective than having programs with singular and clear motivations. Undertaking community development projects with the aim of achieving multiple goals (e.g. community development *and* risk reduction) is unlikely to be effective in achieving any one goal.

Table 7.1 A typology of motivations for corporate community development activities.

Degree of 'voluntariness'	Motivation	Rationale and examples
LOW	Negotiated responsibility	'It's in the agreement: we are required to do it' (compensation and royalty flows, resettlement programs, specified infrastructure)
	Risk management	'If we don't do it, we will get closed down' (restoring justice)
	Social licence to operate	'If we want to maintain community support, we need to do this' (water tanks)
	Corporate social responsibility	'If we want to maintain business reputation and international image, we need to do this' (HIV/AIDS)
	Public relations	'Looks good on our flashy brochures' (some footbridges, aid posts, schools, etc.)
	'The will to improve'	'Leaving behind a better future' (livelihood programs, women's groups)
HIGH	Direct business case	Cost reduction (business development, local training and education)

Source: Banks et al. 2013.

Organisational Structure

The research highlighted a wide diversity of organisational structures adopted to plan and deliver these CCD activities. There is clearly no one model of how corporations can or should deliver community development programs. As with the motivations behind their different initiatives, corporations utilise a variety of mechanisms for this purpose.

At one end sits the Lihir Sustainable Development Plan (LSDP). As previously noted, this program is funded directly by Newcrest, but in many senses it operates at arm's length from the company. The LSDP is the outcome of the second formal round of negotiations between the Lihir community and mine's operator, although the negotiations were concluded before Newcrest took over the operation, after the first round had resulted in a more standard set of agreements bundled together as an 'integrated benefits package'.[4] Much of the design of the LSDP, and perhaps more importantly its vision for community development, has come from the Lihir Mine Area Landowners Association (LMALA).[5] And despite the problem of 'elite capture'—the binding of local elites to corporate interests—the LSDP represents an all-encompassing, locally driven, designed and delivered community 'development' agency that leaves most donor programs far behind in terms of local 'ownership' of development.

Ok Tedi has a long (30-year) history of various development projects and initiatives. The present iteration is relatively recent and complex, in part a legacy of previous agreements and arrangements, as well as more recent initiatives that were envisaged to last beyond the current mine life (see Chapter 8, this volume). At the core sits the new Ok Tedi and Fly River Development Program (OTFRDP), which is gradually assuming a greater proportion of the previous community development roles performed by Ok Tedi Mining Ltd (OTML), and aims to increase its independence from OTML over the next few years. The structure is designed externally, but is now working closely with communities up and down the extensive Ok Tedi Fly River system to be able to respond better to community needs. The relationship between OTFRDP and other stakeholders, including the local and provincial governments, as well as the mine's operator, is still being developed, especially through the implementation of 'mine continuation agreements with local communities', but the intention is that OTFRDP should become the primary mechanism for delivering community development in Western Province by the time the mine finally closes.[6]

---

4   With some additions to the 'standard package', including having the mine's environmental plan included in the agreement. The third version of these agreements was still being negotiated in 2016.
5   Nick Bainton (personal communication, November 2015) has correctly noted that while the LSDP was largely conceived and written by LMALA, it contains an integrated vision for the development of the island with reference to consolidated local development budgets and multistakeholder management structures.
6   Organisational relationships were complicated by the nationalisation of the PNG Sustainable Development Program Ltd in 2013 (see Chapter 8, this volume).

At Porgera, CCD projects and programs have been much more centrally, perhaps conventionally, controlled and directed by the operator's community development department.[7] This has meant that the Porgera Joint Venture has been able to respond more rapidly to threats and opportunities that have arisen in its relationship with local communities, but it has lacked a clear systematic vision and consistent long-term approach to community development. In this sense, there is less independence from short-term, on-site and corporate head office decision-making, and the company's CCD efforts certainly have less of a community-driven flavour when compared to those at Lihir.

The nature of these different institutional arrangements is dependent on a range of factors, such as the various negotiated agreements that have specified particular systems and structures (as at Lihir and Ok Tedi), the history of company–community relationships (as at Porgera) and a mix of corporate, management and local decisions. The situation at the Ramu project, for example, shifted away from an emphasis on the Ramu Nickel Foundation with the arrival of the China Metallurgical Group Corporation as the project's developer. The geographical and cultural context also influences the type and scale of delivery mechanism that is used. The difference between Ok Tedi, with its huge riverine footprint, fragmented and culturally diverse population, and the small contained island environment at Lihir, where the local population has a cultural predisposition to visionary planning, is in part responsible for the very different approaches that have been adopted.

Sustainable Successes Hard to Find

One of the striking features of our survey was the limited number of successful sustainable initiatives that could be identified. Ok Tedi has been operating for more than 30 years, Porgera for more than 20 years, and Lihir since 1997. All, as we have seen above, have committed substantial sums to 'community development' in various guises over this time. And yet there were a limited number of 'successes' that could be pointed to in terms of corporate initiatives. Certainly a lot of new activities and programs had been developed in recent years, but far fewer longer-term projects were regarded by stakeholders as having made a difference to

---

7   The title of this agency has changed several times during the life of the project.

community development.[8] This was reflected in the impression one of the newer members of our team had that 'the companies were only starting community development work in the last couple of years'.

In large part this can be linked to the argument that these 'intentional' corporate development initiatives are responding to, and occur within, a context of the much broader 'immanent' social and economic trends within the surrounding communities. In other words, despite their size and resources, the companies have limited ability (or will) to shape the longer-term trajectory of community social and economic change around their operations. There is also the issue of an almost complete absence of long-term data against which community change, and the effects of corporate activities, can be tracked—a point to which we shall return.

Some projects and activities have clearly had significant outcomes for the affected communities. The ones that are discussed now are projects that in 2009–10 were regarded as having been beneficial to the communities over a longer time period. The Porgera District Women's Association is an institution that with ongoing corporate support has had a significant presence and impact for women at Porgera over a 20-year period. It has coordinated awareness campaigns for women, supported a microcredit scheme, and provided assistance with various livelihood activities (dressmaking and chicken farming, for example). At Lihir and Ok Tedi, there are long-standing health programs and facilities that have been highly effective in terms of health outcomes (Hemer 2005; Bentley 2011; Thomason and Hancock 2011). In economic terms, company assistance with 'landowner umbrella companies' has typically had a chequered path, but at Ok Tedi (Star Mountain Investment Holdings), Lihir (Anitua and its subsidiary, National Catering Services) and Porgera (Ipili Porgera Investments), some landowner companies have developed into major national companies with multimillion dollar businesses (Jackson 2015). These companies have the potential to be major diversified economic assets for the local communities in the post-mine environment, and represent a significant contribution to long-term development that was achieved with the active support of the mining companies involved.

---

8   This may also be a function of a shift from social impact *monitoring* to *reporting*, a point for which we thank Nick Bainton.

We were able to identify a number of instances where the resource company had provided support (mostly logistical) to government or donor projects, particularly health interventions. The Porgera Joint Venture (PJV), for example, had supported at least two Asian Development Bank health initiatives, including most recently an HIV/AIDs program at Porgera, building on studies by the PNG Institute of Medical Research that the PJV had also supported. Such collaboration is seen as particularly beneficial as it combines the developmental focus of the donor with the resources and logistics of the companies involved.

The relationship between the mining companies and the local and provincial governments in and around their operations had varied significantly through time and by location. There were situations where the developmental efforts of the company were closely linked with those of the district administrations and local-level governments, although the latter were usually highly politicised and of limited effectiveness in each of the locations. There were also cases, such as Porgera and Ok Tedi, where a number of attempts, largely funded and driven by the company, were made to develop a district planning framework, yet there was very little effective coordination in delivering development to communities. There was a tension here, though, with the widely observed 'retreat of the state' from these resource enclaves (see Filer 1997). This was partly driven by the pressure on companies to act as a 'pseudo-government' (see Chapter 8, this volume), partly by the huge apparent disparity in resources, capacity and effectiveness between company and local government, yet partly by the desire for companies to work more closely with local actors, including government actors. This latter trend, a reflection of best practice in the aid and development arena, is made more difficult where local government is barely present in the vicinity of the mine sites. This reflects broader concerns with the ways in which governmentality is constructed and practiced in relation to resource extraction (Ferguson 2005; Le Meur et al. 2013).

Corporations are Not Donors

The final issue we discuss here is the difference between these CCD activities and those of bilateral and multilateral donors. It is worth stressing the obvious and stating, as much of the critical literature on CSR does, that these mining companies are not bilateral development donors, and indeed their primary motivation is profit, not community development. Within this context, and to a large extent reflecting the range of motivations identified earlier, we can identify some significant differences between the community development activities of donors and those of these four mining operations.

The first is that community participation in the identification, design and implementation of CCD programs is often limited. While best practice development planning would involve a participatory approach from the start, and would be influenced by notions of development as empowerment, with a focus on building capacity and skills within the target community, the mining companies give little systematic attention to such approaches. Apart from a number of the key projects already identified, such as those at Lihir, very little of this was evident in terms of the practice of these CCD initiatives.

In part, this could be put down to the context within which these CCD programs operated. A significant element of this was the high degree of dependence on the mining company exhibited by local communities, which limits commitment by the community, and ultimately the company, to these community development projects. There is often limited broad-based community support for many of the initiatives because they are often not seen as being driven by local priorities. These priorities are typically dominated by the effects of the highly politically charged nature of the immanent development processes, including in-migration, landownership, representation and identity. This apathy and dependency is a trait frequently maligned by external observers: the lazy, greedy resource owner or landowner is a common stereotype within PNG. However, from the local perspective, given the obvious power, wealth and resource imbalances that exist between the corporation, the state and the local community, a range of culturally derived positions on dependency and reciprocity could be anticipated. Although these are now recognised in the development literature (see Mawdsley 2012), they are probably poorly understood and rarely dealt with by corporate actors.

In terms of other differences from donor development programs, there was an almost total absence of systematic monitoring and evaluation for these corporate programs. This is in contrast to the development industry, for which monitoring and evaluation are central to contemporary practice, and for which there is a specialised branch of the industry. Here we can identify the lack of clear accountability as a factor: lines of accountability for these programs remain closely tied to internal corporate systems rather than reflecting a notion of programs and projects being accountable to the recipient communities, as current aid practice at least normatively preaches. We see this as being important in terms of the success of these corporate programs, since without clear monitoring and evaluation of results—and tying this into institutional learning regarding what makes for successful development—there is little requirement for these programs to be accountable beyond the banal corporate financial accounting for all expenditures. CSR expectations today ensure that the corporations continue to expend resources on their community development sections regardless of their developmental outcomes, underlining the mix of corporate motivations noted earlier. What we do know from the literature is that mixed motivations for bilateral aid—classically trying to mix strategic and developmental objectives—reduces the effectiveness and sustainability of these efforts, and can even further undermine the development of the recipient. Our review here suggests the danger of the same poor outcomes arising from CCD programs that try to achieve a mix of objectives—risk management, development, meeting CSR obligations, and so forth.

Conclusions

Large-scale multinational mining projects are long-term ventures, typically lasting more than 20 years, in contrast to the generally much shorter time horizon of bilateral and multilateral donors. As a result, CCD programs have the potential to provide long-term development assistance to surrounding communities (UNDP 2014). What we have shown here is that these CCD efforts vary across the PNG mining operations, but all are a significant presence for local communities, and sometimes, as in the case of Ok Tedi, much more broadly (see Chapter 8, this volume). Unfortunately, the results of these efforts have typically been piecemeal, and successful sustainable activities are relatively rare. We hope that our review here can provide the start of a discussion about why this is the case.

We would argue that untangling the motivation for CCD programs is likely to improve their effectiveness (see GoA 1997). Mixing motives that include external CSR concerns with risk management, social licence and a desire for 'real' development is always going to lead to contradictions and compromised outcomes.

It is also clear that there was potential for substantial learning and greater cooperation across sectors. The corporations have carried out various forms of community development for decades, tried a variety of formats and approaches, and had a range of successes and failures. They have, as noted earlier, resources and a logistical capacity that is unrivalled in remote parts of the country. At the same time, it is clear from our review that there are elements of current development practice, such as monitoring and evaluation and designing effective participatory approaches, that could be of considerable benefit to the companies and, by implication, the recipient communities. More attention to these 'front-end' and 'back-end' aspects of projects and programs does mean both loosening up control over the design and subsequent direction of the programs, and being more open, transparent and accountable for the failures as well as the successes of these initiatives.

Finally, in terms of the schematic model that was posed at the start of this chapter, there is a need for more attention to the linking and integration of the CCD programs with the financial flows that the mining operations engender. The 'intentional development' of these corporate initiatives and programs occurs within the context of, and is often undermined by, the much broader and rapid 'immanent development' processes that are initiated by these financial flows. Indeed, in many respects CCD is, from the community viewpoint, peripheral to the 'main game': the negotiations and conflicts around the distribution of the economic flows from the operations. We would argue that this will remain the case until companies and communities are able to link these flows to participatory development activities that seek to address, in a positive way, the major 'immanent' shifts and transformations in the communities that are driven by these same financial flows. This may involve corporations being more interventionist in terms of the realm of local politics, for example by negotiating more integrated benefit sharing and community development programs. But failure to do so means not only abdicating responsibility for the 'immanent development' that their economic presence drives; it will also continue to mean that their CCD efforts will produce few sustainable, beneficial development outcomes for affected communities.

References

Bainton, N.A., 2010. *Lihir Destiny: Cultural Responses to Mining in Melanesia.* Canberra: ANU E Press.

Bainton, N.A., C. Ballard, K. Gillespie and N. Hall, 2011. 'Stepping Stones across the Lihir Islands: Developing Cultural Heritage Management Strategies in the Context of a Gold-Mining Operation.' *International Journal of Cultural Property* 18: 81–110. doi.org/10.1017/S0940739111000087

Bainton, N.A. and M. Macintyre, 2013. '"My Land, My Work": Business Development and Large-Scale Mining in Papua New Guinea.' In F. McCormack and K. Barclay (eds), *Engaging with Capitalism: Cases from Oceania.* Bingley (UK): Emerald Group Publishing (Research in Economic Anthropology 33). doi.org/10.1108/s0190-1281(2013)0000033008

Banks, G., D. Kuir-Ayius, D. Kombako and B. Sagir, 2013. 'Conceptualizing Mining Impacts, Livelihoods and Corporate Community Development in Melanesia.' *Community Development Journal* 48: 484–500. doi.org/10.1093/cdj/bst025

Banks, G., R. Scheyvens, S. McLennan and A. Bebbington, 2016. 'Conceptualising Corporate Community Development.' *Third World Quarterly* 37: 245–263. doi.org/10.1080/01436597.2015.1111135

Bebbington, A., L. Hinojosa, D.H. Bebbington, M.L. Burneo and X. Warnaars, 2008. 'Contention and Ambiguity: Mining and the Possibilities of Development.' *Development and Change* 39: 887–914. doi.org/10.1111/j.1467-7660.2008.00517.x

Bentley, K., 2011. *Lihir Social Demographic Health Survey.* Canberra: Centre for Environmental Health.

BGC (Barrick Gold Corporation), 2014. 'Neighborhood Watch: How the Restoring Justice Initiative Addresses Law and Order Challenges in the Porgera District.' Viewed 4 April 2017 at: barrickbeyondborders.com/people/2014/11/neighborhood-watch-how-the-restoring-justice-initiative-addresses-law-and-order-challenges-in-the-porgera-district/#.VIECMdKUfzg

Burton, J., 2014. 'Agency and the "Avatar" Narrative at the Porgera Gold Mine, Papua New Guinea.' *Journal de la Société des Océanistes* 138–139: 37–51. doi.org/10.4000/jso.7118

Cowen, M. and R. Shenton, 1996. *Doctrines of Development*. London: Routledge. doi.org/10.4324/9780203392607

Ferguson, J., 1990. *The Anti-Politics Machine: 'Development', Depoliticization, and Bureaucratic Power in Lesotho*. Cambridge: Cambridge University Press.

——, 2005. 'Seeing Like an Oil Company: Space, Security and Global Capital in Neoliberal Africa.' *American Anthropologist* 107: 377–382. doi.org/10.1525/aa.2005.107.3.377

Filer, C., 1990. 'The Bougainville Rebellion, the Mining Industry and the Process of Social Disintegration in Papua New Guinea.' *Canberra Anthropology* 13(1): 1–39. doi.org/10.1080/03149099009508487

——, 1997. 'The Melanesian Way of Menacing the Mining Industry.' In B. Burt and C. Clerk (eds), *Environment and Development in the Pacific Islands*. Canberra: The Australian National University, National Centre for Development Studies (Pacific Policy Paper 25).

GoA (Government of Australia), Committee to Review the Australian Overseas Aid Program, 1997. *One Clear Objective: Poverty Reduction through Sustainable Development*. Canberra: Committee to Review the Australian Overseas Aid Program ('the Simons Report').

Golub, A., 2006. Making the Ipili Feasible: Imagining Local and Global Actors at the Porgera Gold Mine, Enga Province, Papua New Guinea. University of Chicago (PhD thesis).

——, 2014. *Leviathans at the Gold Mine: Creating Indigenous and Corporate Actors in Papua New Guinea*. Durham (NC): Duke University Press.

Hemer, S., 2005. 'Health Care and Illness in Lihir, New Ireland Province, in the Context of the Development of the Lihir Gold Mine.' *PNG Medical Journal* 48(3–4): 188–195.

HRW (Human Rights Watch), 2011. 'Gold's Costly Dividend: Human Rights Impacts of Papua New Guinea's Porgera Gold Mine.' Viewed 4 April 2017 at: www.hrw.org/en/reports/2011/02/01/gold-s-costly-dividend

Imbun, B.Y., 1994. 'Who Said Mining Companies Take and Do Not Give? The Mining Companies' Role of Social Responsibility in Papua New Guinea.' *TaimLain: A Journal of Contemporary Melanesian Studies* 2(1): 27–42.

——, 2006. 'Cannot Manage without the "Significant Other": Mining, Corporate Social Responsibility and Local Communities in Papua New Guinea.' *Journal of Business Ethics* 73: 177–192.

Jackson, R.T., 2015. *The Development and Current State of Landowner Businesses Associated with Resource Projects in Papua New Guinea.* Port Moresby: PNG Chamber of Mines and Petroleum.

Johnson, P., 2011. 'Scoping Project: Social Impact of the Mining Project on Women in the Porgera Area.' Port Moresby: Porgera Environmental Awareness Committee.

——, 2012. *Lode Shedding: A Case Study of the Economic Benefits to the Landowners, the Provincial Government and the State, from the Porgera Gold Mine: Background and Financial Flows from the Mine.* Port Moresby: National Research Institute (Discussion Paper 124).

Le Meur, P.-Y., C. Ballard, G.A. Banks and J.-M. Sourisseau, 2013. 'Two Islands, Four States: Comparing Resource Governance Regimes in the Southwest Pacific.' In J. Wiertz (ed.), *Proceedings of the 2nd International Conference on Social Responsibility in Mining.* Santiago: Gecamin Digital Publications.

Li, T.M., 2007. *The Will to Improve: Governmentality, Development, and the Practice of Politics.* Durham (NC): Duke University Press. doi.org/10.1215/9780822389781

Mawdsley, E., 2012. 'The Changing Geographies of Foreign Aid and Development Cooperation: Contributions from Gift Theory.' *Transactions of the Institute of British Geographers* 37: 256–272. doi.org/10.1111/j.1475-5661.2011.00467.x

NFHSDP (North Fly Health Services Development Program), 2013. *2011 Annual Report.* Tabubil (PNG): NFHSDP.

OTFRDP (Ok Tedi and Fly River Development Program), 2010. *Business Plan.* Tabubil: OTFRDP.

Owen, J.R. and D. Kemp, 2013. 'Social Licence and Mining: A Critical Perspective.' *Resources Policy* 38: 29–35. doi.org/10.1016/j. resourpol.2012.06.016

Thomason, J. and M. Hancock, 2011. *PNG Mineral Boom: Harnessing the Extractive Sector to Deliver Better Health Outcomes.* Canberra: The Australian National University, Development Policy Centre (Discussion Paper 2).

UNDP (United Nations Development Programme), 2014. *2014 National Human Development Report, Papua New Guinea. From Wealth to Wellbeing: Translating Resource Revenues into Sustainable Human Development.* Port Moresby: UNDP and Department of National Planning and Monitoring.

8. Negotiating Community Support for Closure or Continuation of the Ok Tedi Mine in Papua New Guinea

COLIN FILER AND PHILLIPA JENKINS

Introduction

The Ok Tedi mine has some claim to be an extreme case of a large-scale mine whose closure is beset by politics. Ever since the mine began to operate in 1984, there has been ongoing political debate about whether its operations should continue and, if so, under what conditions. At several moments in its history—in 1985, 1989, 1996, 2001, 2006 and, most recently, 2013—the Papua New Guinea (PNG) Government has been obliged to confront this question directly, and on each occasion it has come down in favour of continuation. Nevertheless, the mining company has already begun to scale down its operations in preparation for closure by 2025, if not before, because most of the ore contained in its original deposit of gold and copper has already been mined.

The political significance of the choice between closure and continuation is primarily a function of the mine's international reputation as an environmental disaster. The choice has therefore been represented as a 'trade-off' between environmental costs and economic benefits. In 2005, at the peak of its output, it was estimated that this one mine was responsible for 15 per cent of PNG's gross domestic product,

25 per cent of the country's export revenues and roughly 20 per cent of the tax revenues raised by the PNG Government (Faulkner 2005). The environmental costs are almost impossible to calculate.

When the PNG Government approved the start of mine construction in 1981, there were plans to store the waste material in containment facilities close to the mine site. But when a landslide halted work on these facilities in 1984, the mine's operator, Broken Hill Proprietary Ltd (BHP), sought to persuade the government that there was no safe method of storing this material without destroying the project's economic viability (Jackson 1993). Once this argument had been reluctantly accepted by the government, the mine proceeded to discharge its waste material into the Alice (or Ok Tedi) River. The effects of this practice were magnified in 1987, as the mine began to produce large quantities of copper in addition to the gold that was its first target, and the volume of waste material increased accordingly. The mine was soon discharging more than 80 million tonnes of tailings and waste rock into the Alice River each year, and a growing proportion of this material was finding its way into the Fly River, more than 100 kilometres south of the mine site. The consequential damage to local ecosystems has been the subject of numerous scientific studies whose content we do not propose to summarise. But as the mine continued to operate, and these studies continued to accumulate, so the terms of the trade-off began to change. The longer the mine continued to operate, the more time it would take for local ecosystems to recover from the damage it had caused, but the longer it continued to operate, the more time and money the operators could spend on finding the best ways and means to limit the extent of this damage.

The trade-off between the economic benefits and environmental costs of the Ok Tedi mine has sometimes been represented as a dilemma for the PNG Government, rather than other stakeholders, or as a trade-off between benefits to the nation as a whole and costs to the people of the mine-affected area—especially the area downstream of the mine where the level of environmental damage has been most acute. But matters are not quite so simple. The PNG Government has certainly experienced an enduring conflict of interest as a result of its triple role as the mine's regulator, tax collector and joint venture partner (Jackson 1982; Pintz 1984). This means that different agencies or ministries within the national government have taken different positions on the question of continuation when the question has been raised. When the mine was first developed, BHP and the other private partners in Ok Tedi Mining Ltd (OTML) may well have discounted the mine's environmental costs in comparison with the profits to be made for their own shareholders, but the environmental costs became reputational costs before the profits from operation of the mine could offset the cost of its construction. BHP stopped operating the mine in 2001 because the company's directors had come to see it as a political and economic liability for which no future profits could compensate their shareholders. But the people of the mine-affected area could not remove themselves from either the costs or the benefits of the mine by means of such a calculated judgment. Throughout the period of its operation, these people have received a growing share of the economic benefits the mine has produced for the nation, even while many have suffered from the growth of its environmental costs. So if the national government has persistently discounted these local costs in the name of a greater national benefit, this does not mean that local stakeholders can equally discount the local benefits of the mine when they protest against the government's neglect of its environmental obligations.

There is another form of government neglect that has served to accentuate the dependency of the mine-affected area on the continued operation of the mine. This consists in the long-standing failure of the Fly River Provincial Government to use its own share of mine-related benefits to provide public goods and services to the people of Western Province as a whole, including that half of the provincial population which officially lives in the mine-affected area (Burton 1998). The PNG Government has generally done very little to discourage mining companies from assuming this type of public function, but the Ok Tedi mine has produced a 'proxy state' which is larger and more complex than others of its kind, and might even hold a world record to match its more familiar global reputation for polluting the natural environment. The size of this proxy state has grown in step with the size of the mine-affected area and the various agreements made to compensate some of its residents for the environmental damage caused by the mine. But its complexity was also enhanced by the specific set of agreements which enabled the mine to keep operating after BHP had stopped operating it.

Under the terms of the *Mining (Ok Tedi Mine Continuation [Ninth Supplemental] Agreement) Act 2001*, BHP Billiton (as it had then become) transferred its 52 per cent stake in OTML to a corporate body that was registered in Singapore as PNG Sustainable Development Program Ltd (SDP). This body was required to act like a cross between a provincial government, an international aid agency and an intergenerational trust fund. So long as the mine continued to operate, SDP would invest two-thirds of its net income from the mine in a 'long term fund' that would be used to mitigate the negative impacts of mine closure for a period of 40 years after the mine had closed and, in the meantime, would put the remaining third of its net income into a 'development fund' to be spent on 'sustainable development projects' in PNG. Sustainable development projects in Western Province would account for one-third of the spending from the 'development fund', while the other two-thirds would be spent in other parts of PNG.

The Ninth Supplemental Agreement also required OTML to establish and fund a body known as the Ok Tedi Development Foundation (OTDF). This was the successor to a body known as the Lower Ok Tedi/Fly River Development Trust, which was established under the terms of a previous agreement made with the PNG Government in 1989. Like SDP, the new body had a broad mission to 'promote equitable and sustainable social and economic development' in the whole of Western Province, but its funding was primarily meant to benefit those mine-affected communities whose members would agree to relinquish their rights to sue BHP or the PNG Government for any additional compensation for the damage caused by the mine. In effect, OTDF became the vehicle through which OTML assumed responsibility for the management of a complex web of compensation and benefit streams earmarked for the people of the mine-affected area under a set of agreements that was already growing before 2001 and has continued to evolve since then. At various points in time, it has also been the vehicle through which OTML has managed the spending of government money earmarked for the people of this area or the rest of Western Province.

While BHP was happy to wash its hands of the Ok Tedi mine in 2001, SDP had a vested interest in keeping the mine open as long as possible so long as it made a profit. By 2010, SDP held 63.4 per cent of the shares in OTML. The balance was held by the PNG Government, half of which was held in trust for Western Province stakeholders, including the Fly River Provincial Government. The SDP stake yielded a net income of roughly US$1.8 billion between the start of 2001 and the end of 2012. If the mine continues to operate until 2022, the 'long term fund' could be worth US$3.7 billion when it closes (Howes and Kwa 2011). However, the profits which grew from the rise in copper prices over the period since 2001 created a predictable dilemma for this arm of the Ok Tedi proxy state. The greater its apparent wealth and power, the greater the incentive for the national and provincial governments to challenge its apparent lack of accountability to anyone except its own board of seven directors, only two of whom were government nominees. In 2013, the PNG parliament passed a bill which made it government property, and thus turned the mine into a wholly state-owned enterprise.

It is not our intention to explore the whole political process that led to this act of expropriation, but rather to understand how 'local communities' have come to be engaged in the political process through which the question of mine closure continues to be addressed. In a paper written many years ago, John Burton argued that OTML (or BHP) was taken by surprise when an environmental protest movement first emerged in the mine-affected area, because 'politics' was a less significant feature of traditional culture in this area than it was in many other parts of PNG (Burton 1997: 33). At the same time, Stuart Kirsch argued that this protest movement 'gave rise to a new generation of political leaders who were able to transcend the traditional limits on power' (Kirsch 1997: 126). Both papers were published in a volume devoted to understanding the sequence of activities that led some local people to sue BHP for damages in an Australian court and eventually win an out-of-court settlement of their case in which BHP promised a substantial compensation package for people living downstream of the mine (Banks and Ballard 1997).

This sequence of activities was well under way by 1991, and the local leaders of the protest movement became international celebrities when the court case received extensive media coverage in 1994. It is a moot point whether they took the lead in mobilising their own local supporters, or whether the leaders themselves were mobilised or engaged by a coalition of national and international advocates for environmental justice and the rights of indigenous peoples (Filer 1997; Kirsch 2007). Kirsch thought that the lawsuit gave rise to new forms of social and political solidarity in the mine-affected area, but Burton noted that leaders of the protest movement won only a small fraction of the local vote in the national election of 1997, and put this down to the persistence of an 'apolitical' culture: 'With few connected traditional political structures above the lineage they had little chance of binding people together into viable political voting blocks, no matter how well-known they were' (Burton 1997: 53).

The question we address in this chapter is the question of how power has been distributed and redistributed between 'local communities' and the other stakeholders involved in making the choice between closure and continuation of the Ok Tedi mine. Our main focus is on the way that this choice has been framed in the period since the legal foundations of the company's operations were amended in ways that placed the representation of mine-affected communities at the heart of this political and economic choice.

To Close or Not to Close: The Context of the Ninth Supplemental Agreement

The out-of-court settlement agreed in 1996 said that BHP and the 'claimants' who had taken it to court would commit to the implementation of any 'tailings option' proposed by the PNG Government provided that BHP regarded it as 'economically and technically feasible' (Banks and Ballard 1997: 216). What actually happened was that OTML commissioned a series of technical reports under the terms of its Mine Waste Management Project and summarised their findings in the form of a 'risk assessment' that was presented to the PNG Government in 1999 (OTML 1999). In this report, the risks were attached to four alternative courses of action that could be taken in light of the fact that OTML had already initiated a 'trial river dredging program' in the lower reaches of the Alice River in order to shift some of the waste material from the river-bed to the river banks. The first option was to keep mining and keep dredging until the 'scheduled end of mine life' in 2010; the second was to stop dredging but keep mining until the point of mine closure; the third was to keep mining and build a new tailings dam by 2001 as an alternative to riverine tailing disposal; and the fourth was to close the mine in 2000.

The report recognised that none of these four options would put a halt to the environmental damage caused by the previous deposition of waste material into the Alice River. It was predicted that the Fly River catchment would continue to sustain such damage for decades after the mine closed, regardless of whether it closed in 2000 or 2010. BHP's preference was to close the mine sooner rather than later, because the first and second options would still cause more environmental damage than the fourth option, while the economic cost of implementing the third option would not produce enough environmental benefits to make it economically worthwhile (Sharp and Offor 2008: 3). Some of the plaintiffs in the previous court case then decided to sue BHP for breach of contract in discounting the feasibility of the third option (Kalinoe 2008: 7), but BHP had already declared that continued operation of the mine would be inconsistent with its own 'environmental values', which had apparently been reconstructed over the course of the previous three years.[1]

This placed the PNG Government in a familiar quandary. This time the government hedged its bets by asking the World Bank to review the risk assessment report and provide a review of the 'independent reviews' already commissioned by OTML. The bank's reviewers agreed that immediate closure was the best option from an 'environmental standpoint', but 'from a social standpoint this would result in a potentially disastrous situation because there is no preparedness for mine closure' (World Bank 2000: 8). They were also concerned by the lack of socioeconomic information in the various technical reports:

> The complex social environment in which Ok Tedi is situated, exacerbated or arguably caused by the fact that Ok Tedi is the only agent of development in the Western Province, has meant that closure has multiple ramifications for the community impacted by its operations and for the government with whom overall responsibility rests for the welfare of its citizens. Thus, much more information is needed regarding the social and cultural history and characteristics of the communities affected by the [mine's] operations and the impacts need to be examined under four major aspects, namely consultation and participation of local communities; compensation; local distribution of benefits and related social, economic and cultural impacts; and employment impacts during and after mine operations. (ibid.: 16–17)

The reviewers therefore suggested that the government should engage in a process of public consultation in order to assess 'the political dimensions of the problem', and allow the mine to operate until a draft mine closure plan had been subjected to a social, as well as an environmental, impact assessment (ibid.: 20, 31).

For its part, OTML was still one step ahead of the government because its managers had already initiated the process of 'community consultation' when the risk assessment report was completed in 1999 (Kalinoe 2008). It was this process that resulted in the six Community Mine Continuation Agreements that were signed by the company's managers and local community representatives and attached as schedules to the

---

1   BHP's desire to wash its hands of the whole operation and avoid a new round of negative publicity may well have been a function of market pressures associated with its imminent merger with Billiton (Kirsch 2007: 309).

Ninth Supplemental Agreement in 2001. These agreements contained a commitment by OTML to spend an additional K180 million to compensate the mine-affected communities over the remaining life of the mine or, more specifically, to compensate those communities whose representatives agreed to withdraw their support for the new lawsuit against BHP Billiton before the end of 2002 (Kalinoe 2008: 7). Fifty-eight per cent of the value of this new compensation package was to be provided in the form of 'development projects', while 16 per cent was to be paid in cash to community members and the remaining 26 per cent was to be held in trust for future generations, rather like the money that was due to accumulate in the 'long term fund' administered by SDP. Each of the six agreements contained a declaration that 'the economic opportunities offered by the Company's Commitments represent to the Communities an acceptable trade-off for the environmental impacts of the future operation of the mine', which was still expected to close in 2010.

Political Constructions and Divisions of the Mine-Affected Area

The problem of negotiating community support for any agreement relating to the closure or continuation of the Ok Tedi mine has grown more complex as the outer limits of the mine-affected area have expanded while the internal social and economic divisions of its population have deepened. The people who were invited to accept the trade-off postulated in the Ninth Supplemental Agreement were not simply the sons and daughters of the people who were regarded as members of mine-affected communities when the Mining (Ok Tedi Agreement) Act was passed in 1976, before the need for such a trade-off was envisaged. That first agreement between BHP and the PNG Government contained a pair of clauses which declared that the developers of the mine would, where 'practicable', give first preference in project employment and local business development to 'the landowners in and other people originating from the Kiunga and Telefomin sub-provinces'. The creation of this 'preferred area', which now comprises the North Fly District of Western Province and the adjoining Telefomin District of West Sepik Province, was not intended as a form of compensation for any sort of environmental damage, but as an extra benefit that was justified by the underdevelopment of this area in comparison with other parts of the country (Filer 2005). The 1980 national census, which was conducted before the start of mine construction, counted 38,440 people resident in this preferred area, nearly all of whom were rural villagers. Less than 3 per cent of these people qualified as 'mine area landowners' who would be compensated for the sacrifice of their customary land to the mine and its associated infrastructure, and only 1 per cent would receive a share of mining royalties because they were customary owners of the mountain that was going to be excavated (Filer 1997: 64).

The boundaries of the 'mine-affected area' were radically transformed when the Lower Ok Tedi/Fly River Development Trust was established in 1990 as a means to compensate communities downstream of the mine whose members were beginning to suffer the effects of riverine waste and tailings disposal. One hundred and two villages containing about 30,000 people were initially identified as beneficiaries of the trust. Nineteen of these villages were located in the 'Ningerum area', immediately south of Tabubil and the 'mine area villages', while another ten were located further south in the 'Alice area', along the lower reaches of the Alice (Ok Tedi) River, where the environmental damage was most acute. The traditional inhabitants of these 29 villages accounted for one-sixth of the trust's individual beneficiaries, but they also accounted for 8 per cent of the 59,337 people recorded as residents of the original 'preferred area' in the 1990 national census. In effect, the mine-affected area had now been divided between an 'upstream area', which contained the other 92 per cent of the people living in the original preferred area, and a 'downstream area', which contained the people entitled to a new form of compensation from the mining company.

But new forms of social division were also emerging within each of these areas. The downstream area was like a corridor passing a set of traditional cultural rooms or spaces whose occupants were now suffering different forms and degrees of damage to their natural environment, while the upstream area was more like a set of concentric circles (or semicircles) whose occupants had more or less access to the material benefits derived from the mining operation. The main axis along which the two areas were divided from each other was the road connecting the towns of Tabubil and Kiunga, which was the first stage in the route by which the products of the mine found their way to the world market. According to the census, these two towns had a combined population of 8,688 in 1990, and thus accounted for 15 per cent of the residents in the original preferred area. Many of the people living in these towns had been recruited from villages in the preferred area and were now dependent on cash incomes generated by the mining operation and its associated service economy. As a general rule, the cost of getting from a rural community to one of the two towns was inversely related to the number of its members who were able to obtain a share of such incomes.

New forms of social division were most acute in that part of the downstream area whose inhabitants were now suffering the most extensive damage to their natural environment. The indigenous inhabitants of the 'Alice area' were divided between two traditional cultures—the Yonggom people on the western side of the river and the Awin people on the eastern side. The Australian colonial administration encouraged the members of both ethnic groups to form larger settlements along the banks of the river in order to facilitate communication (by boat) with what was then the administrative outpost at Kiunga (Jackson 2003). Other members of the Awin group spontaneously formed new settlements along the route of the Tabubil–Kiunga road as mining plans were developed in the 1970s (Jackson 1979), so the Awin people have since been divided between those in the upstream area (along the road) and those in the downstream area (along the river). The history of resettlement created a situation in which the members of Awin and Yonggom communities in the 'Alice area' were divided in the strength of their traditional claims to ownership of the land and resources that were now being damaged by mining operations, and hence to compensation for such damage (Kirsch 1995; Jackson 2003). Meanwhile, another form of social division was created when a large number of Yonggom refugees from the Indonesian province of Irian Jaya (or West Papua) took up residence on the PNG side of the border in 1984, thus creating additional pressure on the land and resources of the Yonggom people who were already living there (Kirsch 1989; King 1997). By 1990, migrants from traditional Yonggom and Awin villages had already established sizeable settlements in the town of Kiunga (King 1993), but the 'Alice people' who remained in their riverside villages suffered a degree of relative deprivation as the Tabubil–Kiunga road became the main economic axis of the North Fly region (Jackson 2003).

When the mine development agreements were renegotiated in 1990, provision was made for the mine area landowners to receive 30 per cent of the royalties collected by the national government, instead of the 5 per cent to which they were previously entitled, while the share allocated to the Fly River Provincial Government was correspondingly reduced from 95 to 70 per cent. This entailed a new degree of economic inequality among the people of the ten 'mine area villages', who then numbered about 2,200. The four villages containing the customary owners of the mine site would get 25 per cent of the royalties, while the village containing the customary owners of the mining town would get 5 per cent. Four villages would continue to receive rental payments for the land containing the tailings storage facility that was never built, but they were now being treated as part of the downstream 'Ningerum area', so the 'mine area landowners' were now understood to consist of about 1,500 people living in the other six villages.

Under the terms of these new development agreements, the Fly River Provincial Government also undertook to spend about half of its own revenues for the benefit of 'landowners in the project lease areas, along the Tabubil–Kiunga road, along the Ok Tedi–Fly River, and in the Fly River Delta area' (Jackson 1993: 71). This meant that people in the eastern part of the original area of preference in Kiunga (now North Fly) District were removed from the 'mine-affected area' within Western Province. Specific proportions of the new benefit package were earmarked for the mine area villages, the 'Alice villages' (including villages in the Ningerum area), the 'Highway villages', and the 'Middle Fly' and 'South Fly' villages. These proportions seem to have reflected the relative size of the respective village populations rather than the extent of environmental damage in each area (Filer 1997: 70). But this division of the mine-affected area was rather less significant than the one that accompanied the establishment of the Lower Ok Tedi/Fly River Development Trust because OTML already had far more capacity to deliver on its undertakings.

OTML adopted its own development program for the 'Highway villages' in 1995, but this was worth a great deal less than the Lower Ok Tedi Agreement that emerged from the out-of-court settlement of the Australian court case in 1997. The beneficiaries of this agreement were the residents of eight Yonggom villages and seven Awin villages in the 'Alice area' whose residents had claims over the land to be affected by OTML's trial dredging program. The distribution of compensation between these villages was to be a function of the size of their populations, the length of their respective riverside frontage, and the extent of the damage caused to the rest of their traditional territories (Banks and Ballard 1997: 207–10; Filer et al. 2000: 60–1). This compensation package was essentially a premium paid on top of the entitlement of all downstream communities to benefit from the development trust and to receive additional compensation under the terms of the Restated Eighth Supplemental Agreement concluded in 1995.

Between 1990 and 2000, the resident population in the downstream part of the mine-affected area increased by almost 100 per cent, from 30,000 to nearly 60,000. This was partly due to a 40 per cent increase in the number of village communities that were included within its boundaries. The national census recorded a similar rate of increase in the combined population of Tabubil and Kiunga, which rose from 8,688 to 16,944, but this was partly due to in-migration, not to any change in the town boundaries. By 2000, the population of the six mine area villages near Tabubil had grown to about 2,300, while almost 50,000 people were living in other rural villages in the upstream part of the original preferred area, including Telefomin District in West Sepik Province. In its broadest definition, the mine-affected area therefore contained about 130,000 people, and accounted for roughly two-thirds of the population of Western Province alone.

A total of 149 village representatives signed the six 'community mine continuation agreements' (CMCAs) that were finalised in 2002. The six mine area villages made up one of the six 'CMCA trust regions'. The other five regions were North Ok Tedi (the Ningerum area), Lower Ok Tedi (the Alice area), the Highway, the Middle Fly and the South Fly (Kalinoe 2008: 19–20). When the CMCAs were reviewed and renegotiated in 2006, the number of trust regions grew from six to nine, as the South Fly region was divided into four parts (see Figure 8.1). There was a slight increase (from 149 to 156) in the number of village representatives who signed the new agreements because of a reduction in the number of villages whose representatives had refused to sign the previous agreements (Menzies and Harley 2012: 4).

An unofficial annual census conducted by OTML showed that the combined population of all the CMCA villages rose from 81,136 in 2007 to 101,413 in 2011. These figures included people who were not normally resident in their home villages but were still recognised (and paid) as members of the affected communities. Most of these absentees were living in urban areas in Western Province—especially the towns of Tabubil, Kiunga and Daru—and some had jobs associated with the mining operation. If the official national census figures from 2011 are reasonably accurate, this means that the 'CMCA people' accounted for 56 per cent of the population of Western Province in that year. The trust deeds and other institutional arrangements negotiated and renegotiated through the CMCA process produced a definition of the mine-affected area as one from which the 'non-CMCA people' in the original preferred area were now largely excluded. However, these villages were combined with all the other non-CMCA villages in Western Province as beneficiaries of what is officially (and rather confusingly) known as the 'non-CMCA region peoples trust account'. No such account has been established for the people of Telefomin District in West Sepik Province.

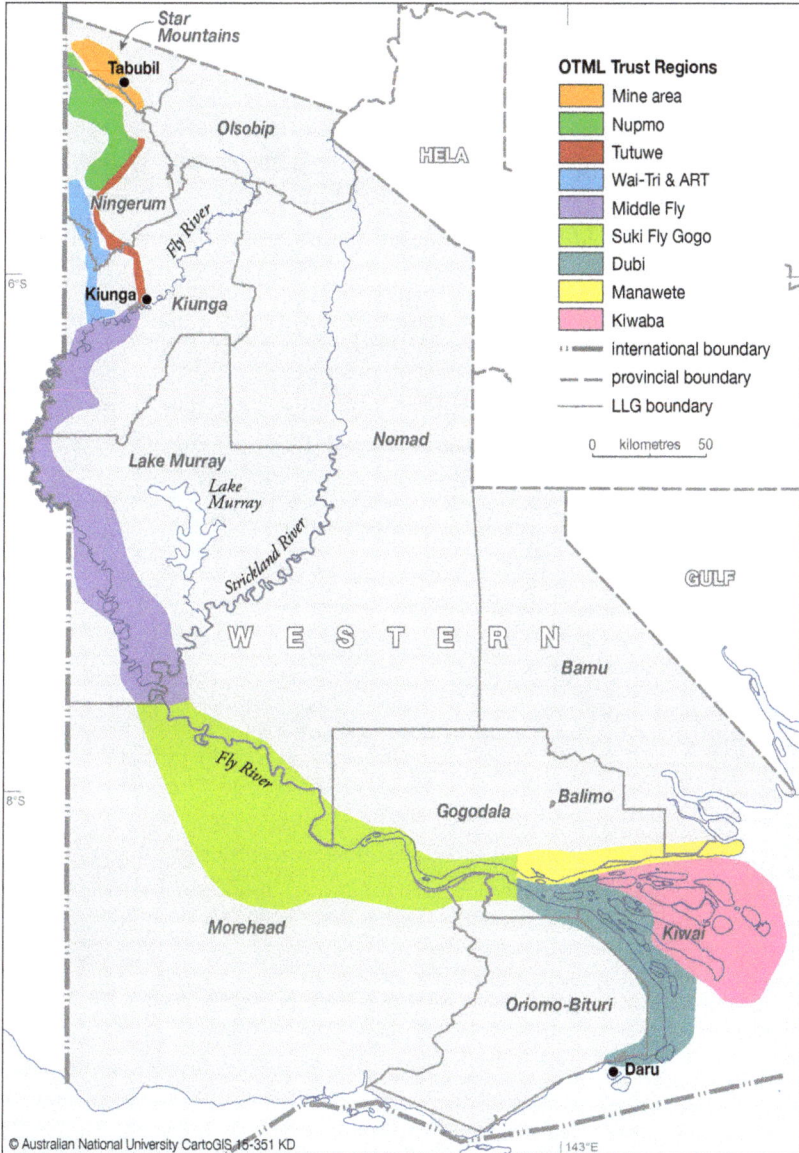

Figure 8.1 CMCA trust regions in Western Province.

Source: CartoGIS, The Australian National University.

By 2007, the three districts of Western Province (North, Middle and South Fly) had been officially subdivided into 14 local-level government (LLG) areas under the terms of the *Organic Law on Provincial Governments and Local-Level Governments 1995*. Kiunga was one of three towns with its own local government, but Tabubil was treated as a single ward (with one local councillor) in the Star Mountains LLG area, which was one of 11 rural LLG areas in the province. The Star Mountains LLG area had 13 wards altogether, including the six 'mine area villages' and six other rural wards, some of which had previously been treated as part of the Ningerum area. If one compares a map of the CMCA trust regions with a map of the rural LLG boundaries in Western Province, it is evident that 9 of the 11 rural LLG areas contain some CMCA trust villages, but some of them contain a lot more than others. The internal divisions of the Ok Tedi proxy state have thus come to diverge from those of the official state, but those of the official state are in many ways less real than those of the proxy state because the institutions of the proxy state deliver most of the economic benefits which matter to the people of the mine-affected area, whatever its outer limits may be.

The institutions of the proxy state also do a better job of counting the number of people in the mine-affected area. According to the official national census, the whole of the Star Mountains LLG area, including the township of Tabubil, had a population of 15,458 in 2011. But an OTML survey of the 'Tabubil Plateau', which accounts for roughly half of that area, counted 23,071 residents in 2009. Of these, 7,721 (33 per cent) were living in the town itself, 4,785 (21 per cent) in the six mine area villages (or council wards) and 10,565 (46 per cent) in a number of peri-urban squatter settlements whose populations seem to have been overlooked in the official census two years later. There is certainly no evidence of mass emigration (or eviction) from these settlements in the intervening period. It is also known that the population of the town may have been under-enumerated in the OTML survey because some of the residents were concealing the presence of 'unauthorised dependants'. So the real population of the Star Mountains LLG area in 2011 (including the six council wards formerly in the Ningerum area) was almost certainly in excess of 25,000. The official census counted 11,260 residents in Kiunga township in 2011, which is probably a more accurate number. The combined population of the two townships in 2011 would therefore have been more than 19,000, which suggests a 20 per cent increase over the urban population enumerated in 2000. But what is almost certainly more significant is the parallel increase in the population of peri-urban settlements, especially around the township of Tabubil, which is the epicentre of the regional mining economy.

It is notable that the population of the six mine area villages was seven times greater in 2009 than it was 30 years earlier, before the start of mine construction. This was not the result of an extraordinary rate of natural increase, but was largely due to the immigration of people from other Mountain Ok (or Min) communities within the original preferred area, including parts of Telefomin District. About two-thirds of the people living in Tabubil town, and one-third of those living in the peri-urban settlements, also seem to have been born in the original preferred area. Some of these would have been 'CMCA people' from the Ningerum, Alice and Highway villages, but the proportion has not been established. In the 2009 survey, more than half of the 3,176 households on the Tabubil Plateau were found to contain at least one wage-earner, and most of the jobs they occupied were directly or indirectly dependent on the mine's continued operation. Even the households without a wage-earner, including 70 per cent of the households in the squatter settlements, had livelihoods that would be hard to sustain if the mine were to close.

The Politics of Community Mine Continuation Agreements

Section 8 of the Ninth Supplemental Agreement states that the 'signature or other execution' of a CMCA by a 'person representing or purporting to represent a Community or clan, or that person's delegate', binds all members of the group to the agreement, even if there is no 'express authority' for that person to represent the members of the group, or some group representatives have not signed the agreement, or some group members are not party to it. The same section then goes on to state that the real or purported group representative who signs or executes a CMCA binds all future members of the group to the agreement, even if they have not yet been born.

These clauses have attracted a good deal of critical academic scrutiny because of their apparent denial of what might normally be regarded as natural justice and human rights (Kirsch 2004, 2007; Kalinoe 2008). The powers they ascribed to individuals who might only be 'purporting' to represent all sorts of other people can partly be explained by the fact that some of these other people were unwilling to accept another big compensation package in return for a promise to desist from legal action against BHP. But the problematic framing of the social relations of representation in the mine-affected area can also be seen as a reflection of the long-standing difficulties faced by OTML staff, and even by 'real' government officials, in their efforts to manufacture consensus or consent in communities without strong leaders.

Interactions between OTML staff and members of downstream communities in the mine-affected area were already characterised by a politics of mutual distrust in the period following the establishment of the Lower Ok Tedi/Fly River Development Trust. The two sides to this relationship were equally unable to believe that the other side had anything other than a set of personal and political agendas that were unrelated to the actual impact of the mine on local livelihoods (Filer 1997: 78–9). In the years preceding the settlement of 1996, matters were complicated by the split between the two branches of the company that were responsible for community affairs and environmental management, each of which was holding separate talks with local villagers. Even after the settlement, company staff were concerned that talk about compensation for environmental damage was covering up the need for local communities to have their own plans and strategies for 'sustainable development' (Simpson 2002: 52). The legal and institutional separation of the OTDF from OTML's core business was seen by company management as an opportunity to address this problem, but the question of compensation was still central to the political process through which community members or their 'representatives' elected to opt in or out of the CMCAs between 2000 and 2002 (Kalinoe 2008: 7). However, the amount of politics in this process was substantially reduced by the absence of any opportunity for community representatives or members to negotiate the contents of the agreements (ibid.: 34).

Each of the six CMCAs made provision for the establishment of one or more community development foundations or trusts administered by OTML through the OTDF, with trustees from each of the villages represented in the agreement.[2] The local trustees were to meet with their administrator once a quarter to decide how best to spend the 58 per cent of the K176 million compensation package that was earmarked for

2   It is not clear how the trustees were selected or appointed, or how many of them had been 'representative signatories' of the CMCAs in 2002.

community development projects rather than cash payments or investments for future generations. It seems that OTML staff made a deliberate effort to avoid the appointment of local government ward councillors as local trustees, and thus created 'a whole new structure of community organisation and governance' (Kalinoe 2008: 10). At the lowest level of this new form of political organisation, the trustees would consult with 'village development committees' made up of 'village elders'. The consultation process has been described as one in which the trustees were 'required to go back to their village community and do patrols, possibly with OTML community relations officials, if not by themselves, and then inform their village community of the decision taken at the last trust meeting' (ibid.: 22). As a result, there was a tendency for them to be regarded by their own constituents as functionaries of the proxy state who were curiously removed from the realm of politics by means of a social contract that ruled out any talk of additional compensation beyond that which the 'communities' had already accepted (Crook 2004: 128).

If the original CMCAs can now be seen, in retrospect, as agreements between OTML and the PNG Government about the way that 'community consent' would be obtained for the continuation of mining operations, they did leave one space for community leaders or members to renegotiate the terms of that consent once they had purportedly given up their right to take further legal action. This was a requirement for the agreements to be reviewed after a period of five years. The company's senior managers decided that this constituted an opportunity for a 'fresh approach' to the business of negotiating community support, 'involving contemporary village leadership, as well as other interested stakeholders not already parties to the original agreement' (Offor and Sharp 2012: 217). At the same time, a fresh approach was needed because OTML's community relations staff were reporting a decline in the level of community support as benefits formerly anticipated had failed to materialise, even while the company's environmental scientists were projecting an increase in the scale and intensity of the environmental damage caused by mining operations (ibid.: 214).

The consultants engaged to design the review process have described the new approach as one that sought to establish an 'informed consensus' based on three 'guiding principles', namely: 'multi-party mediation; interest-based negotiation; and relationship-based communication' (Sharp and Offor 2008: 15; Offor and Sharp 2012: 214). The second and third of these principles were said to be 'culturally appropriate to the PNG setting as they are inherent in the Melanesian style and practice of negotiation in ordinary life' within a 'relationship-based culture' which has 'a firm view as to the longevity of an important relationship for the wellbeing of the community' (Sharp and Offor 2008: 19). An 'interest-based' negotiation was contrasted with one based on people's 'positions' within the political space constructed by previous actions and negotiations (Offor and Sharp 2012: 219).

The review process consisted of six three-day meetings of a working group comprising representatives of 15 groups of stakeholders over a period of 18 months in 2006 and 2007. Aside from the designers of the process, OTML also hired a collection of independent facilitators, advisers and observers to mediate between the different interests at play in the negotiations. The mine-affected communities were now divided between nine regions, rather than six, while separate representation was granted to 'women and youth' and to church organisations from within the mine-affected area, and to a national non-governmental organisation concerned with environmental matters. Community representation in the review process was based on the election of village representatives by village meetings 'supported' by OTML staff, and then on the appointment of three delegates to the working group by the village representatives elected from each of the nine regions (Offor and Sharp 2012: 218). Regional meetings of village representatives were held after the second working group meeting, meetings were held in all villages after the third working group meeting, and both types of meeting were reconvened after the final working group meeting in order to discuss and approve the outcome of the whole review process (ibid.: 221).

That outcome consisted of a compensation package worth K324 million over the seven years between 2007 and 2013—the latter now scheduled as the year of mine closure—and a greater level of community participation in the design and management of development projects funded from this and other sources. Each of the nine 'regional communities' would also have a representative on a body advising the board of the OTDF, which would henceforth operate as the Ok Tedi and Fly River Development Program. This advisory body would not only have the power to vet all development project proposals, but would also appoint three 'associate directors' to the board to sit alongside the two directors representing OTML and the two directors representing SDP and the PNG Government respectively (Offor and Sharp 2012: 228). While OTML would still hold three of the four shares in the OTDF, and SDP would hold one share, one of the OTML shares was supposed to be transferred to a new representative body called the Ok Tedi Impact Area Association in 2012, one year before the anticipated date of mine closure. Despite the fact that there were no women among the village representatives involved in the review process, there was also an agreement that each of the representative bodies involved in implementing the new deal, from the OTDF board down to the level of 'village planning committees', would have to involve some female participation, and that 10 per cent of the benefits should be allocated to a Women and Children's Fund (Menzies and Harley 2012).

Despite the growing level of corporate investment in the production of an 'informed consensus' among the relevant stakeholders, the 'CMCA people' were still defined as the beneficiaries of the new deal, while responsibility for its delivery still fell to the employees of the OTML proxy state. The latter were still imbued with a 'technocratic ethos' that made it hard for them to 'do' community development, while the former were still failing to 'drive' their own community development because of internal wrangles about the structures of participation and representation. A survey of community attitudes to implementation of the new deal, five years after it had been concluded, found that local people still thought that the village planning committees failed to 'faithfully represent community interests' because the members were hoarding information or keeping benefits 'for their personal or family gain' (Menzies and Harley 2012: 7). Most of the committees were found to have only one female member, often a wife or other relative of the male chairman, and some of these chairmen were living in urban areas, not in the villages for which their plans were being made. The authors of this study found that 'intra-village politics' and 'elite capture' were still slowing down the delivery of 'community development' by the 'technocratic' agents of the proxy state, but the bureaucratic agencies of the real state were delivering even less, even though local members of the national parliament now had access to substantial 'constituency development funds'. 'Compounding this, the largely parallel CMCA governance structures are not linked to local government; thus village planning committees do not coordinate with ward development committees, and relations between the mine, Foundation, the PNG Sustainable Development Program, and the provincial government are a topic of continuing concern' (ibid.: 10). So 'politics' was conceived as a problem at every level of political organisation.

The Politics of the Mine Closure Planning Process

Continuation and closure might seem like two sides of the same coin, but do not seem that way when considered as subjects of negotiation between local communities and the real or proxy state. While community representatives were drawn into a conversation about the mine's continuation at the time of the Ninth Supplemental Agreement, they were for many years excluded from a separate channel of communication through which OTML sought government approval for its own mine closure plans.

As we have seen, the World Bank cited the absence of a mine closure plan as one of the reasons for keeping the mine open after BHP decided to wash its hands of the operation in 2000. The need for such a plan had in fact been recognised ten years previously, when the national government made an undertaking to commission a 'long-term economic development plan' for what was then known as Tabubil District—an area which now comprises the Star Mountains and Olsobip LLG areas in the far north of Western Province. The Department of Finance and Planning made some effort to honour this commitment in 1993, with terms of reference that clearly related to the problem of mine closure, but a fiscal crisis consumed the money that would have been used to pay the consultants. By the time the plan was eventually drafted (Jackson 1999), its geographic scope had been made redundant by changes in the definition of the mine-affected area, so in that sense it was true that the company and the government had no plans to deal with the impact of mine closure on communities downstream of the mine.

By that time, the Department of Mining was already working on a new mine closure policy that called for the establishment of mine closure planning committees for each of the big mining projects that were already in operation. The Ok Tedi Mine Closure Planning Committee was established in 2000, and was meant to include representatives from OTML, the Fly River Provincial Government and five national government departments. Although the committee was notionally chaired by the Department of Mining, it was in effect a sounding board for plans made by the mining company. 'While the MCPC [Mine Closure Planning Committee] members may commit to a plan there is no reason to believe, on the basis of past experience, that anyone other than OTML will follow the plan given that some of its recommendations are likely to be unpalatable to provincial politicians, the Administration and mine-affected communities' (Simpson 2002: 49). The likelihood that provincial politicians or members of mine-affected communities would object to the company's plans was somewhat diluted by the fact there were no provincial government or local community representatives on the committee when OTML produced the first of its new plans in 2002 (ibid.: 48).

The Ninth Supplemental Agreement required the company to establish a 'financial assurance fund' that would pay for the implementation of its mine closure plan, once this had been approved by the national government. In the meantime, the company was required to produce draft mine closure plans at two-yearly intervals, each time to be accompanied by 'a report on the social and economic impacts of mine closure and the status of economic programs being undertaken relevant to those impacts by the Company, the Ok Tedi Development Foundation, the Western Province Capacity Building Project, Government and other agencies'. The Western Province Capacity Building Project was itself funded by OTML under the terms of a separate agreement with the national government after the latter had stripped the Fly River Provincial Government of its financial powers in 1999 (Simpson 2002: 46; Kalinoe 2008: 40). In effect, this meant that the long arm of the proxy state had already reached into the field of provincial administration when the planning process got under way, and that helps to explain why the provincial administration was notable for its absence from the Mine Closure Planning Committee.

By 2005, when OTML's managing director brought numerous stakeholders together for a mine closure planning workshop at Tabubil,[3] the provincial government was back in business, and the provincial planner sought to persuade the other workshop participants that the plans of OTML and SDP should henceforth be subordinated to the one that he had recently produced for Western Province as a whole. This looked even less likely than the prospect of collaboration between the two divisions of the proxy state, whose own plans bore no clear relationship to each other. But among the numerous stakeholders at the workshop, there were still none who directly represented the mine-affected communities. The community representatives (or at least some of them) held their own

---

3   One of us (Filer) was a participant observer at this workshop.

meeting at Kiunga, where they showed much less enthusiasm for mine closure than OTML's managing director displayed in his address to the meeting at Tabubil.

> The climax of the [Kiunga] meeting occurred when one of the speakers, using the call-and-response preaching style of public speaking that has become popular, called out to the participants, asking them, 'Do you want the environment or money?' Only a few people responded by saying 'environment'. There was a loud murmuring as the participants discussed the question among themselves. Finally, someone called out in Melanesian Pidgin, '*Tupela wantaim!*' or 'Both of them at once', and the crowd loudly indicated its approval. In other words, they want to simultaneously protect the environment and have access to development opportunities and money. It is almost impossible for them to contemplate early mine closure when so much of the regional economy is dependent on its operation, and when so many of the villages have already been significantly affected by pollution and need the extra income provided by compensation payments to feed their families. They cannot imagine letting the mine close down without gaining some lasting form of economic benefit in return for all the damage that it has done to their environment. (Kirsch 2007: 314)

When OTML came to produce the third of its draft mine closure plans in 2006, the space that might have been occupied by a form of community consultation was already occupied by the CMCA review process, so if continuation and closure really were two sides of the same coin, each side had a distinctive group of stakeholders involved in a different type of conversation.

One thing that set the mine closure planning conversation apart from the mine continuation conversation was the uncertainty about when the mine would actually close. In 2001, at the time of the Ninth Supplemental Agreement, it was expected that the mine would close in 2010. Five years later, the moment of closure had been pushed back by two years. By 2009, one more year had been added to the mine's life expectancy, but the moment of closure was still coming closer. As OTML now faced the prospect of submitting a 'detailed' (and possibly final) mine closure plan to the national government, its new managing director decided that there was no longer room for two separate conversations about the future, but shortly afterwards he announced that consultations with 'Western provincial leaders' had opened up the possibility of postponing the moment of closure by another seven years (Anon. 2009). Two years later, the four politicians representing Western Province in the national parliament issued a joint statement demanding that the mine be closed in 2013, in accordance with the latest plan (Anon. 2011), but OTML countered this demand with the announcement of a Mine Life Extension Plan that would keep the mine in operation until 2022 or even 2025 (Hriehwazi 2011; Anon. 2012a). Local community support for this alternative plan was harnessed through a process of consultation that was understood to constitute the second review of the CMCAs. Representatives from each of the nine CMCA trust regions had signed up to the new deal by the end of 2012, five years after completion of the previous review (Anon. 2012b).

In September 2013, the government of Peter O'Neill enacted a Tenth Supplemental Agreement which deprived SDP of its majority stake in OTML and thus made the mining company a wholly state-owned enterprise. This action was justified by a claim that SDP's board of directors was not accountable to anyone but BHP Billiton. This set the scene for a protracted legal battle over property rights, with a specific focus on control of the 'long term fund' established under the terms of the Ninth Supplemental Agreement. The more immediate effect was to prevent OTML from making any further payments to SDP, which meant that the latter was obliged to shut down its operations and lay off its staff. At the same time, the national government acquired a much bigger financial stake in the continued operation of the mine, so long as it could still make a profit, and the prime minister seems to have thought that he now had more power to persuade the recalcitrant politicians of Western Province that mine continuation could serve their interests as well. As one of his spokesmen put it:

> It is time for the CMCA leaders to work with their newly elected LLG leaders, their provincial government, and national leaders to ensure the new OTML and the new PNGSDP delivers the programs and projects that will go a long way in improving their lives and sustaining that for some time (Martin 2013).

Some of the 'CMCA leaders' objected to the nationalisation of the mine on the grounds that this was not a subject on which they had been consulted when they agreed to the Mine Life Extension Plan and, more importantly perhaps, because the national government was under no obligation to use any of the dividends derived from the shares formerly owned by SDP for the benefit of people in the mine-affected area.[4] However, their right of

---

4   The legislation simply enabled the state 'to restructure PNGSDP and its operations to ensure that PNGSDP applies its funds for the exclusive benefit of the people of Western Province'.

complaint was challenged by the president of the Ok Tedi Mine Impact Area Association, who claimed that his association was now the legitimate representative of all nine trust regions (Anon. 2013). There has not been an end to debate within Western Province on the timing of mine closure, and since there was a sharp fall in OTML's profits between 2012 and 2013, the national government may yet lose interest in extending the life of the mine.

Conclusion

Has the Ok Tedi proxy state operated like an 'anti-politics machine' that has sucked the lifeblood out of the real state in Western Province even at the same time that it has created a form of economic dependency that is inherently unsustainable? Have its engineers and mechanics managed to produce a semblance of community support for mine continuation that conceals the absence of traditional cultural and political institutions through which 'communities' could make any collective decisions about their future? Has the operation now reached a tipping point at which the number of politicians who sense an opportunity to seize control of money held in trust for mine-affected communities is more than a match for the number of technocrats or bureaucrats who have so far been charged with spending it on 'equitable and sustainable social and economic development'?

While we have done our best to summarise the evidence that might seem relevant to questions such as these, we do not think that it provides the answers. Even if we had more evidence, the answers would remain equivocal.

If the Fly River Provincial Government has failed the test of good governance throughout the life of the mining operation, this does not necessarily mean that it would have done a better job of providing public goods and services if the mine had never existed. Perhaps it makes more sense to assert that 'cultural attitudes have taken over the economic management of the province', and that is why the province has displayed a 'limitless capacity for "disappearing" any money given to it for development' (Burton 1998: 173). But this only leads us to another question, which is whether 'political culture' does more to explain the failure of the real state to deliver 'real development' in *this* province, as compared with other provinces, than do various measurable forms of poverty or backwardness that ought to be alleviated by the economic benefits of mining, but may have been exaggerated by the impact of mining on the physical environment. The concept of political culture bears an even bigger load if it is then used to explain the inability of communities or community leaders in the mine-affected area to negotiate a more sustainable or equitable compensation package (Burton 1997). But we might then wonder how that culture is related to the 'relationship-based culture' that has been represented as something in which the mining company needs to participate if it wishes to secure community support for the maintenance of its own operations (Sharp and Offor 2008).

It is still a moot point whether the capacity of the Ok Tedi proxy state to secure this type of community support is a function of its capacity to understand and engage with the political culture of Western Province or its capacity to function as a substitute for the provincial government. The proxy state is clearly not a democratic state, so its engineers and mechanics do not have to satisfy the desires of the local population in order to keep their own jobs; they only have to satisfy their own superiors that they are doing something that is good for both the mining company and its mine-affected clients. From their point of view, 'local politics' or 'community attitudes' are liable to be seen as obstacles to the achievement of this goal, and not as things that need to be appreciated on the way. The social and political landscape of the mine-affected area is thus perceived as a two-dimensional object, as flat as the flood plain of the Fly River, divided and subdivided into separate zones of entitlement to specific benefit streams whose dispensation is ideally free of any political interference.

There are two reasons to regard this as an optical illusion. The first is that the machinery of dispensation is nowhere near as efficient as the mills that churn up the rocks excavated from the remains of Mount Fubilan. The proxy state has been cobbled together from a disparate set of agreements with several government agencies and an assortment of 'community representatives'. With the passage of time it has become a sort of institutional jungle that lacks any mechanical or rational capacity to distribute money, goods or services in ways that are proportional to the impact of the mine or the needs of different groups of mine-affected people.

The second reason is that the growth of this institutional jungle has been matched by the continual enlargement of the 'mine-affected area' and the parallel detachment of a growing proportion of its resident population from livelihoods dependent on the land and natural resources contained within each zone of entitlement. There are still many villagers in the region who continue to make a living from what remains of their customary land, even when that land has been degraded by the impact of the mine, but the institutions of the proxy state may be no more accessible to such people than regular health and education services provided by the government or the churches. As the mining operation has matured, so has the tendency of rural 'community leaders' to gravitate towards the towns where they can mediate or manipulate the distribution of money, goods and services from all quarters. As a result, the superficial distinction between national, provincial and local-level politics, even in the operation of the 'anti-politics machine', has given way to a triangular relationship between elected and appointed leaders who spend various amounts of time in Port Moresby, Kiunga and Tabubil.[5]

Despite the efforts of the proxy state to produce a transparent process of 'community consultation' in the mine-affected area, it is this triangular urban space in which the political choice between mine closure and mine continuation must ultimately be negotiated. This choice is not quite as stark as it sometimes appears. Mine closure is normally a process in which operations gradually wind down before they finally cease altogether. In the present case, the process already began when OTML laid off a substantial proportion of its workforce at the end of 2013, and even if operations continue for another decade or more, the process could last much longer than it normally does because the institutional jungle contains a number of financial mechanisms designed to benefit the people of Western Province and the mine-affected area for many years after the mine itself has finally closed. With an ongoing decline in the proportion of people in the mine-affected area who make their living from wages paid by OTML or its contractors, we can safely anticipate a corresponding increase in the proportion whose primary economic interest lies in the distribution of what is broadly understood to be compensation for environmental damage. This, in turn, would seem to entail the disappearance of the demarcation lines that have previously separated positions of leadership in the real state and the proxy state. The battle for control of SDP can be understood as the start of this political process. Instead of assuming that the proxy state has somehow caused the real state to 'wither away', we may

---

5   Although Daru remains the official capital of Western Province, Kiunga is now the headquarters of the Fly River Provincial Government.

now see that the real state has begun to consume the proxy state as the number of people employed to protect the latter from the former has also begun to fall.

Hitherto, the provincial governor and the three other members of the national parliament elected to represent the people of Western Province have been inclined to call for the mine to be closed sooner rather than later. This in itself does not seem to explain the frequency with which they have then gone on to lose their seats at the next national election, since their replacements have then made the same call. Although sometimes dressed in the language of environmental justice, their choices were more likely based on the resentment of their inability to control the operation of the proxy state and therefore use the promise of mine-related benefits to win over the hearts and minds of their constituents. In recent years, there has been a steady increase in the value of the 'constituency development funds' placed under the direct control of all members of parliament, and provision has recently been made for LLG presidents, who are now directly elected by their own constituents, to be granted their own share of the national development budget. One might suppose that this would give the elected politicians of Western Province less reason to push for control of the proxy state's own 'development budget', and yet the nationalisation of the mine seems to have changed the balance of incentives as well as the balance of power.[6] It remains to be seen what new forms of social solidarity or political accountability might emerge in these new circumstances, but only a bold prophet would predict a more equitable or sustainable set of development outcomes any time soon.

References

Anon., 2009. 'Ok Tedi May Close in 2020.' *Post-Courier*, 30 June.

——, 2011. 'MPs: Ok Tedi Mine Must Close in 2013.' *Post-Courier*, 25 May.

——, 2012a. 'More Life at Ok Tedi.' *Post-Courier*, 8 March.

---

6   The newly elected provincial governor, Ati Wobiro, was formerly employed by SDP to manage its activities in Western Province. He openly supported the expropriation of his former employer but, unlike the re-elected member for North Fly District, he did not voice any public opposition to the Mine Life Extension Plan before he was found guilty of corruption and dismissed from office in 2016.

——, 2012b. 'Ok Tedi Villagers Approve of Mine's Extension.' *The National*, 31 December.

——, 2013. 'Group Claims to Represent Community.' *The National*, 8 November.

Banks, G. and C. Ballard (eds), 1997. *The Ok Tedi Settlement: Issues, Outcomes and Implications*. Canberra: The Australian National University, National Centre for Development Studies (Pacific Policy Paper 27).

Burton, J., 1997. 'Terra Nugax and the Discovery Paradigm: How Ok Tedi Was Shaped by the Way It Was Found and How the Rise of Political Process in the North Fly Took the Company by Surprise.' In G. Banks and C. Ballard (eds), *The Ok Tedi Settlement: Issues, Outcomes and Implications*. Canberra: The Australian National University, National Centre for Development Studies (Pacific Policy Paper 27).

——, 1998. 'Mining and Maladministration in Papua New Guinea.' In P. Larmour (ed.), *Governance and Reform in the South Pacific*. Canberra: The Australian National University, National Centre for Development Studies (Pacific Policy Paper 23).

Crook, T., 2004. 'Transactions in Perpetual Motion.' In E. Hirsch and M. Strathern (eds), *Transactions and Creations: Property Debates and the Stimulus of Melanesia*. New York: Berghahn Books.

Faulkner, K., 2005. 'Ok Tedi Project Update.' Unpublished presentation to Ok Tedi Mine Closure Planning Workshop, Tabubil, 28 October.

Filer, C., 1997. 'West Side Story: The State's and Other Stakes in the Ok Tedi Mine.' In G. Banks and C. Ballard (eds), *The Ok Tedi Settlement: Issues, Outcomes and Implications*. Canberra: The Australian National University, National Centre for Development Studies (Pacific Policy Paper 27).

——, 2005. 'The Role of Land-Owning Communities in Papua New Guinea's Mineral Policy Framework.' In E. Bastida, T. Wälde and J. Warden-Fernández (eds), *International and Comparative Mineral Law and Policy: Trends and Prospects*. The Hague: Kluwer Law International.

Filer, C., D. Henton and R. Jackson, 2000. *Landowner Compensation in Papua New Guinea's Mining and Petroleum Sectors.* Port Moresby: PNG Chamber of Mines and Petroleum.

Howes, S. and E. Kwa, 2011. 'Papua New Guinea Sustainable Development Program Review.' Report to PNG Sustainable Development Program Ltd.

Hriehwazi, Y., 2011. 'OTML Plans to Mine Till 2022.' *The National*, 3 October.

Jackson, R.T., 1979. 'The Awin: Free Resettlement on the Upper Fly River (Western Province).' In C.A. Valentine and B. Valentine (eds), *Going through Changes: Villagers, Settlers and Development in Papua New Guinea.* Boroko: Institute of Papua New Guinea Studies.

——, 1982. *Ok Tedi: The Pot of Gold.* Waigani: University of Papua New Guinea Press.

——, 1993. *Cracked Pot or Copper Bottomed Investment? The Development of the Ok Tedi Project 1982–1991, a Personal View.* Townsville: James Cook University (Melanesian Studies Centre).

——, 1999. 'The Tabubil Outline Longterm Economic Development Plan.' Unpublished report to the PNG Office of National Planning and Implementation and the PNG Department of Mineral Resources.

——, 2003. 'Muddying the Waters of the Fly: Underlying Issues or Stereotypes?' Canberra: The Australian National University, Research School of Pacific and Asian Studies, Resource Management in Asia-Pacific Program (Working Paper 41).

Kalinoe, L., 2008. 'The Ok Tedi Mine Continuation Agreements: A Case Study Dealing with Customary Landowners' Compensation Claims.' Boroko (PNG): National Research Institute (Discussion Paper 105).

King, D., 1993. 'Statistical Geography of the Fly River Development Trust.' Waigani (PNG): Unisearch PNG Pty Ltd for Ok Tedi Mining Ltd (Ok-Fly Social Monitoring Project Report 4).

——, 1997. 'The Big Polluter and the Constructing of Ok Tedi: Eco-imperialism and Underdevelopment along the Ok Tedi and Fly Rivers of Papua New Guinea.' In G. Banks and C. Ballard (eds), *The Ok Tedi Settlement: Issues, Outcomes and Implications*. Canberra: The Australian National University, National Centre for Development Studies (Pacific Policy Paper 27).

Kirsch, S., 1989. 'The Yonggom, the Refugee Camps along the Border, and the Impact of the Ok Tedi Mine.' *Research in Melanesia* 13: 30–61.

——, 1995. 'Social Impact of the Ok Tedi Mine on the Yonggom Villages of the North Fly, 1992.' *Research in Melanesia* 19: 23–102.

——, 1997. 'Is Ok Tedi a Precedent? Implications of the Lawsuit.' In G. Banks and C. Ballard (eds), *The Ok Tedi Settlement: Issues, Outcomes and Implications*. Canberra: The Australian National University, National Centre for Development Studies (Pacific Policy Paper 27).

——, 2004. 'Property Limits: Debates on the Body, Nature and Culture.' In E. Hirsch and M. Strathern (eds), *Transactions and Creations: Property Debates and the Stimulus of Melanesia*. New York: Berghahn Books.

——, 2007. 'Indigenous Movements and the Risks of Counterglobalization: Tracking the Campaign against Papua New Guinea's Ok Tedi Mine.' *American Ethnologist* 34: 303–321.

Martin, M., 2013. 'Landowners Demand Shares.' *Post-Courier*, 26 September.

Menzies, N. and G. Harley, 2012. '"We Want What the Ok Tedi Women Have": Guidance from Papua New Guinea on Women's Engagement in Mining Deals.' Washington (DC): World Bank (J4P Briefing Note 7.2).

Offor, T. and B. Sharp, 2012. 'Turning a Benefit Agreement into Practical Development: A Case Study of a Papua New Guinea Development Foundation.' In M. Langton and J. Longbottom (eds), *Community Futures, Legal Architecture: Foundations for Indigenous Peoples in the Global Mining Boom*. Abingdon (UK): Routledge.

OTML (Ok Tedi Mining Ltd), 1999. 'Mine Waste Management Project: Risk Assessment.' Unpublished report to the PNG Government.

Pintz, W., 1984. *Ok Tedi: Evolution of a Third World Mining Project*. London: Mining Journal Books.

Sharp, B. and T. Offor, 2008. 'Renegotiating a PNG Compensation Agreement: Applying an Informed Consensus Approach.' Canberra: The Australian National University, Research School of Pacific and Asian Studies, Resource Management in Asia-Pacific Program (Working Paper 69).

Simpson, G., 2002. 'Sustainable Development Policy and Sustainability Planning Framework for the Mining Sector in Papua New Guinea— Working Paper 6: Institutional Analysis.' Port Moresby: PNG Mining Sector Institutional Strengthening Project.

World Bank, 2000. 'Ok Tedi Mining Ltd Mine Waste Management Project: Risk Assessment and Supporting Documents.' Unpublished report to the PNG Government.

9. Disconnected Development Worlds: Responsibility towards Local Communities in Papua New Guinea

JOHN BURTON AND JOYCE ONGUGLO

Introduction

Following the publication of the Brundtland Report (UN 1987), diverse global agendas concerned with development, social and environmental sustainability, and the responsibilities of governments and corporations have come to prominence.

A concern arises that, far from being interlocked or even internally consistent, initiatives that play to particular audiences become disconnected from one another and even fail to engage properly with their original objectives. In Papua New Guinea (PNG), where we have been working, it is well known that the state has made a considerable effort to get its planning system in order (GoPNG 2009), while at the same time remaining 'off track', and achieving few of the Millennium Development Goals targets by the end of 2015 (UNDP and GoPNG 2014, Appendix 4). Equally, the Global Mining Initiative and its successor, the Mining, Minerals and Sustainable Development (MMSD) project, which ran from 2000 to 2002, were imagined as marking a new dawn in dealings between mining companies and the communities hosting mining projects. But the MMSD project did not engage with, and its final report

(MMSD 2002) contains no reference to, the global development agenda framed by the Millennium Development Goals (MDGs), despite both being launched in the same year (see Buxton 2012).

A plausible scenario is the creation of separate 'development worlds', each with worthy aims but likely to work at cross purposes because of the different priorities of the stakeholders. Our discussion looks at three of them, centring around the issues of mine-community grievances and the social and economic development of mine-area communities, for evidence that the founding objectives are being adhered to (see Figure 9.1). Our particular focus is the position of indigenous peoples on whose land mining operations are so frequently located. The MMSD project took a special interest in this:

> The post-colonial era has not seen a marked improvement in the status of indigenous peoples or their relationships to the minerals industry in many parts of the world (Danielson 2003: ix).

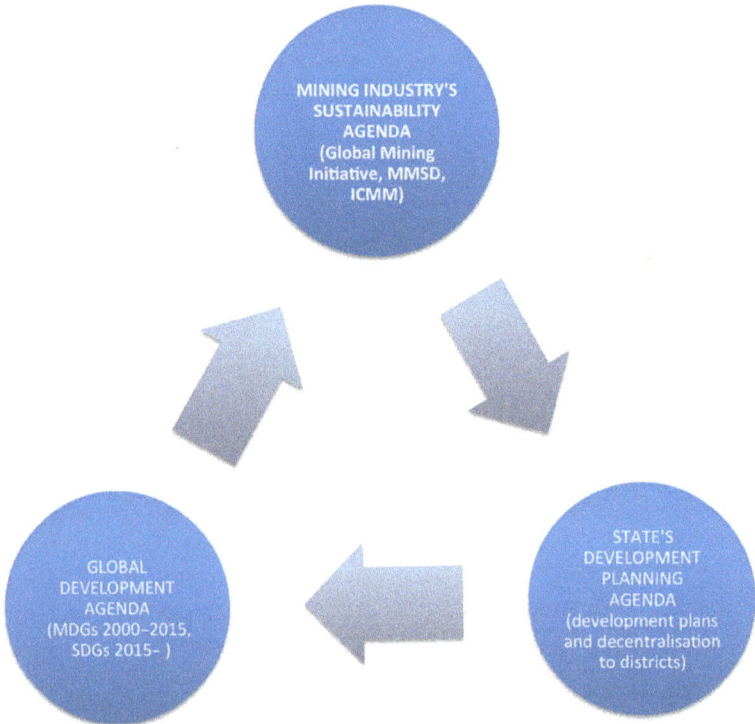

Figure 9.1 Three 'development worlds'.
Source: Authors' diagram.

The loss of two mines in PNG were case studies: the armed conflict that closed the Panguna mine in Bougainville in 1989, majority owned by what is now Rio Tinto; and litigation brought against BHP, the then majority owner of the Ok Tedi mine, in 1994 over environmental damage, causing the exit of the company from this project in 2001 (May and Spriggs 1990; Filer 1990, 1997; Banks and Ballard 1997).

A permanent organisation was born from the MMSD project. This was the International Council on Mining and Metals (ICMM), whose membership in 2015 comprised 21 mining companies—including Rio Tinto and a restructured BHP Billiton—and 35 mining associations worldwide. The ICMM was founded with the goal of improving company performance in the mining and metals industry generally, tied to a set of sustainable development principles (see Box 9.1) and an expanding collection of guidance documents on its website that its members must follow, to make as certain as possible that social and environmental disasters like those seen in Bougainville and at Ok Tedi never happen again.

Box 9.1 The ICMM sustainable development principles.

1. Implement and maintain ethical business practices and sound systems of corporate governance.

2. Integrate sustainable development considerations within the corporate decision-making process.

3. Uphold fundamental human rights and respect cultures, customs and values in dealings with employees and others who are affected by our activities.

4. Implement risk management strategies based on valid data and sound science.

5. Seek continual improvement of our health and safety performance.

6. Seek continual improvement of our environmental performance.

7. Contribute to conservation of biodiversity and integrated approaches to land use planning.

8. Facilitate and encourage responsible product design, use, re-use, recycling and disposal of our products.

9. Contribute to the social, economic and institutional development of the communities in which we operate.

10. Implement effective and transparent engagement, communication and independently verified reporting arrangements with our stakeholders.

Source: www.icmm.com (version in force during the period discussed in the text).

Mining Agreements in Papua New Guinea

Between 2007 and 2014, we have worked in communities around four mining projects in PNG—Porgera and Ok Tedi (Burton), Lihir and Hidden Valley (Burton and Onguglo). All of this work has had the central focus of evaluating the balance between social impacts and the opportunities for development that mining brings, with a special focus on whether the lives of customary landowners have seen any improvement.

In PNG, a mining agreement is the social contract entered into when a tribal people grants permission for mining on its land. The three basic elements of an agreement are:

- Free, prior and informed consent (FPIC)—the decision to allow mining and the negotiation of mining agreements are arrived at in a fair manner, following the principles of FPIC.

- Stakeholder identification—customary owners and other stakeholder groups are properly identified and written into the various sections of an agreement with accuracy, clarity and safeguards to ensure that recognition shifts do not occur over time.

- Agreement governance—processes are specified, and their costs underwritten, that ensure: benefits (royalties, compensation for loss, lease payments, employment, business spin-offs, improvements to local infrastructure, commitments to social programs) will be appropriate and divided fairly among stakeholder groups; beneficiaries receive what the agreement says without hidden transaction costs throughout the mine life; there are appropriate protections for vulnerable people; monitoring and evaluation is carried out to professional standards; and reviews are held following an agreed timetable and to the same standard as used in the original agreement-making process, or better.

A local innovation to try to achieve parts of the above is the 'development forum', first used in the negotiations for the Porgera gold mine in 1988–89 (Golub 2007; Filer 2008). Today, the *Mining Act 1992* lays out the specifications of a forum and sets out a list of parties the mining minister should consider inviting.

At Hidden Valley, a forum was launched on 4 August 2004 (Bonai 2004). The provincial administrator, Manasupe Zurenuoc, praised the cultural appropriateness of the talks, saying that 'in a country such as PNG the Melanesian approach was the secret to success' (Anon. 2004a). However, the participants were not the six sets of communities, labelled Stakeholder Groups A–E, in the impact area that were identified in the company's social impact assessment, the document that should have guided the minister (MCG 2004a). Only Group A—represented by the mine lease landowners' Nakuwi Association—participated in them.

The process ceased to be referred to as a 'forum' after two weeks (Anon. 2004b). Sporadic media reports referred to 'talks' until a year later, when the Hidden Valley memorandum of agreement (MOA) was signed (GoPNG 2005). After the mine construction period, emergency negotiations had to be held with Group B, an omitted stakeholder group made up of the Watut River communities, when their land was impacted by the discharge of waste rock.

On these counts, the process cannot be described as a 'forum' or, for that matter, be said to reflect an inclusive, 'Melanesian' approach. What in fact happened was that decisions were made over the interests of the unrepresented stakeholder communities without their consent.

A surprising inclusion in the MOA was that the six local-level governments (LLGs) surrounding the mine in Bulolo District were allocated royalty shares amounting to 20 per cent of the total. But, here again, the body created to plan the expenditure of funds by LLGs under the *Organic Law on Provincial Governments and Local-Level Governments*, the Joint District Planning and Budget Priorities Committee (JDPBPC), was excluded from the agreement-making process.

In short, the MOA process was deficient: in respect of stakeholder identification because it did not properly represent the parties that should have been involved; in respect of FPIC because of the closed-door nature of the talks; and in respect of agreement governance because it handed money to government entities in a way that bypassed the coordinating body established to guide district development.

This was evident at the time, but it was not until 2015 that any agency reported on the effectiveness of the MOA. This was in the form of research privately commissioned from the PNG National Research Institute (NRI) by the Bulolo District JDPBPC. The NRI's report authors concluded that, while the financial flows to MOA parties were largely as set out by the MOA, the systems in place for managing them were ineffective and their impacts on development were 'minimal' (Sanida et al. 2015: viii, 61). The Nakuwi Association, the 'link between the mine and customary landowners', was described as 'defective' and its business subsidiary had not submitted a tax return for ten years (ibid.: 48). This is not a surprise: the pathologies can be traced back to the 1980s, when a previous business subsidiary delivered little to its community owners (Burton 2003: 215).

These things flow on to agreement governance as a whole—the technical work of seeing that what agreements say is actually implemented. The former Department of Mining noted the 'isolation of the development forum from the process of planning for sustainable development' more than a decade ago (GoPNG 2003, para 2.3.1.10), and we can widen this to say that the more the state leaves most of the work of agreement-making to local parties, the less likely it is that attention will be paid to agreement governance, frustrating the broader national and international objectives of poverty reduction.

Reporting Sustainability Performance

A major change since the millennium is that members of the ICMM undertake to report transparently on their performance each year (Principle 10). Member companies must publish a 'sustainability report' that is compliant with the Global Reporting Initiative (GRI), while non-member companies are encouraged by peer example to do so voluntarily. All should seek to continually improve their environmental and social performance, and obtain better development outcomes for communities in mining areas.

In practice, few mining companies follow the ICMM guidelines. In the case of 33 mining companies with interests in PNG in 2013–14 (see Figure 9.2), 27 (82 per cent) did not refer to sustainability on their websites or publish a sustainability report on their PNG operations.[1]

---

1   The 33 companies are mining companies that had detectable interests in PNG in the 2014 reporting year; there may be 50 or more companies altogether. Of the 33, Barrick, Anglo American and Glencore were ICMM members. In 2011, Anglo American and Teck held 11.1 per cent and 6.8 per cent respectively of the shares of Nautilus Mining, the developer of the Solwara deep-sea mining project in PNG (Nautilus Minerals 2011). Neither mentioned this in annual sustainability reporting. Teck has since divested but Anglo American retained 5.95 per cent in 2015. Anglo American invested in a Star Mountains exploration project in 2014, but did not mention this in its report for that year (Anglo American 2015).

Of the remainder, six (18 per cent) did sustainability reporting (Table 9.1). Of these, four (12 per cent) were to the standard required by ICMM or better, one (Ok Tedi Mining Limited) was non-compliant, and one (PanAust) was in the process of acquiring a PNG operation; so, while it reported on operations in Laos and Thailand, it did not do so for its (Frieda River) operation in PNG for the part of the first year in which it was involved.

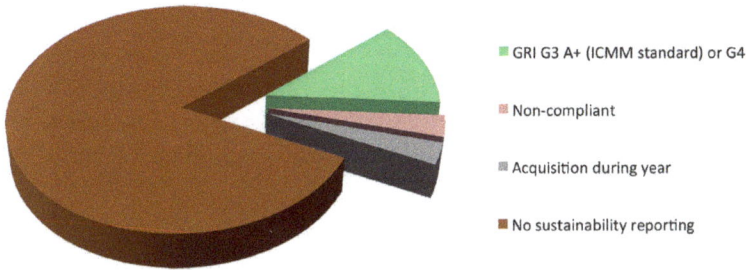

■ GRI G3 A+ (ICMM standard) or G4

■ Non-compliant

■ Acquisition during year

■ No sustainability reporting

Figure 9.2 Sustainability reporting by 33 mining companies with PNG interests in 2013–14.

Source: Authors' diagram.

Table 9.1 Disclosures in relation to mining on the lands of indigenous peoples by six mining companies undertaking sustainability reporting in PNG.

Company	Year	Operation	Global Reporting Initiative standard	Discloses operations on lands of indigenous peoples?
Barrick	2013 (cy)	Porgera	G3	In Australia, North America and Chile, but not Porgera
Glencore	2014 (cy)	Frieda (to August 2014)	G3	Yes
PanAust	2014 (cy)	Frieda (from August 2014)	G4	Yes
Newcrest	2014 (fy)	Lihir, Hidden Valley, Wafi	G3	Yes
Harmony	2014 (fy)	Hidden Valley, Wafi	G4	Yes
Ok Tedi Mining Limited	2013 (cy)	Ok Tedi	G4 (non-compliant)	There is a reference to a page about free, prior and informed consent and the Community Mine Continuation Agreements, but indigenous peoples are not mentioned on it

For latest reporting period available: cy = calendar year (1 January–31 December); fy = financial year (1 July–30 June).

Sources: BGC (2014a); OTML (2014); Glencore (2015); NML (2015); PanAust (2015).

Indicators of Heightened Importance to Indigenous Peoples

As noted already, all mining in PNG takes places on the customary land of tribal landowners. What content is there in GRI reporting that might help raise the alarm over incipient conflicts or the grievances that landowners may have over mine operations?

The GRI framework requires that organisations address a standard set of indicators—79 at the G3 standard and 91 at the newer G4 standard—plus an additional 11 Mining and Metals Sector Supplement (MMSS) indicators for mining companies (GRI 2010). A subset of the indicators are of heightened importance to local communities and deal with human rights, grievances and resettlement: HR1, HR9, MM5, MM6, MM7 and MM9 (and HR12 in the G4 standard) (see Table 9.2). Four indicators specifically refer to indigenous people. An immediate difficulty is that not all of the six reporting companies acknowledge that they run mines in PNG on the territories of indigenous peoples.

Digression: Who Are Indigenous Peoples?

What factors define an individual or group as being 'indigenous'?

The special status of indigenous peoples arose in the context of the colonial occupation of tribal lands, though scant progress was achieved prior to the advent of the post-war international institutions. Thus Chief Deskaheh (1873–1925) led a delegation of Six Nations Iroquois, the largest First Nation band in Canada, to Geneva in 1923, helped by the Bureau International pour la Défense des Indigènes, but was blocked from making a formal presentation at the League of Nations by Britain, the United States and Canada, all worried by the wider implications (Smith 2000; Niezen 2003: 31–5). Once the United Nations (UN) came into existence, the focus in respect of subject peoples in their own lands was decolonisation. Tribal peoples within larger settler nations still struggled for self-determination, arguably bringing into being 'indigenous peoples' as a category distinct from other subject peoples by virtue of being ethnic minorities. Definitional issues of this nature, however, were resolved by the UN Special Rapporteur José Martínez Cobo in a multivolume study in the 1980s (Box 9.2). This study informed two subsequent international agreements: the International Labour Organization's Convention No. 169, adopted in 1989, and the UN Declaration on the Rights of Indigenous Peoples, adopted by the General Assembly in 2007.

Box 9.2 Definition of Indigenous people by the UN Special Rapporteur José R. Martínez Cobo.

379. Indigenous communities, peoples and nations are those which, having a historical continuity with pre-invasion and pre-colonial societies that developed on their territories, consider themselves distinct from other sectors of the societies now prevailing in those territories, or parts of them. They form at present non-dominant sectors of society and are determined to preserve, develop and transmit to future generations their ancestral territories, and their ethnic identity, as the basis of their continued existence as peoples, in accordance with their own cultural patterns, social institutions and legal systems.

380. This historical continuity may consist of the continuation, for an extended period reaching into the present, of one or more of the following factors:

a.   Occupation of ancestral lands, or at least of part of them;

b.   Common ancestry with the original occupants of these lands;

c.   Culture in general, or in specific manifestations (such as religion, living under a tribal system, membership of an indigenous community, dress, means of livelihood, life-style, etc.);

d.   Language (whether used as the only language, as mother-tongue, as the habitual means of communication at home or in the family, or as the main, preferred, habitual, general or normal language);

e.   Residence in certain parts of the country, or in certain regions of the world;

f.   Other relevant factors.

381. On an individual basis, an indigenous person is one who belongs to these indigenous populations through self-identification as indigenous (group consciousness) and is recognized and accepted by these populations as one of its members (acceptance by the group).

Source: Martínez Cobo (1987).

In another example of disconnection between the 'development worlds', the mining industry (Danielson 2003; ICMM 2010) and global development agendas (Caruso et al. 2003; Anaya 2011) have both embraced the need for special care when indigenous peoples are involved, but the Pacific island states as a bloc—PNG, Solomon Islands, Vanuatu, Fiji, Nauru, Palau, Marshall Islands, Tonga, Tuvalu, Kiribati—have not. They were absent from the UN General Assembly at the time of the vote on the Declaration on the Rights of Indigenous Peoples.

More pointedly, a PNG Department of Mineral Policy and Geohazards Management presentation explained it would be 'wrongful' for a legislative amendment to give effect to the Declaration on the Rights of Indigenous Peoples within the Mining Act, because PNG was not among the parties that had adopted it (GoPNG 2012). This creates a potential consultation gap in respect of mining on tribal lands. The development forum has come into play in PNG prior to the granting of mining leases but this has given rise to the paradox of state institutions handing most of the work of agreement-making to local parties—empowering indigenous landowners—yet propped up by an official position that, logically, denies their existence.

On Lihir Island, the mining lease is wholly on 'ancestral land' and royalty is calculated and paid on the basis of its customary ownership by matrilineal clans. Indeed, ICMM's *Good Practice Guide* (2010) was developed by experts from the Centre for Social Responsibility in Mining at the University of Queensland with the collaboration of the mine's operator (see also Macintyre 2007).

In 2010, the mine was acquired by Newcrest Mining Limited. The company was specific in using one criterion, that of 'non-dominant sectors of society' (paragraph 379 in Box 9.2), to rule that its PNG operations did not involve indigenous people:

> On the basis of this definition, it is only Newcrest's Australian operations where issues of specific Indigenous rights arise. Community relation issues with regards to Indonesia and Papua New Guinea operations are reported elsewhere in this report. (NML 2011: 57; see also NML 2012)

Barrick became the operator of the Porgera mine in 2006. Its position on indigenous people reads as follows:

> Barrick endorses the ICMM Position Statement on Indigenous Peoples and Mining, which includes specific commitments and obligations related to Free Prior and Informed Consent (FPIC). Barrick is committed to implementing the Position Statement globally. (BGC 2013: 29)

But though successive sustainability reports show that Barrick considers communities in Chile, Australia and North America to be indigenous, it does not include the Ipili people of Porgera, even though land issues are handled through Ipili agents empowered under a section of PNG's Land Act that deals with customary land (Golub 2007).

In the two cases, a single one of the special rapporteur's criteria was used to trump the full set. In reality, all landowners in PNG fit the overall definition: they occupy their ancestral lands; they are distinguished by specific customs setting them apart from their neighbours; they speak their own language as the habitual means of communication within the community; they reside in 'certain parts of the country', namely on their tribal territories; and they transmit these 'to future generations ... in accordance with their own ... legal systems', not state ones.

In the 2012 reporting year, perhaps in response to a draft of this paper circulated after the Nouméa conference in November 2011 (see Chapter 1, this volume), Newcrest (NML 2013: 62) began counting its PNG landowners as indigenous people. Barrick did not follow suit.

Sustainability Reporting Continued

A good test of whether the mining industry agenda is on track with its founding objectives is to weigh up whether company disclosures can provide an alert for the kinds of mine-community grievances and conflicts that can lead companies to 'lose the mine'.

We evaluate the companies in Table 9.2 on the way they reported the previously mentioned indicators in 2009–10, before the Nouméa conference, and again in 2013–14. Two of the companies began using the newer GRI G4 standard in the second period, which meant that a new indicator (HR12) was added for them, relating to the handling of grievances. In both periods, the most recent available report was examined, depending on the reporting cycles of the organisations.

The significance of claiming not to deal with indigenous people may be seen in the manner in which sustainability reporting is curtailed. Table 9.2 shows that in the earlier period none of the companies acknowledged the presence of indigenous people, making four indicators unreportable. In the later period, two now did so, but the commonest answer to the indicators about disputes was to say there were none. This is not really believable in PNG: there is always something to report, whether a minor grievance or something more serious.

Table 9.2 Seven Global Reporting Initiative indicators of heightened importance to local communities and indigenous peoples, for the Lihir, Hidden Valley, Porgera and Ok Tedi mines, 2009–10 and 2013–14.

HR1	Percentage and total number of significant investment agreements that include human rights clauses or that have undergone human rights screening.	
	2009–10	LGL: not addressed, Newcrest: not addressed, Harmony: indicator not reported, Barrick: globally reported–cannot disaggregate Porgera, OTML: (non-GRI).
	2013–14	Newcrest: 'human rights [not yet screened] in investment agreements', Harmony: indicator not reported, Barrick: uses TRACE International anti-corruption screening system, OTML: indicator omitted.
HR9	Total number of incidents of violations involving rights of indigenous people and actions taken.	
	2009–10	LGL: ● + 'No incidents', Newcrest: ●, Harmony: ●, Barrick: ●, OTML: (non-GRI).
	2013–14	Newcrest: ● 'No incidents [involving the Company] … The Company, on occasion, has been impacted by intra-communal tensions in the surrounding communities', Harmony: ● indicator not reported, Barrick: ●, OTML: ● indicator omitted.
MM5	Total number of operations taking place in or adjacent to indigenous peoples' territories, and number and percentage of operations or sites where there are formal agreements with indigenous peoples' communities.	
	2009–10	LGL: ●, Newcrest: ●, Harmony: ●, Barrick: ●, OTML: (non-GRI).
	2013–14	Newcrest: ● correctly reported. Harmony: ● correctly reported, Barrick: ●, OTML: ● discussion omits indigenous people.
MM6	Number and description of significant disputes relating to land use, customary rights of local communities and indigenous peoples.	
	2009–10	LGL: ●, Newcrest: ● + 'Union of Watut River Communities published a list of claims', Harmony: ●, Barrick: ●, OTML: (non-GRI).
	2013–14	Newcrest: ● '2 significant disputes' (at Lihir only), Harmony: ● 'There were no disputes during the year', Barrick: ●, OTML: ● correctly reported.
MM7	The extent to which grievance mechanisms were used to resolve disputes relating to land use, customary rights of local communities and indigenous peoples, and the outcomes.	
	2009–10	LGL: ●●, Newcrest: ●, Harmony: ●, Barrick: ●, OTML: (non-GRI).
	2013–14	Newcrest: ● Lihir–correctly reported, Hidden Valley–'total of 276 grievances were received across Hidden Valley, Wafi-Golpu and exploration operations', Harmony: ● indicator not addressed–see HR12, Barrick: ●, OTML: ● correctly reported.

MM9	Sites where resettlements took place, the number of households resettled in each, and how their livelihoods were affected in the process.	
	2009–10	LGL: not addressed, Newcrest: 'one instance of resettlement', Harmony: indicator not reported, Barrick: not addressed, XStrata: 'In 2009, we resettled 116 households in four locations', OTML: (non-GRI).
	2013–14	Newcrest: correctly reported, Harmony: indicator not reported, Barrick: resettlement under review, OTML: correctly reported.
HR12 (GRI G4 only)	Number of grievances about human rights impacts filed, addressed and resolved through formal grievance mechanisms.	
	2013–14	Harmony: 'illegal strike activity occurred in March 2014, when six days were lost due to grievances', OTML: indicator not reported.

LGL = Lihir Gold Limited, OTML = Ok Tedi Mining Limited. LGL merged with Newcrest in mid-2010.

Indigenous people: ● = recognised ● = not recognised ● = unclear.

Sources: LGL (2010); HGMC (2010, 2014); BGC (2011, 2014a); NML (2011, 2015); OTML (2012a, 2012b, 2014).

Lihir

Lihir Gold Limited (LGL) was one of the reporters saying it had 'no incidents' in 2009 (LGL 2010: 115), but in reality the mine had a shutdown over landowner grievances:

> Australia's second-biggest listed gold producer, has been forced to shut its mine in Papua New Guinea because of landowner disputes, the firm said on Monday (Anon. 2009).

A mine shutdown is a material incident to report under HR9 and MM6. In Table 9.2, it can be seen that we score LGL for this year as both ● and ● on the MM7 indicator. LGL was one of the companies that denied the presence of indigenous people but (confusingly) had grievance policies in place that were tailored to suit dealings with indigenous people (LGL 2010: 69). The subsequent owner, Newcrest, has prevaricated about whether incidents are 'indigenous' or not. In 2012–13, it reported 'no incidents' under HR9 (NML 2014a, vol. 2: 12) when landowners shut down mine operations 'by placing the traditional gorgor plant around the mine site' (Nalu 2012). The use of *gorgor* shows that it was a dispute where an indigenous people pressed their case using a customary rite. It was also serious enough to affect the share price (Fitzgerald 2012). On both counts it was an HR9 incident and should have been reported as such. In 2013–14, Newcrest again said there were 'no incidents' on this indicator, but added:

> The Lihir operation is subject to a number of agreements with local landowners, which cover (among other things) compensation and other benefits, including commercial opportunities related to land ownership. The Lihir agreements are subject to periodic review, with an ongoing review at the time of this report. In the context of the current review, the Company and landowners have outlined matters that they would like addressed within the review, including the implementation of projects under the agreements, compensation payments (including distribution of benefits among landowners), the impacts of the mine and the impact of certain practices on the mine operations. These matters will be advanced as part of the ongoing Lihir agreements review. (NML 2015, vol. 2: 19)

This was a long-winded way of avoiding saying anything of substance. In reality, an impasse had been reached in reviewing agreements signed in 1995 and 2007. The Lihir Mine Area Landowners Association (LMALA) was reported as being 'angry' (Anon. 2014a) and had described a deal between the provincial and local-level governments 'as illegal' (Anon. 2014b). In mid-year, the company reported 'receiving threats' from LMALA to shut the mine (Anon. 2014c). In August and October 2014, a 'rebel group' erected roadblocks at the North Kapit stockpile. On the one hand, Newcrest said that it only dealt with LMALA, but on the other it facilitated a K6 million finance deal with the dissident group (Anon. 2014d).

To not report the incidents was to follow a very pedantic reading of HR9 on the grounds that it was unproven that the mining company had violated indigenous rights. Delayed (or cancelled) Bougainville Copper Agreement reviews were a key factor leading to conflict at Panguna in 1988–89 (see May and Spriggs 1990; Oliver 1991; also Chapter 12, this volume), and observers with long memories also recall that the present governor of New Ireland Province, where Lihir is located, said in 1986 that Lihirians were living in a 'sociological time-bomb' (Filer 1990: 78). To fudge the difficulties faced on Lihir must be judged imprudent.

Hidden Valley

At Hidden Valley in 2014, Harmony reported 'no disputes' under MM6 (HGMC 2014: 187), but elsewhere in the same report the company said that 'illegal strike activity occurred in March 2014, when six days were lost due to grievances' that were then handled by the Employee Representative Council (ibid.: 69). The question is, did the strike have anything to do with customary rights?

The strike was about terms and conditions for semi-skilled workers and, after it turned violent, three employees were taken to hospital (Anon. 2014e; Keslep 2014). This was a significant incident, and Harmony correctly reported it under 'labour disputes and strikes', as well as flagging that this was relevant to the GRI's new HR12 indicator. However, Annex B of the Hidden Valley MOA, the mine's 'employment and training plan', clearly sets out that all unskilled positions are reserved for landowners—the bona fides of the latter being vouched for by the Nakuwi Association—and that landowners would have the highest priority for recruitment and training as heavy equipment operators, maintenance workers and drill and blast operators (MCG 2004b, sections 2.4, 3.2, and 3.5.1). This means that the complainants were local landowners whose jobs came into existence through their connection to the land needed for mining. They were correct to link their grievances to the principal mining agreement, whether or not they were right about the substance of them.

The strike was illegal only if legislation allowed employment contracts to outlaw industrial action, but as far as is known this was not the case. On the other hand, anyone involved in issuing threats or using violence 'to the person or property of another' in or near a workplace is likely to have been committing offences under the section 508 of the Criminal Code. It was incorrect to call the strike itself illegal.

Newcrest mentioned that it had handled 276 grievances at Hidden Valley and Wafi (a nearby exploration site) in 2013–14, 'tracking them to ensure that a resolution is ultimately reached' (NML 2015, vol. 1: 43). But this drew a veil over whether there had been a strike or a shutdown. In reporting the MM4 indicator ('total strikes and lock-outs exceeding one week'), it said in one place that there had been none in 2013–14 (ibid., vol. 2: 9), but in another that there had been one at Hidden Valley (ibid., vol. 3: 30). Cryptically, it noted in a market release that there was 'an 8% reduction in mill throughput' in the first quarter of the year (NML 2014b: 5) but it did not say why.

Porgera

In the 2009 reporting year, the Porgera Joint Venture published a 354-page Porgera-specific environmental monitoring report (PJV 2010), but cast a veil over 'landowner concerns' that had prevented environmental sampling at various times and locations in the impact area. Actually, we know these were physical stand-offs between company staff and landowners, and therefore reportable 'incidents'.

But 'concerns' were not the half of it. Warlords were directing conflicts among well-armed factions in various parts of the Porgera Valley, including in and around the leases. Figures given in a policy document for local consumption, 'Restoring Justice', show that there were 70 killings during 2007 (Kikala et al. 2008: 5). If the population of the valley was about 35,000, this represented a spike in the homicide rate touching 200/100,000—a rate not reported globally outside war zones (see Burton 2014).

Barrick buried this news in passages such as 'sustainable development ... may mean ... developing a greater capacity to plan for and manage the changes of modern life, including new law and order issues' in an optimistic 60-page publicity report (BNL 2009). This stood in stark contrast to the realism of the 'Restoring Justice' document (Kikala et al. 2008), which preceded it by about three months, and to raw reality six weeks later when a police and defence force sweep called 'Operation Ipili' was launched 'to flush out the warlords' (Eroro 2009; Muri 2009a).

The police action targeted a section of the community living inside the boundaries of the mining lease at Porgera, which included about 7,000 recognised landowners at the time. Approximately 300 houses were destroyed in the raid, which received widespread coverage internationally and was investigated by at least four civil society groups. Barrick's *2009 Responsibility Report* (BGC 2010) remained silent on the matter. The company received the maximum possible rating in the ICMM's annual 'member performance table'[2] and its quality assurer wrote that the report gave 'a fair representation of Barrick's activities over the reporting period' (Bureau Veritas 2010).

Both Amnesty and Human Rights Watch took Barrick to task during the second half of 2010 (Amnesty International 2010; HRW 2011), and after this it was no longer possible to issue blandishments. Barrick's next Responsibility Report (BGC 2011) addressed their findings, conceding abuses by the company's security force. There was further coverage under the title 'Human Rights and Security at Porgera' on the company website.

---

2   www.icmm.com/en-gb/members/member-reporting-and-performance.

This was promising, but it still fell short of the GRI requirements in respect of mining on the land of indigenous peoples (HR9, MM6), and there was no mention of whether, if effective grievance mechanisms (MM7) had been in place, complaints such as those affecting the security force could have been avoided.

In respect of the resettlement indicator (MM9), Barrick started a program to move customary owners to land away from the mining lease in 2006. The program was cancelled in late 2007 but Barrick said no more than that it 'works together with host governments to manage the resettlement of people that may be affected by our planned operations in a manner consistent with local laws and international best practice' (BGC 2007: 15, 2008: 15). For two years after cancelling the program, Barrick said nothing about it (or in answer to MM9), but then said that '[r]esettlement activities continued at Porgera … in 2010' (BGC 2011: 72).

None of this commentary would give any reader a clue that the issue of resettlement had been a running sore for years (e.g. Anon. 2002; Muri 2009b) compounded by the (local) mismanagement of benefit streams that might have made mine-area villages pleasant to live in (Johnson 2012). At the time of the Nouméa conference, the answer to the question of what progress had been made at Porgera in respect of the ICMM objectives is that a litany of blunders at Porgera had a 20-year history of obfuscation and denial (Bonnell 1994; Filer 1999; Filer et al. 2008; Wiessner 2010; Burton 2014). Since the conference, the mine's oversight body, the Porgera Environmental Advisory Komiti, has published a report on the plight of women in Porgera (Johnson 2011), and Barrick has issued a 'framework of remediation initiatives' to help women abused by company guards (PJV and BGC 2012). This went some of the way towards completing the requirements of a grievance mechanism and has commendably undergone a recent review (BGC 2014b). It should not have been preceded by years of denial.

Ok Tedi

A joint venture led by BHP operated the Ok Tedi copper and gold mine from 1984 to 2001. After litigation by landowners (Banks and Ballard 1997), BHP divested its majority shareholding to a holding company, PNG Sustainable Development Program Ltd (SDP) in 2001. SDP's share was in turn nationalised in 2013 and the operator, Ok Tedi Mining Limited (OTML), is now a state-owned enterprise (see Chapter 8, this volume, for further detail).

OTML has historically disclosed information to the public in a patchy manner. In 2011, Howes and Kwa noted that 'OTML has been promising to use the internationally recognized Global Reporting Initiative (GRI) standards for some time, but has failed to do so' (Howes and Kwa 2011: 39). Its next two annual reviews were claimed to be 'prepared in accordance with the Global Reporting Initiative' (OTML 2012b: 3, 2013a: ii), but neither reported any indicators. No environment report to date (most recently OTML 2013b) has mentioned the GRI at all or reported against the more than 30 environmental indicators (34 from 2013). This is a serious omission, given Ok Tedi's place in the history of mining and the environment.

In its *Annual Review 2013*, general standard disclosures are finally reported and three of the indicators of Table 9.2 are correctly reported (OTML 2014: 125–9). This represents progress, as is also demonstrated by the voluminous documentation on the 'community mine continuation agreement' (CMCA) process to be found on its website,[3] as published by the facilitators who led the first round of negotiations in 2006–07 (Sharp and Offor 2008; Offor and Sharp 2012), and as further reported by World Bank observers (Eftimie 2010; Popoitai and Ofosu-Amaah 2013).

In large part, the CMCA outcomes were expressed in terms of new benefits for the environmentally affected communities: spending on infrastructure, community projects and annual payments in cash. They are an intensification of activities in the Ok Tedi impact area handled by SDP and the OTML-funded Ok Tedi Development Foundation (OTDF) since 2001. But how can we tell if the infrastructure is in place, the projects work, and the CMCA communities are satisfied with their benefit shares?

SDP produced annual reports during its 12-year lifespan. These show that total spending on development projects (throughout PNG) was US$578 million, in line with SDP's articles of association. Reports up to 2012 discuss projects and the company's strategy, but the word 'evaluation' does not make an appearance until the year of the Howes and

---

3   www.oktedi.com/our-corporate-social-responsibility/mine-continuation-consultation.

Kwa report (2011). Thereafter, the 'design and rollout of an appropriate monitoring and evaluation system' for all projects was flagged for 2012. The government then announced it would nationalise the mine (Elapa 2013), and after the single-page chief executive officer's message, the entirety of the final report is taken up by a tabulation of all the projects undertaken by SDP over 12 years (SDP 2013: 3–31). No evaluation of them was published.

As OTDF did not have a website until 2011, its activities prior to that year are unclear. Annual reports for 2011, 2012 and 2013, detailing projects undertaken, are available. A monitoring and evaluation policy was introduced in 2013 (OTDF 2014: 14), but no results had been published by the time of writing.

Discussion

Being able to talk about social and environmental sustainability is not the same as being able to show 'a marked improvement in the status of indigenous peoples' around mining projects.

In terms of the three worlds of development model (Figure 9.1), starting with the mining industry's sustainability agenda, our discussion of projects in PNG points to poor—if any—use of the new reporting tools. Devised both to warn of difficulty in a company's social and environmental performance, and to provide internal feedback, the evidence that managements understand their purpose is not convincing. Corporate media offices, the presumed source of sustainability reports, leave gaps and recast alarming accounts of protests, strikes and conflicts in impact areas in more soothing prose: 'threats' to close a mine dissolve into an 'ongoing agreements review'; a violent melee with tribal workers is diluted into 'grievances' handled by a review process; and 'warlords' on a killing spree around a mine become 'landowner concerns'. This is unlikely to alert managements, communities or the public in general *before* a future social or environmental crisis is triggered, any better than the failed systems in place prior to the Bougainville and Ok Tedi crises two decades ago.

The world of the state's development planning agenda intersects with this situation to the extent that the biggest stakeholder in mining is always the state. At the national level, royalties, taxes and duties are collected, planning objectives set, and funds channelled to the lower levels of government for implementation. But at all mining projects in PNG, observers find themselves struck by the failure of the state to use its windfall incomes to run essential services, let alone to progress its long-term objectives. Now that the Ok Tedi mine has been nationalised, we may speculate as to whether the livelihoods of the impact-area communities, and the quality of information bearing on their situation, will take a turn for the better or the worse. The state's performance in managing the Tolukuma mine, acquired in 2008, is not propitious: information on the website of Petromin, the state-owned enterprise of which the mine is a subsidiary, was not updated during the seven years of national ownership, and no financial report has been published since 2012. The mine closed in 2015.

A paradox is that it is often visitors from other parts of government who blow the whistle on service delivery failures. The director of PNG's National Research Institute went to Porgera and said afterwards that 'the local primary schools were closed … the high school was closed … the communities are missing out' (Howes 2012). His researcher showed that it was 'impossible to see where the money has gone' because local governance institutions did not produce financial reports (Johnson 2012: xii, 89–90). As noted, National Research Institute researchers found that development impacts around Hidden Valley were 'minimal' (Sanida et al. 2015). And, rather dramatically, during a parliamentary committee tour of Western Province, home of the Ok Tedi mine, an MP burst into tears when visiting a broken-down health centre (Alphonse 2015).

The world of the global development agenda barely intersects with this. The *2014 National Human Development Report* (UNDP and GoPNG 2014) is a valiant attempt to gauge the impact of the country's resource incomes on poverty reduction. But since it concluded that PNG was 'off track' to achieve the Millennium Development Goal (MDG) targets (ibid.: Appendix 4), the answer can be given that, despite a substantial economic boost from the extractive industries, only weak gains have been made in human development. The report did not attempt a lower-level audit of progress in mining areas.

If MDG8—'global partnership for development'—or the post-2015 replacement SDG17—'partnerships for the goals'—were working properly, then the three different development worlds would have become integrated long ago, but they have not. When the Sustainable Development

Goals[4] were launched at the UN on 25 September 2015, the UN's expert panel declared a 'One World: One Sustainable Development Agenda' (UN 2013: 4; see also UNDP and GoPNG 2013). On the evidence of what we have seen, there is a great deal of work to be done even to reach the starting line.

References

Alphonse, A., 2015. 'MP Breaks Down in Shock, Shame.' *Post-Courier*, 25 May.

Amnesty International, 2010. *Undermining Rights: Forced Evictions and Police Brutality around the Porgera Gold Mine, Papua New Guinea.* London: Amnesty International.

Anaya, J., 2011. 'Report of the Special Rapporteur on the Rights of Indigenous Peoples: Extractive Industries Operating within or near Indigenous Territories, A/HRC/18/35.' New York: United Nations General Assembly.

Anglo American, 2015. *Sustainable Development Report 2014.* London: Anglo American plc.

Anon., 2002. 'Local Clans Call for New Site.' *Post-Courier*, 13 March.

——, 2004a. 'Goldfield Vows Support.' *Post-Courier*, 6 August.

——, 2004b. 'Talks on Mine Work.' *Post-Courier*, 20 August.

——, 2009. 'Lihir Gold Says Papua New Guinea Mine Shut on Disputes.' Reuters, 26 January.

——, 2014a. 'Discussions to Continue on Lihir Mine MOA.' *The National*, 4 February.

——, 2014b. 'Owners Oppose Signing of Deal.' *The National*, 13 February.

——, 2014c. 'Newcrest Confirms Threats.' *The National*, 18 June.

——, 2014d. 'Stop Work Resolved.' *The National*, 17 October.

---

4   sustainabledevelopment.un.org/?page=view&nr=1021&type=230&menu=2059.

——, 2014e. 'Employees Strike Halts Mine Operation.' *Post-Courier*, 21 March.

Banks, G. and C. Ballard (eds), 1997. *The Ok Tedi Settlement: Issues, Outcomes and Implications.* Canberra: The Australian National University, National Centre for Development Studies (Pacific Policy Paper 27).

BGC (Barrick Gold Corporation), 2007. *Responsible Mining: 2006 Responsibility Report.* Toronto: BGC.

——, 2008. *Responsible Mining: 2007 Responsibility Report.* Toronto: BGC.

——, 2010. *Responsible Mining: 2009 Responsibility Report.* Toronto: BGC.

——, 2011. *Responsible Mining: 2010 Responsibility Report.* Toronto: BGC.

——, 2013. *Responsible Mining: 2012 Corporate Responsibility Report.* Toronto: BGC.

——, 2014a. *Responsible Mining: 2013 Corporate Responsibility Report.* Toronto: BGC.

——, 2014b. 'The Porgera Joint Venture Remedy Framework: *Olgeta Meri Igat Raits*.' Toronto: BGC.

BNL (Barrick Niugini Ltd), 2009. 'Porgera: The Mine, Its People and the Future.' Port Moresby: BNL.

Bonai, S., 2004. 'Ministers to Launch Wau Projects.' *Post-Courier*, 4 August.

Bonnell, S., 1994. 'Dilemmas of Development: Social Change in Porgera, 1989–1993.' Brisbane: Subada Consulting Pty Ltd for Porgera Joint Venture.

Bureau Veritas, 2010. 'Independent Assurance Statement.' In BGC (Barrick Gold Corporation), *Responsible Mining: 2009 Responsibility Report.* Toronto: BGC.

Burton, J., 2003. 'Fratricide and Inequality: Things Fall Apart in Eastern New Guinea.' *Archaeology in Oceania* 38: 208–216. doi.org/10.1002/j.1834-4453.2003.tb00546.x

——, 2014. 'Agency and the "Avatar" Narrative at the Porgera Gold Mine, Papua New Guinea.' *Journal de la Société des Océanistes* 138–139: 37–51.

Buxton, A., 2012. *MMSD+10: Reflecting on a Decade of Mining and Sustainable Development.* London: International Institute for Environment and Development (Sustainable Markets Discussion Paper June 2012).

Caruso, E., M. Colchester, F. MacKay, N. Hildyard and G. Nettleton, 2003. *Extracting Promises: Indigenous Peoples, Extractive Industries and the World Bank.* Moreton-in-March (UK): Forest Peoples Programme. Baguio City (Philippines): Indigenous Peoples' International Centre for Policy Research and Education (Research Report 3).

Danielson, L., 2003. 'Foreword.' In *Finding Common Ground: Indigenous Peoples and Their Association with the Mining Sector.* London: International Institute for Environment and Development and World Business Council for Sustainable Development.

Eftimie, A., 2010. 'Engendering Mines in Development. A Promising Approach from Papua New Guinea.' Washington (DC): World Bank, Oil, Gas and Mining Policy Division (East Asia and Pacific Region—Social Development Notes).

Elapa, J., 2013. 'Marriage Over, Says PM.' *The National*, 21 March.

Eroro, S., 2009. 'Porgera up in Flames.' *Post-Courier*, 30 April.

Filer, C., 1990. 'The Bougainville Rebellion, the Mining Industry and the Process of Social Disintegration in Papua New Guinea.' In R.J. May and M. Spriggs (eds), *The Bougainville Crisis.* Bathurst: Crawford House Press. doi.org/10.1080/03149099009508487

——, 1997. 'The Melanesian Way of Menacing the Mining Industry.' In B. Burt and C. Clerk (eds), *Environment and Development in the Pacific Islands.* Canberra: The Australian National University, National Centre for Development Studies (Pacific Policy Paper 25).

—— (ed.), 1999. *Dilemmas of Development: The Social and Economic Impact of the Porgera Gold Mine 1989–1994*. Canberra: Asia-Pacific Press (Pacific Policy Paper 34).

——, 2008. 'Development Forum in Papua New Guinea: Upsides and Downsides.' *Journal of Energy and Natural Resources Law* 2(6): 120–149.

Filer, C., J. Burton and G. Banks, 2008. 'The Fragmentation of Responsibilities in the Melanesian Mining Sector.' In C. O'Faircheallaigh and S. Ali (eds), *Earth Matters: Indigenous Peoples, Corporate Social Responsibility and Resource Development*. London: Greenleaf Publishing. doi.org/10.9774/gleaf.978-1-909493-79-7_11

Fitzgerald, B., 2012. 'Land Fight Shuts Lihir Gold Mine.' *The Australian*, 29 August.

Glencore, 2015. *Sustainability Report 2014*. Baar (Switzerland): Glencore plc.

Golub, A., 2007. 'From Agency to Agents: Forging Landowner Identities in Porgera.' In J.F. Weiner and K. Glaskin (eds), *Customary Land Tenure and Registration in Australia and Papua New Guinea: Anthropological Perspectives*. Canberra: ANU E Press (Asia-Pacific Environment Monograph 3).

GoPNG (Government of Papua New Guinea), 2003. *Sustainable Development Policy and Sustainability Planning Framework for the Mining Sector in Papua New Guinea*. Port Moresby: Department of Mining.

——, 2005. 'Memorandum of Agreement Relating to the Hidden Valley Gold Project between the Independent State of Papua New Guinea; and the Morobe Provincial Government; and Morobe Consolidated Goldfields Limited; and Nakuwi Association Inc.; and the Wau Rural Local Level Government; and the Watut Rural Local Level Government; and Wau Bulolo Urban Local Level Government.' Port Moresby: Office of the State Solicitor.

——, 2009. *Papua New Guinea Vision 2050*. Port Moresby: National Strategic Plan Taskforce.

——, 2012. 'Presentation to Members of Parliament: Comments on the Proposed Mining (Amendment) Bill 2012.' Port Moresby: Department of Mineral Policy and Geohazards Management (PowerPoint presentation).

GRI (Global Reporting Initiative), 2010. *Sustainability Reporting Guidelines & Mining and Metals Sector Supplement.* Amsterdam.

HGMC (Harmony Gold Mining Company Ltd), 2010. *Sustainable Development Report 2010.* Randfontein: HGMC.

——, 2014. *Integrated Annual Report 2014.* Randfontein: HGMC.

Howes, S., 2012. 'Thomas Webster on Visas, Porgera, PNG Elections and the Resource Boom.' Devpolicy blogpost, 31 May. Viewed 11 June 2015 at: devpolicy.org/thomas-webster-on-visas-porgera-png-elections-and-the-resource-boom20120531/

Howes, S. and E. Kwa, 2011. *Papua New Guinea Sustainable Development Program Review.* Canberra: The Australian National University, Crawford School of Public Policy, Development Policy Centre. Port Moresby: Constitutional and Law Reform Commission.

HRW (Human Rights Watch), 2011. 'Gold's Costly Dividend: Human Rights Impacts of Papua New Guinea's Porgera Gold Mine.' Viewed 11 June 2015 at: www.hrw.org/en/reports/2011/02/01/gold-s-costly-dividend.

ICMM (International Council on Mining and Metals), 2010. *Good Practice Guide: Indigenous Peoples and Mining.* London: ICMM.

Johnson, Penny, 2011. 'Scoping Project: Social Impact of the Mining Project on Women in the Porgera Area.' Canberra: Emic Consultancy for Porgera Environmental Advisory Komiti.

Johnson, Peter, 2012. 'Lode Shedding: A Case Study of the Economic Benefits to the Landowners, the Provincial Government and the State, from the Porgera Gold Mine.' Port Moresby: National Research Institute (Discussion Paper 124).

Keslep, S., 2014. 'Disgruntled Mine Workers Protest.' *Post-Courier*, 11 March.

Kikala, P., I. Temu and P. Ipatas, 2008. 'Restoring Justice: Law and Justice Sector Partnerships in Enga Province, Papua New Guinea.' Port Moresby: Department of National Planning, Barrick Gold (Australia Pacific) and Enga Provincial Government.

LGL (Lihir Gold Limited), 2010. *2009 Sustainability Report.* Port Moresby: LGL.

Macintyre, M., 2007. 'Informed Consent and Mining Projects: A View from Papua New Guinea.' *Pacific Affairs* 80: 49–65. doi.org/ 10.5509/200780149

Martínez Cobo, J.R., 1987. 'Study of the Problem of Discrimination against Indigenous Populations by the Special Rapporteur Mr José R. Martínez Cobo. Volume V. Conclusions, Proposals and Recommendations, E/CN.4/Sub.2/1986/7/Add.4.' New York: United Nations Economic and Social Council, Commission on Human Rights, Sub-Commission on Prevention of Discrimination and Protection of Minorities.

May, R.J. and M. Spriggs (eds), 1990. *The Bougainville Crisis.* Bathurst: Crawford House Press.

MCG (Morobe Consolidated Goldfields Ltd), 2004a. 'Hidden Valley Project Social Impact Assessment.' Lae: MCG.

——, 2004b. 'Hidden Valley Memorandum of Agreement: Annex B. Employment and Training Plan.' Lae: MCG.

MMSD (Mining, Minerals and Sustainable Development Project), 2002. *Breaking New Ground: Mining, Minerals, and Sustainable Development.* London: Earthscan.

Muri, D., 2009a. 'Crackdown: Porgera Security Ops Set to Kick Off.' *The National*, 6 April.

——, 2009b. 'Barrick Urged to Resettle Villagers.' *The National*, 9 April.

Nalu, M., 2012. 'Lihir Mine Shuts Down.' *The National*, 28 August.

Nautilus Minerals, 2011. *2011 Annual Report: Making Headway.* Toronto: Nautilus Minerals.

Niezen, R., 2003. *The Origins of Indigenism: Human Rights and the Politics of Identity*. Berkeley: University of California Press. doi.org/10.1525/california/9780520235540.001.0001

NML (Newcrest Mining Limited), 2011. *Newcrest Sustainability Report 2010*. Melbourne: NML.

———, 2012. *Newcrest Sustainability Report 2011*. Melbourne: NML.

———, 2013. *Newcrest Sustainability Report 2012*. Melbourne: NML.

———, 2014a. *Newcrest Sustainability Report 2013* (3 volumes). Melbourne: NML.

———, 2014b. *Quarterly Report for the Three Months Ended 31 March 2014*. Melbourne: NML.

———, 2015. *Newcrest Sustainability Report 2014* (3 volumes). Melbourne: NML.

Offor, T. and B. Sharp, 2012. 'Turning a Benefit Agreement into Practical Development: A Case Study of a Papua New Guinea Development Foundation.' In M. Langton and J. Longbottom (eds), *Community Futures, Legal Architecture: Foundations for Indigenous Peoples in the Global Mining Boom*. Abingdon (UK): Routledge.

Oliver, D., 1991. *Black Islanders: A Personal Perspective of Bougainville 1937–1991*. Melbourne: Hyland House.

OTDF (Ok Tedi Development Foundation), 2014. *Annual Report 2013*. Tabubil (PNG): OTDF.

OTML (Ok Tedi Mining Ltd), 2012a. *Annual Review 2010*. Port Moresby: OTML.

———, 2012b. *Annual Review 2011*. Port Moresby: OTML.

———, 2013a. *Annual Review 2012*. Port Moresby: OTML.

———, 2013b. *Annual Environment Report FY2013*. Port Moresby: OTML.

———, 2014. *Annual Review 2013*. Port Moresby: OTML.

PanAust, 2015. *Shaping Our Future: 2014 Business Review & Sustainability Report* (2 volumes). Brisbane: PanAust Ltd.

PJV (Porgera Joint Venture), 2010. *Environmental Monitoring: 2009 Annual Report. January–December 2009.* Mt Hagen (PNG): Environment Department, PJV Communities and Environment.

PJV (Porgera Joint Venture) and BGC (Barrick Gold Corporation), 2012. 'Framework of Remediation Initiatives in Response to Violence against Women in the Porgera Valley.' Port Moresby: PJV and BGC.

Popoitai, Y. and W. Ofosu-Amaah, 2013. 'Negotiating with the PNG Mining Industry for Women's Access to Resources and Voice: The Ok Tedi Mine Life Extension Negotiations for Mine Benefit Packages.' Washington (DC): World Bank Institute.

Sanida, O.O., A.A. Mako and C. Yala, 2015. 'A Review and Assessment of the Benefit-Sharing Arrangements of Large-Scale Mining Activities in Wau-Bulolo, Papua New Guinea.' Port Moresby: National Research Institute (Development Perspectives Paper 1).

SDP (PNG Sustainable Development Program Ltd), 2013. *Annual Report Summary 2012.* Port Moresby: SDP.

Sharp, B. and T. Offor, 2008. 'Renegotiating a PNG Compensation Agreement: Applying an Informed Consensus Approach.' Canberra: The Australian National University, Research School of Pacific and Asian Studies, Resource Management in Asia-Pacific Program (Working Paper 69).

Smith, D.B., 2000. 'Deskaheh (Levi General).' *Dictionary of Canadian Biography* (Volume XV, 1921–1930). University of Toronto. Viewed 11 June 2015 at: www.biographi.ca/en/bio/deskaheh_15E.html

UN (United Nations), 1987. Our Common Future: Report of the World Commission on Environment and Development. New York: UN General Assembly.

——, 2013. *A New Global Partnership: Eradicate Poverty and Transform Economies through Sustainable Development.* New York: UN Publications.

UNDP and GoPNG (United Nations Development Programme and Government of Papua New Guinea), 2013. 'The Future We Want: Voices from the People of Papua New Guinea.' Port Moresby: UNDP and Department of National Planning and Monitoring.

———, 2014. *From Wealth to Wellbeing: Translating Resource Revenue into Sustainable Human Development.* Port Moresby: UNDP and Department of National Planning and Monitoring.

Wiessner, P., 2010. 'Youths, Elders, and the Wages of War in Enga Province, Papua New Guinea.' Canberra: The Australian National University, State, Society and Governance in Melanesia Program (Discussion Paper 2010/3).

10. Gender Mainstreaming and Local Politics: Women, Women's Associations and Mining in Lihir

SUSAN R. HEMER

Introduction

In 2004, Lihirian women staged a silent protest march in Londolovit town with banners calling for greater involvement in discussions surrounding the review of the agreement (integrated benefits package) between the operators of the Lihir gold mine and the local community. They presented the chairman of the review committee with a petition arguing that women had been left out of 'all aspects of the planning, decision-making and programming for development issues' (*Lihir i Lamel* 2004: 5). This protest was led by the Petztorme Women's Association, and was unprecedented in the history of Lihir. In part, this chapter aims to understand the strategies that have been employed by women's associations in Lihir to advance the position of Lihirian women and to call for greater political representation, and to appreciate why some have been more successful than others.

The Lihir gold mine, operated by Newcrest Mining Ltd, is located on the main island of the Lihir group of islands in New Ireland Province, Papua New Guinea (PNG). Construction began in 1995, and production in 1997. At the time of writing, active mining was set to continue until 2031, with production from stockpiles for a further ten years or so.

Annual production has been in the realm of 600,000 to 800,000 ounces. Prior to mining, the Lihir Islands were home to 7,000 Lihirians, whose livelihoods consisted of subsistence production and a small amount of income from cash crops and remittances. Since mining began, the population has grown to include some 16,000 Lihirians, about 4,000 informal migrants, and a further 4,000 migrant employees of the mine and its contractor companies. Vast changes have occurred on the islands, including the construction of a township and a ring road, a medical centre and improved education facilities, as well as increased consumption of alcohol and store-bought goods.

It is unsurprising that with these changes there have been both positive and negative impacts on men and women, and on the relationships between them. Briefly, both men and women have benefited from greater access to health, education and transportation services in Lihir. Paid work is now available for many on the island group through the mining company or its contractors, though this is dominated by men, particularly in more senior roles. Money has also become available through the payment of compensation and royalties, largely paid to men, but onward distribution of this money through lineages has been problematic, and women rarely have any access to it. With money, alcohol consumption has increased dramatically, leading to increases in violence, including domestic violence, and to motor vehicle accidents. Women continue to bear a heavy load of subsistence agriculture, child-rearing and household work, and this remains the case even when women have paid employment. It is mostly Lihirian men who have moved into positions of authority in negotiations with the mining company, who have positions in the local-level government (LLG), the Lihir Mining Area Landowner's Association and village development committees, as well as being the ones much more likely to hold more highly paying positions in the mining company, to have their own business or to be on the boards of businesses (Macintyre 2003a; Bainton 2010; Hemer 2013).

Two women's associations on Lihir, Petztorme and Tutorme, have aimed to improve the position of women, but have developed quite different strategies to do so. Petztorme has drawn upon international understandings and conventions on gender and development to work for women's development at the international, national and local level, and to argue for greater representation in local-level politics and decision making. Tutorme, on the other hand, has remained focused on the local level and has gained support from both the LLG and the company managing the mine. Through an analysis of this case, I argue that gender mainstreaming has not been effective at the local level in Lihir, and that instead women continue to gain their status from their role as guardians of the future through children, youth and health. I begin with an analysis of the broader framework of literature on gender and mining to contextualise the Lihir case.

Gender and Mining

Male domination of the mining industry around the world is well documented, and the Pacific is no exception. In recent years, however, a number of studies have demonstrated and discussed women's involvement in the industry, with a key aim being to challenge the hegemonic notion of mining as masculine (e.g. Eveline and Booth 2002; Gier and Mercier 2006; Lahiri-Dutt and Macintyre 2006; Moretti 2006; Lahiri-Dutt and Robinson 2008; Lahiri-Dutt 2012).

Mining has also been characterised by its negative impacts, particularly on women. Oxfam Australia (2009) notes that the impacts of mining are not gender neutral. There is a great deal of literature that describes and analyses the impacts of mining on women as miners, as wives and as community members. Women have often been excluded from employment or, as miners, have suffered from poor working conditions, less pay and fewer opportunities for advancement than men (Lahiri-Dutt and Robinson 2008; Sharma 2010). Mining has also been discussed in terms of its negative impact on women's psychological well-being as wives and mothers, in terms of isolation from friends and family in remote locations (Sharma 2010), and also on family life through long working hours and shift work (Chase 2001). As community members, women have been viewed as bearing the brunt of the negative impacts of mining, with little input to negotiations between mining companies and local communities. In the Melanesian context, women have little say in negotiations or control over the compensation that flows into communities (Byford 2002; Macintyre 2002; Oxfam Australia 2009; Wainetti 2013). While Scheyvens and Lagisa (1998) aim to show how women resist mining and logging activities, their paper documents much more thoroughly the lack of empowerment in economic, social, psychological and political domains in regard to mining in Lihir and logging in Solomon Islands, than the ways in which women resist these activities.

It is in the context of these negative impacts that there have been calls for gender mainstreaming in mining. Gender mainstreaming is the 'process of assessing the implications for women and men of any planned action, including legislation, policies and programmes in all areas and at all levels ... [so that] women and men benefit equally and inequality is not perpetuated' (UN 1997: 27). Lahiri-Dutt (2006) has argued that gender mainstreaming is the right direction for mining projects, yet there have also been critiques of the concept, arguing that women face structural disadvantages that gender mainstreaming is unable to overcome. The common approach is still to add women as a concern rather than have gender as an integral part of planning and processes. As Macintyre (2011: 30) states:

> Whatever gender mainstreaming might be in academic terms, by the time it gets into aid projects or workplace policies it has become 'add women and stir' with nobody prepared to actually do this.

There are concerns that 'gender mainstreaming', with its emphasis on 'gender', draws the focus away from women's specific disadvantages (O'Neill 2004; Macintyre 2011). Moreover, it has been suggested that mainstreaming universalises notions of gender that can disempower women even further (Rimoldi 2011).

In recent years, a number of analyses have begun to document and move beyond the 'negative impacts of mining' to understand women as agents who can strategically draw upon mining to improve their lives. Mahy (2011) aims to move beyond the dichotomous representations of women as either indigenous blameless victims of their husbands' sexuality in mining areas or as blameworthy migrant sex workers, to understand the livelihood choices and strategies of sex workers and community women. Lahiri-Dutt (2012) seeks to understand women's agency in mining, whether this be in terms of how women draw upon mining as one of a number of livelihood strategies, or how they act as political agents in protests around mining. Lahiri-Dutt argues that there is a need to appreciate 'the enormous evidence of women's agency—in their productive roles in mines and at home, and in their resistance to exploitations of mining' (2012: 193). It is these differing forms of strategy and agency—the ways in which Lihirian women are actively navigating local politics to position themselves to benefit from mining—that I explore in this paper. In what ways have they succeeded and how are they positioned as a result?

Petztorme and Tutorme: A Tale of Two Associations

Early in the history of the Lihir gold mine there was recognition of the need for an overarching women's association to provide a forum for women's participation in decision making and development programs. So in 1991, with the assistance of the mining company's community relations department and a female consultant, Suzy Bonnell, the Petztorme Women's Association was formed (Membup 2003). Petztorme, meaning 'work together' in the local language, was founded on the church structures of the *Katolik Mamas* ('Catholic Mothers') and United Church Women's Fellowship already in place in Lihir (Membup and Macintyre 2000). The majority of Lihir villages are predominantly Catholic, with seven United Church villages and a handful of other denominations present in the islands (Hemer 2011). Each village in the island group had its own women's group, with a president, vice-president, secretary and treasurer. The Petztorme executive mirrored this structure, with a president, vice-president, secretary and treasurer and the presidents of the women's groups in the four local churches (Catholic, United, Pentecostal and Seventh-Day Adventist). Petztorme also had a general council comprising representatives of each council ward on Lihir.[1]

Since 2014, Petztorme has altered its structure to require financial membership rather than simply encompassing all women on Lihir. Groups of 5–20 women choose to become financial members of Petztorme by paying a small fee (K50 in 2014). There were 35 member groups in 2014, from nine of the 15 council wards. This financial membership seems to promote more active presence and membership at meetings, and more engagement with the programs of the association.

In the early years of mining, Petztorme worked with women who were being relocated from two villages in order to allow mining to go ahead. It also began a number of income-generating projects: a nursery, market and can crusher in order to provide incremental self-help development for women. The difficulties with this approach, and the sectarian divisions in the association, have been well documented by Macintyre (2003b), who argues that most women's projects in PNG fail due to issues with leadership once external advisers or funders withdraw.

---

1   Lihir is divided into 15 wards for the purpose of the local government council system. Most wards comprise two or three villages.

By 2000, when I began work in the mining company's community relations department, Petztorme was still being supported by the 'women's section' of this department to successfully undertake a number of projects and programs with local women. While the nursery and can crusher had failed, the market was still functional, and was earning Petztorme a good ongoing revenue. It was at this time that a new project began, emerging out of a cultural exchange program between expatriate and Lihirian women, and the desire for training in sewing skills. A training centre was set up, with an expatriate seamstress able to train Lihirian women in sewing skills (see Hemer 2016; Macintyre 2003b).

This training centre came to be known as the Tutorme Training and Sewing Centre, and functioned out of a small building provided by the mining company.[2] It was managed by an advisory committee of expatriate and Lihirian women, including myself as secretary. The seamstress became the key trainer, with her salary provided by the mining company as well. Sewing training was taken up with gusto, with some 200 women paying to take classes in sewing by 2002. Many had completed a full sequence of training, and a select number had found employment with Tutorme. In 2001, Tutorme moved to a new building, once again provided by the mining company, and had taken on a number of commercial sewing projects, such as providing curtains for dormitories and embroidering company logos on uniforms. It was also during this year that Tutorme was registered as an association, much to the concern of Petztorme women, who argued that having two associations for women in Lihir was likely to be a source of confusion and friction.

Even at this early stage, sectarian divisions within the two associations, as well as between them, threatened the continued viability of Tutorme. In 2002, the mining company provided Petztorme with a building of its own, at least partially to placate the Petztorme executive members who were critical of their lack of input into the direction in which Tutorme was heading. While this temporarily appeased most women, and provided a much-needed office space within the township, tensions between Tutorme, now managed through a financial advisory committee heavily composed of expatriates, and the Petztorme executive continued.

---

2   Tutorme means 'stand together' (*tu torme*), rather than 'tutor me'.

From 2002, Tutorme continued for some time as a relatively successful sewing centre. Then, with the departure of the expatriate manager in 2005, its continued viability was threatened. In about 2008, the building that housed Tutorme was condemned and, while the sewing machines remained onsite and the intention was to relocate these to a new building, in reality Tutorme as a sewing centre is now defunct. This building has since been refurbished for use as a general conference and meeting space, and named the Lihir Meri Developmen Senta ('Lihir Women's Development Centre'). Its use is not restricted to any particular women's group, or to women in general, though any revenue from use of the centre is allocated to women.

In 2009, Tutorme re-emerged on the Lihir scene as the Lihir Tutorme Women's Association, a general women's association working for women's development. This re-emergence was in response to concerns in the mid-2000s about the lack of programs for Lihirian women, and Petztorme's lack of financial transparency. There was a failed attempt to change the executive committee of Petztorme, and eventually it was decided to bypass it by creating a new association.

Local Initiatives in Women's Development

Both Tutorme and Petztorme have run a number of successful initiatives on Lihir. In its new incarnation Tutorme has been functional for a number of years, and in that time has focused on education and awareness programs for women and youth in villages, covering topics such as law and order issues, education and adult literacy. One of its recent programs was to hire a theatre group to provide entertaining and engaging education on HIV/AIDS. Tutorme has drawn upon the Nimamar Rural LLG, links with the mining company (formerly Lihir Gold Limited, now Newcrest), and the Lihir Sustainable Development Plan (LSDP),[3] both for small annual budgets and for material assistance and transport to implement these initiatives.

---

3   The LSDP arose out of the review of the integrated benefits package, the agreement between the Lihir community and the mining company which specifies the benefits to be provided to Lihirians on a five-yearly cycle. The LSDP is a plan for the development of Lihir using mining revenue, and is also an organisation with a budget to carry out that plan.

Petztorme has also been involved in women's health programs in Lihir. In the early 2000s, the association was a key partner and supporter of initiatives by the Lihir Medical Centre and the mining company's women's section as they carried out programs on nutrition in villages around the island group. Petztorme was also heavily involved in supporting the programs of the medical centre's maternal and child health clinic, including the key nurse in her work on family planning. There were concerns in the early 2000s about population and family planning issues, and Petztorme organised and supported a workshop on this issue, with Petztorme members taking information back to their village groups. This interest in health has continued since that time, and programs on healthy families and villages were scheduled for 2014.

Petztorme continued to manage the market in the township until its 'temporary closure' in 2011, and this provided a source of revenue to fund more projects.[4] In late 2003, it launched a women's microcredit scheme with loans of K500–700 charging 10 per cent interest and, in 2004, employed a woman to coordinate this project. This seems to have been problematic, however, as the credit scheme was halted in 2005 and had not been relaunched at the time of writing. Also in 2004, Petztorme was successful in gaining grants from the LLG of K10,000 per ward for women's projects, hence a total of K150,000. This money was to be used for vegetable gardens, bakeries, vanilla plants and small market houses to provide ongoing sources of money to women in villages. More recently, Petztorme has provided a route for providing information to women about financial opportunities provided by organisations like the New Ireland Savings and Loan Society and the Nationwide Microbank, and more generally on financial management and literacy.

On a more political note, in the early 2000s Petztorme pushed for women's representation in local government, leading to the appointment of two women's representatives to the Nimamar Rural LLG. This LLG is made up of 15 elected members, one each for the 15 wards in Lihir. The two female representatives are additional to this body, and were chosen by Petztorme to be the conduit for women's views to be taken to the LLG and for information from the government to be brought back to Petztorme

---

4   The temporary closure came about due to concerns over the cleanliness of the market and continued upkeep of the area, as well as debates over the position of migrants selling goods at the market and forcing prices too high. While 'temporary', it proved very difficult to resolve these issues and the market was closed for at least 12 months.

and hence to Lihir village women. Also, as noted at the beginning of the chapter, Petztorme successfully lobbied for the inclusion of women in negotiations between the community and the mining company, and an additional woman was added to the Joint Negotiating Committee in the 2000–07 negotiations to bring the total to two. However, in the more recent negotiations beginning in 2012, Lihirian women have again struggled to get representation on the Joint Negotiating Committee despite Petztorme executives asking for this. In this respect, it can be seen that Petztorme has been more overtly concerned than Tutorme with the processes of local politics.

The initiatives of Petztorme and Tutorme at the local level have generally focused on aspects of everyday life understood as key concerns for women. These include gardening, cooking and sewing, and sometimes the sale of fruit and vegetables, or cooked food and sewn items, to make small amounts of money. Health is also seen as another key concern for women, as it is women who oversee the health and nutritional needs of their families. However, the claims by Petztorme for greater involvement and representation in local politics and negotiations signal a more global understanding of gender relations.

Petztorme: Going National and Global

In the two decades since the inception of Petztorme, the association has begun to choose to work beyond the local arena of Lihir. At the national level, members of Petztorme participated in the two 'women in mining' conferences held in Madang in 2003 and 2005 and spoke about the experiences of Lihirian women (Membup 2003). Petztorme has also provided a link between initiatives at the national level and women in Lihir. Members of Petztorme attended information sessions in Port Moresby about small grants from the United Nations Development Programme (on climate change and the environment) and the Women in Mining National Action Plan, then conveyed what they learned to women in Lihir, and planned to apply for grants under these programs.

One member whom I shall call Mary has also been sponsored by the Centre for Environmental Research and Development (CERD), a PNG non-governmental organisation largely funded by Oxfam Australia, to be a member of the Mine Affected Women's Foundation. Mary was able to attend the 2005 women in mining conference, as well as the 2007

Pacific women and mining conference (Oxfam Australia 2009), and was a signatory to the 2007 declaration by CERD opposing any new mines in PNG. Mary has also been able to raise Lihirian women's concerns over mining at a global level through overseas travel, interviews and publications. In 2003, she travelled to Portugal for a meeting of the World Bank's Extractive Industries Review, as well as making a written submission to it in 2004. Mary also travelled to London to raise environmental concerns at Rio Tinto's annual general meeting in 2003, when Rio Tinto still had a controlling interest in the Lihir mine. She was interviewed by Friends of the Earth International, and her comments about environmental disruption in Lihir are recorded in a 2003 publication as well as on their website.[5]

One Petztorme executive member is also the co-author of a chapter in an internationally published book that documents concern about the social and environmental risks of large-scale mining projects (Moody 2005). The Petztorme executive as a whole has been active and able to raise Lihirian women's concerns about the impacts of mining on Lihirian women and youth in particular, and also on the Lihirian environment, to national and global levels. At the national level in PNG, understandings of women in mining have been shaped by global discourses on gender.

Gender Mainstreaming and the National Action Plan

The PNG Government developed its *Women in Mining National Action Plan* as a result of national conferences on 'women in mining' held in 2003 and 2005 (GoPNG 2007). These conferences were themselves part of an institutional strengthening project funded by the World Bank. The aim of the National Action Plan is to draw attention to the issues affecting women in mining areas in the country, and to then set the direction for addressing these issues. The plan draws upon key international conventions and obligations including the Millennium Development Goals, the Beijing Declaration, and the Convention on the Elimination of All Forms of Discrimination against Women. In this respect, it highlights

---

5   In order to maintain confidentiality, I am not including the reference for these comments here.

the importance of gender mainstreaming for the PNG Government in aiming to achieve women's empowerment and equality. As outlined in the National Action Plan, gender mainstreaming in mining is the:

equitable distribution of the resources, opportunities and benefits of the development process and addresses gender inequalities in the mainstream of organizational policies, plans and programs, not just as separate, ad-hoc activities (GoPNG 2007: 3).

There is a clear recognition, then, that gender mainstreaming in mining aims to move beyond approaches that simply add a recognition of women to supposedly neutral plans and programs.

The National Action Plan lists eight goals which are, very briefly, to raise women's education and literacy; to improve access to reproductive health services; to prevent or control communicable diseases, including tuberculosis and sexually transmitted infections; to increase women's participation in economic, political, cultural and social life; to ensure that women's associations function beyond mine closure; to ensure sustainable livelihoods; to mitigate or avoid environmental degradation; and to promote security and peace in communities. Each of these goals has related objectives, strategies and targets. Many of these involve awareness raising, and numbered or named targets such as, for example, a certain number of training sessions held, water tanks at six health facilities, or 10 per cent of women multiskilled by 2012. It is clear that such an ambitious plan would require high-level coordination, and this is acknowledged in the implementation framework, yet there is little discussion of how this would actually occur.

The plan acknowledges that women should be involved 'in the development of policies, design and management of programs at all stages and levels of decision making' (GoPNG 2007: 3). This then becomes part of the fourth goal of the plan, which is to increase women's participation in the economic, political, cultural and social life of their communities. One of the objectives of this goal is for women's participation in local political processes, such as landowner associations and mine closure committees, as well as in employment. This is one of the few goals that have an objective that aims for equality, such as 50 per cent participation in landowner associations. Most of the objectives, even in this goal, aim for an addition of women to existing committees, or for awareness raising and 'gender sensitivity' amongst males. Yet once again there is little discussion in the plan of how participation of women in local political processes will be achieved, other than by awareness and sensitivity training.

Most of the goals in the plan, despite its acknowledged context, have little to do with gender mainstreaming and equality, whether this be in education, health or agriculture. Most of the targets (such as a 50 per cent increase in female student enrolment) are set without any discussion of the communities' needs and strengths prior to the plan, which would vary between the different mine sites around the country. Overall, the aim of the plan is for improvements to the current conditions that women face, rather than for gender mainstreaming or equality.

There is little evidence of the National Action Plan having any impact in Lihir. Despite the number of Lihirian women attending the women in mining conferences, no one I spoke to in 2011, 2012 or 2014 mentioned the existence of the plan. One of the leaders of Tutorme did say that their emphasis on awareness programs arose out of one of the conferences, which was possibly a reference to the plan, since awareness programs were a key feature of the activities listed in it. Yet there is no explicit discussion of the plan itself. This suggests at the least that women are unlikely to be using the plan to leverage changes in Lihir. Of the eight goals mentioned in the plan, some were already being pursued, funded and organised through relevant organisations such as the Lihir Medical Centre or the LLG, but no one carrying out these activities seemed to be aware that they were fulfilling the goals of the National Action Plan, so it seems quite likely that the goals were being pursued by accident rather than by design.

There were few calls for gender mainstreaming in Lihir. Where there was some activism was in terms of calls for some level of representation for women in local politics and on bodies negotiating over the benefits of mining, as noted at the beginning of this chapter. More recent years have seen sustained activism against gender violence. Women also called for more business opportunities and for the need for access to sources of funding. In the realm of employment there was little activism around equity of access or rates of pay.[6] The ways these issues became part of local practice, however, was in terms of adding women in to existing

---

6   Where there was activism in relation to employment, it tended to be in terms of Lihirians' access to employment and conditions relative to other Papua New Guineans, rather than being concerned with differences between men and women.

processes—the 'add women and stir' approach (Macintyre 2011: 30)—rather than significantly challenging the processes or practices themselves. Hence more global notions of gender mainstreaming and equality had little currency in Lihir, where other understandings of gender relationships held sway—in particular, relationships between men and women based in nurturance, respect and work.

Women's Status in Lihir: *Sio, Ertnin* and *Pniez*

Lihirians are members of both lineages and named clans with matrilineal descent. Land tenure is said to be based on matrilineal inheritance, but is complex and has cognatic tendencies. As argued by Macintyre (2003b), in pre-colonial times it is unlikely that women participated in exchanges in their own right, and even today they assert little control over land or ritual events. Despite this, a woman could develop the reputation and status of a 'big woman' (*wok tohe*) on rare occasions, and more commonly it was known that the renown of 'big men' (*a tohe*) rested largely upon their wives and sisters (Bainton 2010: 83–5).

Women's status is best understood through the concept of *sio*, meaning the respect shown to someone or that someone has earned. *Sio* is most commonly used in reference to big men, to indicate that people respect them by demonstrating respectful behaviour and providing them with shell money (*a le*) with which they could participate in exchanges. Thus the respect shown to a big man would allow for the further development of the reputation of the clan through ritual exchanges.

Women did not receive *sio* in the form of shell money, but gained it through the bearing and raising of children, and through the production of garden food and pigs that could be exchanged for shell money. The key term used to refer to these processes is *ertnin*, a term that indicates a productive future-oriented practice, often meaning nurturance (Hemer 2013). Hence a woman would carefully nurture her children, gardens and pigs. As a mother she would aim to have many children, but space them carefully so they would all grow to healthy maturity. A woman who was highly productive, with large gardens full of yams and pigs grown fat to adulthood was one who was highly valued by her husband, and one who could inspire the jealousy of her husband's male rivals. Sometimes she would be the subject of sorcery attacks intended to inhibit her capacities. Such a woman was one of high status—one who had earned *sio*.

Sio was also demonstrated through appropriate behaviour between classes of kin. For example, sisters were understood to be mutually supportive. A brother and sister should not speak of sexual matters in one another's presence, nor should a sister walk past her sleeping brother's head. Male and female cross-cousins should refrain from using each other's names, but were expected to joke. Male cross-cousins were expected to have a highly supportive relationship that was not the site for jealousy or anger. It was understood that often behaviour did not necessarily meet these ideals, and was the focus for angry confrontations and relationship tension. Yet conformity with these expectations was a demonstration and practice of respect, and in itself could earn respect. Many of these expectations underpin both the practice of gender relations and the interpretation of appropriate and respectful conduct.

While it is understood that both benefits and respect can be gained either through hard work or through positions or associations, women generally gained their *sio* not through their positions or associations, but through their own hard work (*pniez*). In the context of mining there has been some backlash against benefits received through positions, like the royalties gained by a person as a member of a particular lineage. This backlash is particularly strong where these royalties are then not distributed throughout the lineage, as it seems that the receipt of these benefits relies on relational qualities that are simultaneously denied. This has meant that the moral value of *pniez* as the source of benefits has become more acutely understood. This point is key to understanding the criticisms made of some Lihirian women.

Women's Associations and Women's Status

In order to carry out many of its programs, Petztorme has drawn upon Lihir women's voluntary labour. This has been one of the weak points of the organisation, in the sense that Lihirian women already have many other things that draw upon their time. As many men and young women have gained employment with the mine, the volume of work carried out by village women has increased, meaning that there is even less time available for voluntary communal work (Macintyre 2003b). Given that much of the work for Petztorme drew upon existing skills and understandings of women's work, such as growing plants or vegetables for sale or providing health and nutrition information, it has not been the basis for gaining additional respect nor a source of challenge to conventional notions of women's status. Although it constitutes an extra obligation for women who are already very busy, this work at the local level has still been seen as generally appropriate for women.

Similarly, Tutorme in its first incarnation extended this understanding of appropriate women's activities through its sewing training. Most of the women who sought training were able to use these skills to provide for their families in the village setting: only a handful went on to gain employment with Tutorme. For those who did, like women who work for the mine, their earnings are expected to be used to provide for their immediate family or for the customary work and feasting of their lineage. The new incarnation of Tutorme likewise draws upon women's time and work to provide further awareness and health education for women and youth in the village. At this local level, then, the strategies of both Petztorme and Tutorme confirm and bolster women's status based in their work and the nurturance of families, gardens and pigs.

The activities of Petztorme at the national and global level were perceived quite differently. For most women, the travels of women like Mary or other executives of Petztorme were seen not as working for the broader benefit of Lihirian women. Rather, these were seen as a personal benefit—a chance for them to experience the wider world and gain a name for themselves. Their efforts to raise the concerns of Lihirian women at the national and international levels, and their experience with government and non-governmental organisations, have not translated to additional *sio* for them at the local level. While the ability to speak up about the concerns of women, particularly in front of men, is valued by other women, it is generally not highly valued by Lihirian men, and does not appear to be a source of *sio*. Mary, for example, has instead been the subject of much criticism about 'doing nothing' for Lihirians while benefiting from her position in the Petztorme executive. Her actions are not perceived as morally valued work. Thus Mary has been caught in the disjunction between national and global discourses on gender and mining on the one hand, and local Lihirian notions of the appropriate work and place of women on the other.

The existence of two women's associations, and divisions both within and between them, has been a continual focus of criticism on the part of Lihirian men. Men have continually pointed to the mismanagement (or lack of management) of programs, equipment and buildings as evidence of Lihirian women's inability to work cooperatively or organise development initiatives. There has been ongoing insistence that Lihirian women need to develop 'one voice' before they should seek to have an input to negotiations over benefits from the mine. Lihirian women have largely accepted this insistence and justification for their lack of representation in local political and economic matters, and have continually attempted to develop models for reconciliation or cooperation. This was ongoing even as recently as 2016, with men insisting that women reconcile to have 'one voice' in order to contribute to the agreement review process, and women concurring. There is often little reflection by women on the lack of cooperation among men or among the various groups that represent them. There is also very little comment on the different conditions for other associations, often with almost exclusively male membership, which do not rely on voluntary labour. These include the Nimamar Rural LLG and the Lihir Mining Area Landowners' Association, in which men are salaried, and the boards of various businesses that pay men but require little work. Thus far, the differences among women's groups have proven too difficult to resolve despite lengthy efforts to achieve reconciliation, and women remain largely excluded from local politics and decision making.

Lihir Women's Status and Gender Mainstreaming

In a context where value is placed upon women's nurturance of children, gardens and pigs, gender mainstreaming can appear as a threat to these traditional sources of status. Women, through Petztorme and Tutorme, have appealed to other women, and to both expatriate and local Lihirian males, in their roles as guardians of the future to effect changes at the local level. This has been an effective strategy in gaining acceptance of programs about health or sewing, to make changes at the market run by Petztorme, or to call for changes to the sale of alcohol. Such activities fit well with the broader understanding of women's position and positively contribute to their status or standing (*sio*). In particular, Lihirian men have been supportive of such approaches, and hence there has been funding support for programs such as these by the Nimamar Rural LLG and the Lihir Sustainable Development Plan.

Yet when Lihirian women aim or argue for greater access or representation, particularly in the realms of politics and employment, they undercut those understandings of women's status. Instead, women then become the subject of criticism and jealousy from other women as they are seen to be benefiting themselves rather than performing morally valued work. It also appears that, as they draw less upon those nurturance roles, they have to build their status in new ways. This roughly translates as being seen to be 'doing something', often in various forms of leadership, which may include proposing and organising successful meetings, training opportunities and projects from which Lihirian women can benefit. Yet such a definition of success in raising their status leaves them open to criticism similar to that which many Lihirian men also face. Thus Mary and other members of the Petztorme executive over the past ten years have been subject to charges of misuse of funds or association property, and of responsibility for a lack of real development for Lihirian women. As argued by Macintyre (2003b), the local model of successful leadership is one where leaders must distribute largesse; if they are unable to do so then their leadership, and hence their standing in the community, is contested.

If gender mainstreaming, as defined internationally, implies that women should be the equals of men in their access to employment, politics, and hence sources of status, then such a model can be problematic in a country such as PNG. Rimoldi has pointed out the danger in a discussion of the Leitana Nehan Women's Development Agency in Bougainville:

> By taking on the vocabulary and orientation of international development or welfare agencies (and affiliating with them), this once very grassroots organisation gradually came to believe that it was dependent on aid money for existence. In a sense it looks like a 'buy out' of women's influence and authority intrinsic to their traditional standing in Bougainville society. (Rimoldi 2011: 191)

Rimoldi goes on to question what precisely 'mainstream' means, and how gender in Bougainville may be shaped in response to these international concerns (ibid.: 191–2).

Onyeke has argued, in the context of PNG, that it is necessary to go beyond notions of equality to aim more broadly for respect:

> [T]he idea of using political or leadership gender equality to measure the value and dignity of PNG women is far too simplistic and overly optimistic (Onyeke 2010: 12).

This respect should be based on the dignity of the person rather than due to their roles or achievements. This accords with Walby's (2005) argument that, where gender mainstreaming implies women gaining equality on terms set by male norms, there is a risk that women's sources of authority can be undermined. This is clearly the case for Lihir, where a woman's status continues to be tied to her capacity to contribute through the nurturance of children, gardens and pigs, and in a lesser way through the contribution of her earnings to her family and lineage. Similarly for Onyeke, mothering—and by extension the 'social act of caring and nurturing'—should be the model and basis for respect of women (Onyeke 2010: 13). It is possible that these activities of nurturance may come to be understood as of lesser value by Lihirian men and women if arguments for equality become more commonplace in Lihir.

Conclusion

In moving away from analyses that emphasise the negative impacts of mining on women, it is possible to productively examine the strategies and choices made by women in the context of mining developments. The two women's associations on Lihir have followed different strategies in an attempt to advance women's positions: both Tutorme and Petztorme emphasise traditional forms of women's status, yet Petztorme has appealed to a broader audience and more global understandings of gender and mining. Strategies adopted at the local level by both associations have aimed to consolidate and strengthen traditional sources of women's status through work on the nurturance of children and youth, with attention to human health and a productive environment. Yet such a focus does little to challenge Lihir women's position relative to men and their lack of local political representation, as well as the real conditions of disadvantage that they face.

Petztorme's efforts at the national and global levels to raise the concerns of Lihirian women about the impact of mining, and to call for a place in local political representation and decision making, clearly resonate with some of the aims for equality and access that characterise gender mainstreaming. To date their claims fall well short of the 'equitable distribution of resources, opportunities and benefits', and equal participation in policies, plans and programs, that are core to gender mainstreaming (GoPNG 2007: 3). Even in this partial form, the strategies of Petztorme have not been met with much enthusiasm or acclaim on Lihir, and speak to the difficulty of improving the conditions for women and their representation in local political processes.

At present, it is uncertain how successful each of the strategies adopted by Lihirian women's associations will be. It may well be the case that the strategy of Tutorme, which places greater emphasis on the nurturing roles of women, will be much more successful in this local context. Rather than directly challenging men and arguing for equality, it highlights the strengths of the unique contributions that women can make to social life. Yet it is clear that women should have a voice and a place at the negotiation table, as well as equitable access to any opportunities and benefits afforded by mining within their community. It would appear that gender mainstreaming, as understood at the national and international levels, is unlikely to prove to be a workable model, at least for Lihirian women. To judge by the local perception of Petztorme's efforts at the national and global levels, it is quite possible that their strategy may work against Lihirian women in the long run by undercutting their traditional sources of status, which would place them at an even greater disadvantage.

References

Bainton, N.A., 2010. *The Lihir Destiny: Cultural Responses to Mining in Melanesia.* Canberra: ANU E Press (Asia-Pacific Environment Monograph 5).

Byford, J., 2002. 'One Day Rich: Community Perceptions of the Impact of the Placer Dome Gold Mine, Misima Island, Papua New Guinea.' In I. Macdonald and C. Rowland (eds), *Tunnel Vision: Women, Mining and Communities.* Fitzroy: Oxfam Community Aid Abroad.

Chase, J., 2001. 'In the Valley of the Sweet Mother: Gendered Metaphors, Domestic Lives and Reproduction under a Brazilian State Mining Company.' *Gender, Place and Culture* 8: 169–187. doi.org/10.1080/09663690120050779

Eveline, J. and M. Booth, 2002. 'Gender and Sexuality in Discourses of Managerial Control: The Case of Women Miners.' *Gender, Work and Organization* 9: 556–578. doi.org/10.1111/1468-0432.00175

Gier, J. and L. Mercier, 2006. *Mining Women: Gender in the Development of a Global Industry, 1670–2005.* New York: Palgrave Macmillan. doi.org/10.1007/978-1-349-73399-6

GoPNG (Government of Papua New Guinea), 2007. *Women in Mining National Action Plan 2007–2012*. Port Moresby: Department of Mining.

Hemer, S.R., 2011. 'Local, Regional and Worldly Interconnections: The Catholic and United Churches in Lihir, Papua New Guinea.' *Asia Pacific Journal of Anthropology* 12: 60–73. doi.org/10.1080/144 42213.2010.535844

——, 2013. *Tracing the Melanesian Person: Emotions and Relations in Lihir*. Adelaide: University of Adelaide Press.

——, 2016. 'Sensual Feasting: Transforming Spaces and Emotions in Lihir.' In S. Hemer and A. Dundon (eds), *Emotions, Senses, Spaces*. Adelaide: University of Adelaide Press.

Lahiri-Dutt, K., 2006. 'Mainstreaming Gender in the Mines: Results from an Indonesian Colliery.' *Development in Practice* 16: 215–221. doi.org/10.1080/09614520600562488

——, 2012. 'Digging Women: Towards a New Agenda for Feminist Critiques of Mining.' *Gender, Place and Culture* 19: 193–212. doi.org/ 10.1080/0966369X.2011.572433

Lahiri-Dutt, K. and M. Macintyre (eds), 2006. *Women Miners in Developing Countries: Pit Women and Others*. Aldershot: Ashgate.

Lahiri-Dutt, K. and K. Robinson, 2008. '"Period Problems" at the Coalface.' *Feminist Review* 89: 102–121. doi.org/10.1057/fr.2008.5

Lihir i Lamel, 2004. 'The Silent March for Equality.' Unpublished petition.

Macintyre, M., 2002. 'Women and Mining Projects in Papua New Guinea: Problems of Consultation, Representation, and Women's Rights as Citizens.' In I. Macdonald and C. Rowland (eds), *Tunnel Vision: Women, Mining and Communities*. Fitzroy: Oxfam Community Aid Abroad.

——, 2003a. 'The Changing Value of Women's Work in Lihir.' Paper presented at the conference on 'Women in Mining: Voices for Change', Madang, Papua New Guinea, 3–6 August.

——, 2003b. 'Petztorme Women: Responding to Change in Lihir, Papua New Guinea.' *Oceania* 74(1–2): 120–133.

——, 2011. 'Modernity, Gender and Mining: Experiences from Papua New Guinea.' In K. Lahiri-Dutt (ed.), *Gendering the Field: Towards Sustainable Livelihoods for Mining Communities.* Canberra: ANU E Press (Asia-Pacific Environment Monograph 6).

Mahy, P., 2011. 'Sex Work and Livelihoods: Beyond the "Negative Impacts on Women" in Indonesian Mining.' In K. Lahiri-Dutt (ed.), *Gendering the Field: Towards Sustainable Livelihoods for Mining Communities.* Canberra: ANU E Press (Asia-Pacific Environment Monograph 6).

Membup, J., 2003. 'The Status of Women Affected by Mining in Lihir.' Paper presented at the conference on 'Women in Mining: Voices for Change', Madang, Papua New Guinea, 3–6 August.

Membup, J. and M. Macintyre, 2000. 'Petzstorme: A Women's Organisation in the Context of a PNG Mining Project.' In B. Douglas (ed.), *Women and Governance from the Grassroots in Melanesia.* Canberra: The Australian National University, State Society and Governance in Melanesia Program (Discussion Paper 00/2).

Moody, R., 2005. *The Risks We Run: Mining, Communities and Political Risk Insurance.* Utrecht: International Books.

Moretti, D., 2006. 'The Gender of the Gold: An Ethnographic and Historical Account of Women's Involvement in Artisanal and Small-Scale Mining in Mount Kaindi, Papua New Guinea.' *Oceania* 76: 133–149. doi.org/10.1002/j.1834-4461.2006.tb03041.x

O'Neill, P., 2004. 'Rethinking Gender Mainstreaming (Or, Did We Ditch Women When We Ditched WID?)—A Personal View.' *Development Bulletin* 64: 45–48.

Onyeke, D., 2010. 'Real Men Don't Hit Women: The Virtue of Respect As a Strategy for Reducing Gender-Based Violence in Papua New Guinea.' *Contemporary PNG Studies* 13: 1–16.

Oxfam Australia, 2009. *2007 Pacific Women and Mining Conference.* Carlton: Oxfam Australia.

Rimoldi, E., 2011. 'Force of Circumstance: Feminist Discourse in a Matrilineal Society.' *Asia Pacific Journal of Anthropology* 12: 180–194. doi.org/10.1080/14442210903289348

Scheyvens, R., and L. Lagisa, 1998. 'Women, Disempowerment and Resistance: An Analysis of Logging and Mining Activities in the Pacific.' *Singapore Journal of Tropical Geography* 19: 51–70. doi.org/10.1111/j.1467-9493.1998.tb00250.x

Sharma, S., 2010. 'The Impact of Mining on Women: Lessons from the Coal Mining Bowen Basin of Queensland, Australia.' *Impact Assessment and Project Appraisal* 28: 201–215. doi.org/10.3152/14615 5110X12772982841041

UN (United Nations), 1997. 'General Assembly Fifty-Second Session: Report of the Economic and Social Council for 1997.' Viewed 2 March 2016 at www.un.org/documents/ga/docs/52/plenary/a52-3.htm

Wainetti, U., 2013. 'Responding to Fast Changing Community Aspirations.' Paper presented at the conference on 'Mining for Development', Sydney, 20–21 May.

Walby, S., 2005. 'Gender Mainstreaming: Productive Tensions in Theory and Practice.' *Social Politics* 12: 321–343. doi.org/10.1093/sp/jxi018

11. Migrants, Labourers and Landowners at the Lihir Gold Mine, Papua New Guinea

NICHOLAS A. BAINTON

Introduction

The Papua New Guinea (PNG) landscape is punctuated by large-scale resource extraction projects that have created concentrated nodes of hyper-intensive capital development that draw people from near and far. Existing patterns and processes of uneven development have found new expression in these enclave spaces that have reshaped the national 'metageography' of development (Sidaway 2007). These new centres attract hopeful migrants who seek economic opportunities and access to services that remain absent throughout many rural districts in contemporary PNG. The movement of people around resource development projects, or what the International Finance Corporation calls 'project-induced in-migration' (IFC 2009), or what Glenn Banks has more aptly described in the PNG context as 'rural-to-resource migration' (Banks 2005: 135), is a common social phenomenon in developing country contexts. While the drift towards resource development projects throughout PNG is motivated by similar aspirations as rural-to-urban migration (Strathern 1975; Clunies Ross 1984), and bears some resemblance to rural-to-rural migration, there are several distinguishing spatial and temporal features.

Migration has been a central feature of PNG's human landscape for centuries, as part of the maintenance of kin relations, marriage, trade and exchange or displacement from warfare and, more recently, in response to environmental hazards (Connell 2012). In many cases, it is regulated by customary norms and reciprocal obligations over extended periods of time. In contrast, rural-to-resource migration typically occurs at a greater pace over a more compressed period of time. There is often an increased level of in-migration during project construction, but it can vary over time as operations expand and contract. It is strongly influenced by existing regional development and settlement patterns that are usually concentrated around project areas. While most migrants draw upon regional networks and existing social relations with project area communities, the process is also formulated as an opportunistic strategy whereby migrants with previously limited social connections to the project area forge new relationships in order to legitimise their presence. For many migrants, the intention is to stay as long as the resource project remains in operation, or as long as it remains socially and economically feasible. In addition to unprecedented forms of social and environmental change generated by large-scale resource development (Bainton 2010; Golub 2014; Kirsch 2014), in-migration places enormous pressure upon host communities, project operators, and local government authorities. If it is readily acknowledged that resource projects provide many of the pull factors that influence regional migration patterns, it is also evident that host communities play a major determining role in shaping in-migration and settlement patterns. As such, in-migration can create complex and overwhelming consequences, and the question of so-called 'migrant management' is a vexed issue for all parties.

This chapter presents material collected through a long-term demographic monitoring program that covers the Lihir Islands in PNG's New Ireland Province, where the Lihir gold mine is situated. The scale of in-migration around the Lihir project is smaller than that associated with some resource projects on the PNG mainland, but an awareness of this difference has only served to sharpen local concerns over the growing presence of outsiders in Lihir. Elsewhere I have explored Lihirian cultural and political responses to migrants, which are largely characterised by the emergence of a strong sense of ethnic and cultural difference that Lihirians regularly articulate in relation to outsiders, the transformation of guests into strangers, and the codification of Lihirian identity (Bainton 2009). In different ways, each response attempts to manage or exclude outsiders in order to contain mine-related benefits within Lihir, and these responses can be seen as part of the social relations of compensation (see Chapter 1, this volume). In this chapter, I look more closely at the migrant population. I propose that in-migration, or rural-to-resource migration, is best understood as a socially embedded phenomenon that arises from the strategic opportunism exercised by both migrants and host communities or, in other words, as the socioeconomic result of a dialectical resourcefulness. In Lihir, this is best exemplified through the complex and often contradictory intersection between migrant labour and landowner business development, the strategies that migrants pursue in order to secure access to land, accommodation and livelihoods, and the precarious nature of their lives. I conclude with some reflections on the practical implications for the management of this phenomenon.

Resource Enclaves, Project Corridors and Migrant Groups

The legal recognition of customary landownership in PNG ensures that resource companies are drawn into intense and continuing engagement with local landowners or their representatives over access to land and resources and the operational activities that occur once leases have been granted. Benefit-sharing agreements outline the conditions for the payment of statutory compensation and royalties to lease-area landowners and the provision of a broader set of benefits such as employment opportunities, service delivery and economic opportunities, which may also filter out to the wider host community (Filer 2012). Even when resource development projects require the relocation of village communities, landowners often remain within close vicinity of the project. Notwithstanding trips abroad or the minority elite who establish themselves in urban centres like Port Moresby, it is rare for local landowners to permanently migrate away from project areas and their related disturbances.[1] If this reflects the difficulties that people may face in securing access to land in other areas, it also reflects their desire to maintain claims over new forms of development and the fear that, by relocating elsewhere, they might relinquish any rights

---

1   The PNG context tells a rather different story compared to other mining-dependent countries like Peru, where local communities are known to migrate *away* from the project area, while more skilled migrants settle around the mine (Bury 2007). Thanks to Glenn Banks for bringing this point to my attention.

over this new 'inheritance'. In some locations, such as Porgera, it also reflects a fear among landowners that other people may come and take up residence on their land if they do not stay to protect their interests. Social pressures increase as people from neighbouring villages, and from other valleys, islands and provinces similarly gravitate towards these areas in search of the 'good life'. As a result, the nature and scale of rural-to-resource migration often irreversibly restructures regional networks and relationships. Local communities may well feel dispossessed, marginalised or displaced, not only by the project, but from the sheer number of people arriving on their lands (Banks 2009: 49).

Without overstating the differences, it is possible to draw some basic distinctions between migrant 'types' around resource development projects in PNG. The bulk of the migrant population is often composed of semi-skilled rural people seeking employment with local companies or the developer, or indirect engagement with the resource economy as entrepreneurs or market sellers, or access to better services. And in some locations the glitter of artisanal gold mining is simply too much to resist. These migrants differ from the small number of people posted to project areas to work in essential government services where accommodation is provided, and skilled professionals and tradespeople who come to manage or work for local businesses. These migrants are distinct from the employees of the developer or major contractor companies, who are often engaged on fly-in/fly-out arrangements from their home and stay in camp accommodation during their work rotation.[2]

The Wau-Bulolo region of what is now Morobe Province hosted the first major mining operations in PNG in the 1920s and drew in large numbers of migrants, many of whom were recruited from the Sepik region through the government's labour recruitment program. Pre-existing tensions surrounding the presence of outsiders working on plantations and in administrative roles in Bougainville during the period after World War II were compounded when mining operations started in that region in the early 1970s. The presence of thousands of outside labourers, termed 'redskins' by local Bougainvilleans due to their comparatively light skin colour, was a major contributing factor that led to widespread dissatisfaction around the operation of the mine (Nash and Ogan 1990; also Chapter 12, this volume). By contrast, the Misima gold mine in

---

2   For an overview of fly-in/fly-out employment in the PNG mining sector, see McGavin et al. (2001). See also Storey (2001) and Markey (2010).

Milne Bay Province is one of the few operations in PNG that was not impacted by in-migration, partly as a result of its remote location and the lack of transport options, the limited economic opportunities for contractors, and the policy adopted by Misima Mines Ltd to maximise local recruitment.

At the Ok Tedi mine daily life for the previously isolated Wopkaimin people has certainly been affected as thousands of people from cognate Mountain Ok language groups and from further afield have moved into the area and settled around Tabubil town. The prospect of instant riches sparked a gold rush at Mt Kare and lured vast numbers to that inhospitable landscape (Ryan 1991; Vail 1995). But nothing compares to the Porgera gold mine, where tens of thousands of people have been drawn to the Porgera Valley by the allure of mine-related development and an artisanal gold mining boom (Callister 2008). The pre-mining population has increased from roughly 2,500 people recorded in 1957 (Meggitt 1957: 33), to approximately 10,000 in 1990 (Filer 1999: 3) and an estimated 45,000 in 2012 (BGC 2012).[3] Many people have exploited a combination of customary ties and local Ipili people's notions of incorporation that make it possible for people from outside the valley to come and live there (Golub 2007). In-migration was further encouraged as Ipili people strategically allowed 'outsiders' to come and reside in the valley in order to increase the number of their allies and supporters for greater security and to expand the network of economic possibilities that they might exploit. The socially embedded nature of in-migration makes it difficult for local community members to now 'manage' or stop the influx of outsiders. Ipili people fear retribution for any attempt to remove outsiders, particularly from their neighbours who feel entitled to benefit from the mining project. Mounting pressure to resettle the large population residing in congested settlements on lease areas close to mining activities presents an intractable problem for all parties: the complex ties between landowners and migrants have made it impossible for the Porgera Joint Venture to forge a clear distinction between these household units and to establish eligibility for relocation packages (Kemp and Owen 2015). At the same time, as the company and local authorities contemplate the eventual closure of the mine, they face difficult questions around the provision of post-mining services that will adequately support a large community that is heavily concentrated around the project area with

3   In the absence of firm census figures, estimates of the population resident in the Porgera Valley have been the subject of some debate. A recent report for the Restoring Justice Initiative suggests that the population in the valley may be as high as 73,000 people (Robinson 2014).

little incentive to move elsewhere. The question of relinquishing the mining lease at the end of its term may become seriously problematic in light of the difficulties surrounding the safe decommissioning of the mine site when hundreds of so-called 'illegal miners' (BGC 2010) continue to enter the lease area and tunnel and sift through stockpiles and pit walls in search of gold. These miners will doubtless expect to have access to remaining gold traces as their main form of livelihood in the post-mining era.

Whereas mining operations often produce more concentrated impact zones, and benefit streams tend to accumulate in quite discrete geographic locations, oil and gas projects are temporally different and generate more spatially distributed impacts.[4] PNG's new liquefied natural gas project, operated by ExxonMobil, sources gas from various fields and wells, incorporating these locations into a wider productive landscape as gas is piped to export facilities on the outskirts of Port Moresby. This gas network covers the lands of many culturally and linguistically different groups and will reconfigure relationships at a much larger scale as a more diverse group of people gravitate towards wells and proposed pipeline corridors in anticipation of economic opportunities (Goldman 2009). Instead of coalescing around a single project site, the extended network of facilities has provided migrants with a multitude of possible destinations, and this could mean that a much greater number of communities will experience social and demographic pressures without the development offsets that are sometimes found in mining enclaves.

Lihir and Its Gold Mine

The Lihir Islands are situated off the east coast of mainland New Ireland and are divided into 15 local-level government wards. The Lihir gold mine, which commenced operations in 1997 and is currently operated by Newcrest Mining Ltd, is located in Ward 2 on the main island of Aniolam (Figure 11.1). While the mining lease areas occupy slightly more than 12 per cent of Aniolam Island, the social and economic

---

4   Of course, some projects, like the Ok Tedi mine, have generated far-reaching downstream impacts. In other locations, mining projects may develop infrastructure across wide geographic zones, including ports, pipelines, power stations, roads and railways that increase both the possibilities and complexity of in-migration patterns. Work elsewhere, particularly in Africa and South America, where oil and gas pipelines thread across parts of those continents, provides useful points of comparison (e.g. Thibault and Blaney 2003; Lavachery et al. 2005; Smith 2005; Finer et al. 2008).

footprint of the operation has encompassed the entire island group and all villages have experienced substantial social, economic, cultural and political change.[5] As with other resource development projects in PNG, unmanaged in-migration has generated major social impacts and numerous operational risks.

Lihirians historically lived in scattered hamlets along the coastal strips. Land and other property rights are generally held by matrilineal descent groups, although this pattern is changing due to greater emphasis being placed on the nuclear family unit and fathers looking to secure land and resources for their children or pass on permanent houses to their sons. Matrilineages traditionally occupied their own hamlets, and male leaders were responsible for maintaining the men's house of the lineage. Men's house sites were a primary locus for the social reproduction of lineage groups through the performance of customary feasts and the exchange of pigs and shell money. Mining has introduced greater amounts of cash and commodities that are now incorporated into these events, thereby boosting the local ceremonial economy and reaffirming the symbolic importance of the men's house (Bainton and Macintyre 2016).

Lihirians have strong cultural links throughout New Ireland with people from the Namatanai, Tanga and Tabar districts, and with people from West New Britain. Historical migration patterns, both within Lihir and the broader region, tended to be more short term. People travelled to other areas to engage in feasting and exchange activities and to maintain relations established through marriage and clan ties. Reciprocal notions of hospitality, coupled with comparatively less pressure upon land, meant that if people chose to stay on for extended periods of time they could be granted access to land and resources for subsistence. Some people also migrated to pursue work on plantations or in urban areas.

---

5   The total land mass on Aniolam Island is 19,960 hectares. The combined area of the special mining lease, the lease for mining purposes, and the mining easement is 2,527 hectares—about 12 per cent of the island.

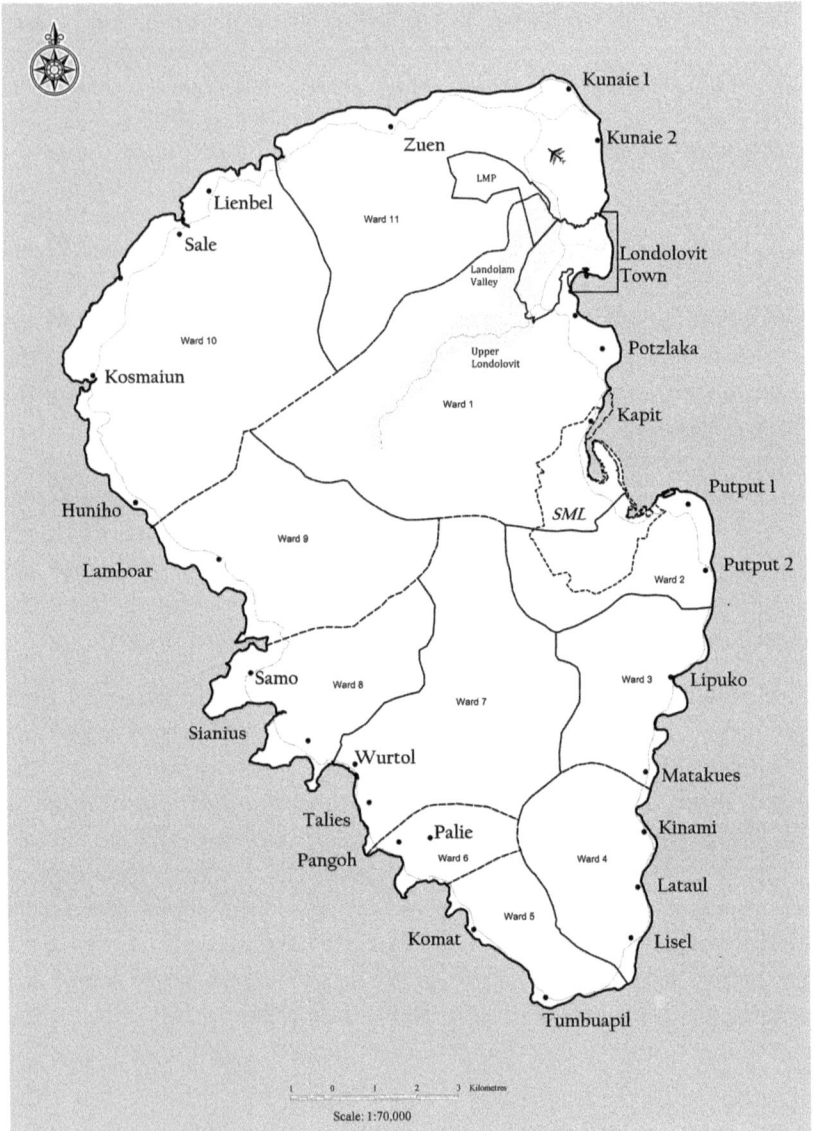

Figure 11.1 Villages and council wards on Aniolam Island.
Source: Newcrest Lihir Lands and Geographic Information System unit.

At the crudest level, resource development has divided Lihir into 'affected' and 'non-affected' areas. This distinction is crystallised in the 1995 'integrated benefits package' (IBP) agreement and is based upon whether village land has been disturbed for mining purposes.[6] Those Lihirians who own land within the mining lease areas receive royalties, compensation and other economic benefits. The wider community benefits in more indirect ways such as employment, service provision and community development projects. Development of the mine in the Louise Caldera required the relocation of two coastal villages, Putput and Kapit (Owen and Kemp 2014). The physical relocation of Kapit village and the subsequent resettlement programs have been complex and protracted, and are still incomplete. Due to a lack of customary land within the vicinity of the original village of Kapit, the community was divided as households were relocated around Lihir to areas where they have clan connections. Few of these relocated families have strong land tenure rights in these new locations, further compounding their vulnerability. The people of Putput were resettled as a relatively cohesive community to nearby land where they had customary land rights. Putput village is largely composed of Lihirians who claim ownership over land within the 'special mining lease' (SML) area and is consequently positioned at the top of the new socioeconomic hierarchy. Other nearby villages, including Zuen, Kunaie and Londolovit, also contain many Lihirians who claim ownership over land within the SML and the 'lease for mining purposes' (LMP) area. These villages were not relocated to make way for the mining operation, but because portions of their land were alienated for the LMP, the airport or the mining easements, these villages are also classified as affected areas and have received specific benefits and compensation.

Most Lihirians maintain high expectations for mine-derived development, and many lease-area landowners feel entitled to an exclusive claim over economic benefits and opportunities that arise from the project. At

---

6   The 1995 IBP was signed by the developer, the landowners and the government. It was organised into four chapters which represented the fourfold classification of compensation presented by the landowners during negotiations—compensation for destruction, development, security and rehabilitation. It contains the compensation and relocation agreements and three memoranda of agreement (MOAs) between the landowners and the three tiers of government, as well as the environmental monitoring plan and provisions for trust funds. The MOAs establish the distribution of royalties and define the responsibilities and undertakings of the parties in relation to the development of the mine and its benefits for the community. The MOAs were included in the agreement so that the landowners could see the entire package of compensation and benefits they would receive from the mining operation. The IBP agreement was revised in 2007 (Bainton 2010) and is the subject of a current review process.

the same time, the technical or legal distinction between affected and non-affected areas fails to capture the extent to which the so-called non-affected villages have also experienced the social impacts arising from the operation. Consequently, the unequal distribution of the costs and benefits associated with the project has generated considerable tensions and divisions throughout Lihir.

Demographic Monitoring

The social impact monitoring program for the Lihir gold mine, managed by the company's social performance department, has maintained a consistent focus on demographic monitoring. The current program has its origins in the baseline population survey and landowner identification work, and the earlier mapping of Lihirian men's houses (Filer 1992; Burton 1994). In 1994, John Burton developed the Lihir 'village population system' (VPS) database built on his customised genealogy software. An in-house team of research officers continues to work closely with a group of village-based informants to collect quarterly population updates for the VPS database, and since 2003 demographer John Vail has provided the team with technical support and produced annual population reports. The VPS now comprises a genealogical database of the Lihirian population and a database for the non-Lihirian population. The reason for this separation reflects Lihirian concerns over maintaining the 'purity' of Lihirian data to ensure that non-Lihirians are not mistakenly ascribed Lihirian status, which might allow them to access mining benefits. These two databases are linked through a third geographic information system database that combines aerial imagery and ground-level surveys, and contains information on the location of residential houses and their inhabitants.

Research officers and village informants have maintained relatively accurate data on the Lihirian population, partly due to the quality of the original genealogical survey, the ease of identifying Lihirians, and because many Lihirians tend to reside close to their natal village. On the other hand, it has proven more difficult to maintain accurate data on the migrant population because it is more mobile and less integrated into Lihirian society. For this reason, a residential household mapping approach has been required. Demographic datasets collected through a variety of other sources, including birth and death records from the health centres, attendance at maternal and child health clinics, school enrolment records and employment applications, are cross-referenced for inclusion in the VPS.

Overall, the longevity of this program has started to allow for a more sophisticated analysis of changes in fertility, mortality and life expectancy in the Lihirian population, and a more detailed understanding of the migrant population and its residential patterns. The VPS has proven to be an important corporate risk management tool, as it allows for a detailed level of information on local demographic trends and broader stakeholder connections. It also provides the basis for understanding the customary ownership of mining lease areas, the lack of which has been the source of considerable social tension and financial cost at other operations in PNG. In the absence of reliable government census data, the VPS has been of great value to local stakeholders, including the landowner association and the local-level government, and it has assisted with community development planning and land dispute resolution. While the VPS has many positive applications, and is maintained with the express endorsement of the landowner association and the local-level government, in its capacity as the curator or manager of this genealogical and personal identity database, the company has a clear duty of care to safeguard against its improper use and ensure its preservation for future generations (Burton 2007).

Demographic Shifts in the Mining Era

Over the past century the Lihirian population has increased from around 3,600 in the 1920s to 5,500 in 1980, and approximately 9,890 in 1995.[7] By the end of 2013 the Lihirian population was 16,095 (Figure 11.2). These shifts partly reflect a combination of improved community health

---

7   It was estimated that in the early 1980s around 500 Lihirians were absent from the islands. Many of these Lihirians were married to non-Lihirians and later returned to Lihir with their families to seek better living conditions. By the late 1990s, this group of returnees comprised some 2,000 people. Many of the women in this group were of childbearing age, which contributed to natural population increase.

facilities, increased life expectancy and lower infant mortality rates.[8] The bulk of the Lihirian and migrant population now resides in a broad arc across the northeast of Aniolam Island, which is mainly a result of people moving closer to services and employment opportunities.

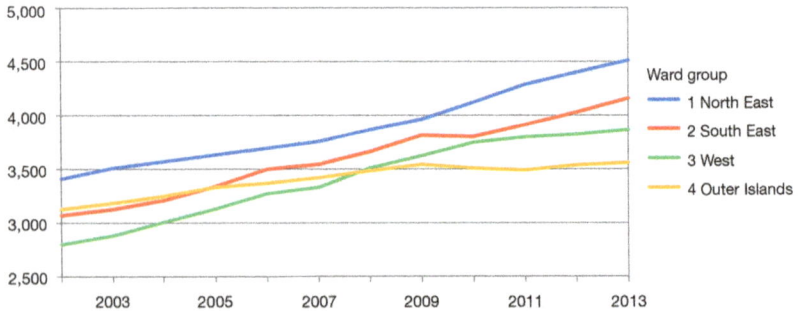

Figure 11.2 Lihir population by ward group, 2002–13.
Source: Newcrest Lihir Social Impact Monitoring unit.

The expanding migrant population in Lihir can be broadly divided into three categories: those who claim some form of existing or historical social connection; those who have forged relatively recent relations with Lihirians; and those people with very tenuous social connections in Lihir.[9] In local Tok Pisin terms, Lihirians tend to regard people from the first two groups as *ol wantok* (literally 'one talk'), a flexible and contextually contingent relational category that can be extended to relatives, people who share the same language or ancestry, people from the same province or region,

8   In 1997, John Burton concluded that single motherhood had been a common feature in Lihir since Independence, a period for which accurate data was available (Burton 1997a). Datasets for 1975–97 indicated that one in every seven families was headed by a woman, and one in every nine births, or about 10.8 per cent of all births, was to a single mother. Inquiries as to the whereabouts of the father were met with the habitual reply *em karim pikinini long rot* ('the child was conceived during her wanderings'). More recent data suggests a further increase, with 36 per cent of births recorded to single mothers in 2014. We might conclude that this increase is a result of rapid social changes; however, more research is required to understand whether these mothers are indeed single, or if this reflects an increase in de facto relationships that Lihirians may not report or 'recognise' as legitimate marriages for religious or cultural reasons. At this stage, it is impossible to conclude whether an increase in the number of single mothers has contributed to population growth. Similarly, more work is required to understand whether the growth rate has actually slowed over the past decade as more women have become empowered through access to employment and incomes.
9   This does not include temporary visitors from the surrounding parts of New Ireland Province, who mainly come for specific purposes for short periods of time and later return to their home village or town, or the fly-in/fly-out workforce associated with the mining operation.

or simply friends.[10] In other instances, Lihirians regard incorporated migrants as their *bisnis* ('business'), a term that Lihirians use to denote both their extended relatives and exchange partners and the practice of small-scale enterprise. In their relationships with migrants, these social and economic meanings often converge. Alternatively, migrants with little or no incorporation in Lihir, or those from the third category, are typically glossed as *ol autsait lain* ('outsiders') or *ol weira* (effectively 'strangers'). The latter term is deliberately derogatory and may be applied to migrants from neighbouring islands or remote mainland districts as a way of positioning them in terms of perceived social connections or their 'rights' over mine-derived development.

Although in-migration has peaked throughout the mining era, during the colonial period large numbers of labourers were recruited from other parts of PNG to work on Lihir plantations (Bainton 2008: 297). Not all of these labourers were able to return to their home districts, and several of them are now quite elderly and find themselves especially dependent upon their landowner patrons and still work as labourers in order to survive (Figure 11.3). Rather than an instant influx, in-migration has increased in step with the development and expansion of the mining operation and the local economy. Migrants first arrived from the surrounding New Ireland and New Britain region, drawing upon established networks and relationships and settling in hamlets as incorporated guests in similar ways to some of the early informal settlements found in other urban areas in PNG (Numbasa and Koczberski 2012). While the number of migrants was comparatively low during this early period of mining operations, there was already a level of concern among some Lihirians about the prospect of further in-migration and the negative impacts that this might have. Women in particular were rather vocal about the prospect of migrant women seeking to marry Lihirian men in order to claim access to mine-related benefits.

---

10   Tok Pisin is the lingua franca in PNG and used in conversation between Lihirians and migrants. Lihirians generally communicate among themselves in the Lihirian language (but see Bainton 2015).

Figure 11.3 Former plantation labourers from Sandaun Province.
Note: Those shown are (from left to right) Thomas Waipano, Paulus Make, Nelson Sira, Camillus Mawe and Albert Esiuke.
Source: Photo by John Burton.

During this earlier phase there were fewer economic opportunities. The mining company was the main source of employment and non-Lihirians were predominantly engaged on fly-in/fly-out rosters. Nevertheless, migrants still arrived in search of employment, moving between villages and attaching themselves to other migrants. In-migration was further encouraged by a small number of Lihirian men with prior business experience outside Lihir who later returned to establish small companies. Some of these men were shrewd businessmen, refused to employ relatives, and remained aloof in order to avoid demands from kin that would compromise their business pursuits (see Martin 2007). As local business development increased, more migrants were employed by private or local companies contracted to the mining operation, not all of which had formal provisions for accommodation. Many migrants came to sell goods at the town market, being regularly supplied by people from their home areas with betelnut and other produce for sale. These people were soon the target of Lihirian resentment as they were seen to be making money that Lihirians felt entitled to. In a typical chain migration sequence, these migrants were joined by other relatives seeking similar opportunities, although subsequent migrants were often far less incorporated, particularly unaccompanied young males.

By the early 2000s, greater numbers of people were arriving from the Highland and Momase regions. One Lihirian man in particular provided initial support for migrants to settle in the previously unoccupied interior area between Londolovit and Kunaie villages, in the hamlets of Kul and Bombel. He was one of several outliers who bolstered personal support and renown through the patronage of migrant settlers. Before long, these interior hamlet areas were predominantly occupied by migrants. The growth of the migrant population is closely linked to the expansion of the local business environment, which has created a more diverse range of economic opportunities. Lihir is easily accessed by daily flights and the regular movement of dinghies from mainland New Ireland. The number of boats travelling to Lihir has increased as more money has spread throughout the region, and the advent of cheap mobile phones has enabled migrants to regularly contact relatives and friends to arrange their passage to Lihir or maintain a constant supply of market goods. Some of the more savvy entrepreneurs now utilise courier services to import goods and artefacts for sale. These strategies ultimately enable them to remain in Lihir for longer periods of time.

There is now a distinct clustering around the affected-area villages in the local government wards that are closest to the mine site (Figure 11.4).[11] The number of people from New Ireland, and from the broader Islands, Momase and Highlands regions, has increased (Figure 11.5). Migrants outnumber Lihirians in Ward 1, and it is likely that they outnumber Lihirians in Ward 11 as well. The number of migrants residing in Putput village, in Ward 2, is comparatively lower, but the impact of their presence is compounded by existing land pressure and other rapid social changes related to resettlement. In these areas the migrant population is weighted towards younger mobile males from the Islands Region in the 20–35-year age range, but there are also many young families who seek employment and access to services and a small number of elderly migrants.

---

11   Due to the likely under-enumeration of the non-Lihirian population in the affected-area wards, an attempt has been made to reach a more conclusive *estimate* of the non-Lihirian population in these areas. Figure 11.4 presents positively enumerated figures. The adjusted population figures for non-Lihirians are derived from certain assumptions based upon known residency patterns, the number of mapped residential structures, and average household numbers. See Vail (2011) for technical details for calculating the adjusted non-Lihirian figures.

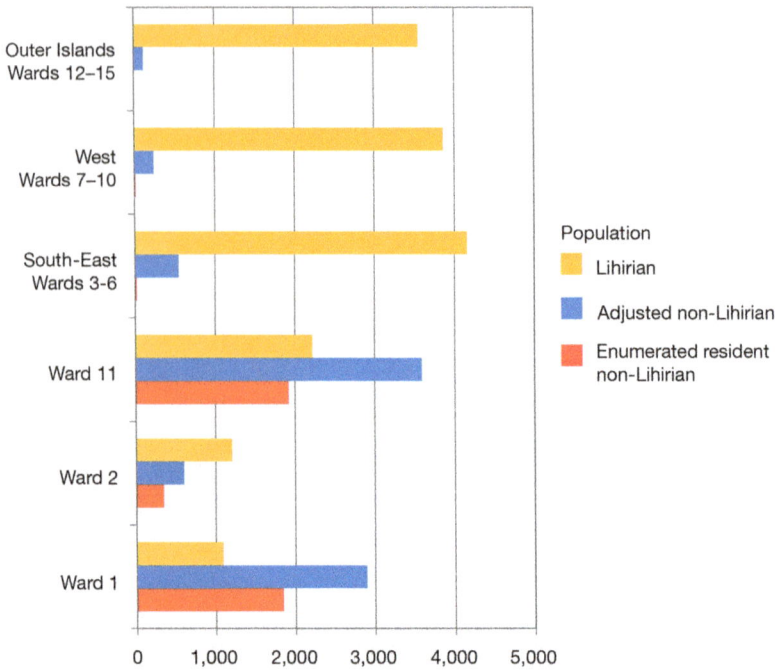

Figure 11.4 Enumerated Lihirian and non-Lihirian population figures, and adjusted non-Lihirian population figures, December 2013.
Source: Newcrest Lihir Social Impact Monitoring unit.

One of the most attractive services for migrant families is the well-resourced Lihir medical centre that provides vital primary health care to mine employees and the residents of Lihir. This has led to a form of 'medical tourism' as people come to access services that are not available elsewhere in the province, or many other parts of the country for that matter.[12] While there have been major improvements in the health of the Lihirian community, successful health initiatives must also reach the migrant population, which is a considerably more difficult task.[13]

---

12   The Namatanai hospital, which is located in the nearest town on the mainland of New Ireland, also in the electorate of the former Minister for Mining, Byron Chan, has remained chronically dilapidated for many years. Redevelopment of the Namatanai hospital would certainly help to redirect in-migration and reduce the burden upon the Lihir medical centre, which has seen patient numbers more than double between 2005 and 2012, from approximately 36,000 to around 80,000 visits per year. In 2012, more than half of these patients were recorded as being contractors or their dependents, and the overall majority were recorded as non-Lihirians. More recently there has been a slight reduction in patient numbers due to the decline of contractor opportunities. However, there is still a need to integrate health service provisions across the district to meet future demands.

13   For more information on recent health programs see Bentley (2011) and Mitjà et al. (2011, 2012, 2013, 2015).

Scheduled village health programs typically exclude the settlement areas, which is partly due to political decisions by Lihirian leaders to restrict the distribution of benefits. But, as health workers well understand, the settlement areas are only a short walk from the medical centre and the environmental conditions in these locations present major health risks for residents, which in turn places greater pressure upon health services.

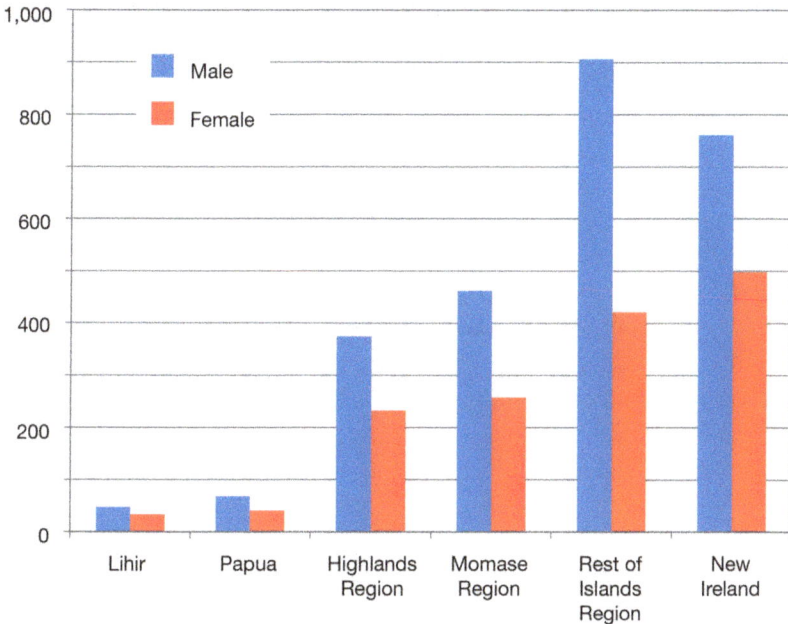

Figure 11.5 Birthplace of enumerated resident non-Lihirians, December 2013.

Source: Newcrest Lihir Social Impact Monitoring unit.

As the affected-area villages have been steadily urbanised, settlement patterns have remained regulated by customary land boundaries and more recent village planning exercises. On the other hand, in the interior hamlet areas of Kul and Bombel, and more recently the upper Londolovit and the Landolam valleys, settlement is more sporadic and opportunistic, and reflects more temporary arrangements. Although most of the migrants residing in these areas have limited social incorporation in Lihir, and have been the target of wider Lihirian frustration in relation to in-migration, many of these migrants remain in Lihir at the sufferance or direct sponsorship of specific landowners. However, as we shall see, many of these landowners have also lost control as patrons.

329

The Contractor Labour Force

At its most recent peak in 2012, the mining company engaged a direct workforce of approximately 2,400 people and more than 4,100 contract labourers (NML 2013). In the earlier years of the operation, the contract labour force was considerably smaller, but as the project has expanded this has created more opportunities for local companies. Following the initial construction boom, the operation of the mine brought the need for reliable service provision and generated opportunities for local companies. These companies varied in size and composition and gained a range of business contracts including grass cutting, rubbish collection, gardening, transport services, construction, electrical and maintenance work and other mine-related services. From the outset, Lihirian landowners have been more interested in gaining business contracts for these services rather than performing the actual work themselves. This has opened up avenues for migrant labourers, many of whom have settled in the interior valleys and affected-area villages. By 2012, some 650 businesses were registered in Lihir, although certainly not all of these companies had contracts with the mine or any other kind of current operation. The largest and most successful of these business operations is the Lihirian-owned Anitua group of companies, which includes several companies that offer security, drilling, retail, catering and mining services. In 2012, the Anitua group employed over 2,000 people, with at least 45 per cent of its workforce coming from Lihir, 51 per cent from other parts of PNG, and the remainder from overseas.

Most of the companies in Lihir are small, locally owned businesses seeking contracts with the mining operation. Many of these companies provide labour hire services, sourcing Lihirian and non-Lihirian labourers for various work packages. Local businessmen not only recruit migrants on Lihir but regularly travel to other parts of PNG to find labourers for current or anticipated contracts. In 2008 the mining company commenced a major upgrade of the processing plant facilities, which included K40 million in seed capital to landowner companies, and provided millions of kina in commercial opportunities to local businesses. In order to maximise the contracting opportunities and to organise the distribution of the seed capital, Lihirian leaders amalgamated many of the smaller companies in a set of larger 'specific issues companies' that represent the landowners from particular lease areas (such as the plant site, the pit, the town and airport), collapsing the clan-based and geographical distinctions that operate elsewhere. This also had the peculiar effect of turning 'issues' into commercial entities, which has precluded the resolution of certain issues and guaranteed further 'compensation' through commercial contracts. Despite the anticipated economies of scale to be achieved through the consolidation of small and often unsustainable landowner companies, some landowners have found this umbrella model unsatisfactory. Not all of the new umbrella companies have operated effectively or regularly distributed dividends to the landowners they represent. Moreover, the distribution of contracting opportunities has remained uneven, which in turn has increased the existing divisions between landowners over access to business opportunities.

According to the environmental impact statement (EIS) for the plant upgrade, it was anticipated that the construction phase of this upgrade, completed in 2012, would require approximately 500 additional labourers (CNS 2009). Landowners and migrants took advantage of the opportunities for contract labour, and it now appears that a far greater number of hopeful migrant labourers have made their way to Lihir than what was originally projected in the EIS.[14] The number of companies and labourers seeking contracts in Lihir has now outstripped the available opportunities. The proliferation of companies has arisen for several interconnected reasons: the perceived low start-up costs; the modern status associated with owning a business and employing people; a ready supply of migrant labour; and the expectation among landowners for business contracts based upon the belief that, because the mine is operating on their land within the lease areas, they are entitled to exclusive business opportunities, which is best characterised by the local catchcry 'my land, my work' (Bainton and Macintyre 2013). In this sense, landowner business development is perhaps better understood as another form of compensation or rent, or a kind of 'tax' levied against developers for access to resources. This kind of 'rentier capitalism' is now a common feature of the PNG resource sector and many landowners seem to conceptualise business development in terms of 'risk-free entrepreneurialism', which requires developers to carry the cost and exposure of local business development so that landowners might collect their due entitlement (Jackson 2015). If this appears to outside observers like a form of dependency or unsustainable development, from a local landowner

---

14 See Banks (2013) for a discussion on the social and environmental impact statement for the Lihir plant upgrade.

perspective it is more likely perceived as a way of securing a guaranteed income for life, or at least exploiting an immediate opportunity.[15] In the Lihir context, the relationship between landowners, migrants and the mining company is thoroughly shaped by local forms of patronage and the art of 'rentier leadership' (Burton 1997b).

Housing the Labourers

Two main scenarios for employee accommodation were tabled during the negotiation phase of the mine: a fly-in/fly-out arrangement and a full-scale mining town that would house workers and their families. Lihirian leaders wanted to maximise the opportunities for urban development and the associated economic spin-offs but feared the negative impacts that might accompany an influx of migrant workers. The end result was a combination of both options. The mining company utilised a dormant plantation on the Marahun plateau to construct a mining camp for fly-in/fly-out employees. A purpose-built mining town was also developed adjacent to the camp with around 70 houses for expatriate and senior national employees who reside in Lihir with their families, and 35 houses for government employees. A local commercial service centre was constructed in the nearby lower Londolovit area (also previously used as a plantation), and a ring road was cleared around Aniolam Island linking all villages to the town and the camp. Prior to the establishment of Londolovit town, the nearby Potzlaka government station and the Palie mission station in the southwestern corner of the island were the only places that might be considered 'public' spaces, and were the main link to the outside world. The previous lack of transport options, compared with today, also ensured far less internal and outbound travel.

According to the terms of the 1995 mining development contract and the 2007 revised IBP agreement, non-Lihirian employees and contractors were to be engaged on fly-in/fly-out arrangements. Local Lihirian employees travel to work from their home villages, and it was originally hoped that this would generate less disruption to daily village life. The intention was to somehow 'contain' the large and predominantly male non-Lihirian workforce within the confines of the mining lease areas in order to limit

---

15 Thanks to Richard Jackson for bringing this point to my attention.

the social footprint of the operation. This strategy is premised upon the developer being the main source of employment. However, as the number of landowner and joint venture companies operating in Lihir has increased, there has been a much higher demand for employee accommodation.

While some of the contracts gained by local companies include provisions for employee accommodation, this does not always ensure that these companies fulfil their obligations to their employees, including adequate accommodation, standard employee entitlements, or even regular meals. Some of the larger companies have built basic 'camp' accommodation, but in many cases non-Lihirian employees are forced to seek other arrangements. Not all of the Anitua group employees are housed within the mining camp, which is partly due to the size of the camp, but also because many of the lower-paid and semi-skilled positions do not include provisions for camp accommodation. For most migrants in Lihir, the primary intention is to gain employment or secure some form of livelihood. Access to land remains a secondary consideration since long-term residence is not the ultimate goal.

Since 1995, more than 1,000 permanent timber kit houses have been provided to the wider Lihirian community through the 'village development scheme' that is part of the IBP agreement (Figure 11.6). Lihirians have recognised a flourishing rental market, and many rent their new houses to migrant workers and their families. As more local businesses source migrant labour, there is a shortage of suitable accommodation. In the villages of Zuen, Kunaie, Londolovit and Putput, landowners can charge K1,000–3,500 (approximately A$450–1,500) per month to rent a wooden kit house. Most houses in these villages are connected to electricity and have access to rainwater tanks or mains water. The high rental rates do not always reflect real landowner expenses, especially given that most houses are provided under the benefits package. Landowners have seized upon the inflated resource economy and the demand for housing, and these rates could be seen as another type of 'tax' levied against migrants for access to Lihirian economic development.

Figure 11.6 Village development scheme house.
Source: Photo by Nicholas A. Bainton.

For many migrant tenants, this is rarely an unproblematic rental arrangement. In some cases, landowners remain on the same block of land in a semi-permanent or bush material house and share water tanks and sometimes illegally connect electricity to their houses. These arrangements can create tensions between tenants and landlords, especially when tenants expect exclusive use of the house and any amenities. Some tenants complain that landlords demand additional contributions or advance payments on rent for customary feasting or other forms of daily consumption.[16] This situation is only compounded by the absence of rental contracts.

---

16   For examples of these additional demands in other contexts, see Curry and Koczberski (2009), Allen (2012) and Koczberski et al. (2012).

Settlers, Squatters and Opportunistic Businessmen

In Lihir the term 'rent' is applied loosely to cover a variety of informal arrangements that provide migrants with temporary access to housing or land upon which they may erect housing or grow food gardens.[17] As yet, migrants have not utilised this land for agro-economic development, and there are few de facto changes to customary land tenure because Lihirians refuse to extend long-term use rights to migrant families. Effectively, landowners retain and exert their land rights: insecure or temporary use rights might be granted, and informal agreements established, but customary land is by no means alienated.

A technical legal distinction can be drawn between squatters, as people who enter and live on unoccupied land without permission, and settlers, who are invited to stay on land for a long period with express or tacit approval. In Lihir, these categories are often blurred and highly contextual. For instance, each of the 18 migrant settlement clusters identified in a social mapping exercise in the Landolam Valley area in 2011 were found to be sponsored by specific landowners (Burton and Onguglo 2011). Migrants living in these clusters exist in patron–client style relationships with individual landowners or local lineages. Some of these migrant settlers rent semi-permanent housing, while others are granted access to land upon which they erect makeshift homes. Settlers may pay from K20 to K100 on a semi-regular basis for access to land, and make additional contributions that help to sustain the relationship. Landowners often collect another form of 'rent' by asking a group of residents to 'raise funds' for an upcoming customary event. At the same time, many of these landowners also claim that they are overwhelmed by additional migrants who were not originally given permission to settle on their land regardless of any collective 'rental' payments that might have been made on their behalf. Rapid chain migration has meant that many landowners have lost control as patrons. Consequently, Lihirian political leaders often characterise all migrants as squatters, conveniently homogenising them as 'illegal trespassers'. In some cases the label is clearly correct, as people are squatting on customary or leased land without permission. In other cases,

---

17 This is similar to Koczberski et al.'s (2009: 35) observations in the PNG oil palm sector. See also Koczberski et al.'s (2017) discussion on informal land markets in PNG.

migrants have ongoing economic and social relationships with individual landowners, complicating any simplistic image of migrants as unlawful tenants or parasitic vagrants.

In order to maintain use rights over land and access to residences, migrants frequently maximise new opportunities for social incorporation with Lihirian lineage groups. This is commonly expressed by migrants as: *Mipela kamap bisnis wantaim sampela papagraon bilong Lihir* ('We are incorporated with some Lihirian landowners'); or *Mipela pas wantaim dispela lain* ('We are connected to this lineage group'). This relationship may be confirmed by Lihirians with expressions like: *Ol i pas wantaim klen bilong mi* ('They are connected to my clan'); *Ol i kam ananit long lain bilong mi* ('They are incorporated under our lineage'); or *Em ol bisnis bilong mipela* ('They are our "business"'). In the context of landowner–migrant relations, both the social and economic meanings of *bisnis* are emphasised. As migrant families become partially incorporated into Lihirian clans, as a sort of subgroup of the landowning lineage, the social (and economic) relationship is foregrounded and frames property rights. Migrants remain indebted to their hosts, who retain control over property and grant temporary forms of access. These relations are forged and maintained through contributions to funerals and large feasting and exchange activities (sometimes involving several thousand kina for the purchase of pigs), and more regular gifts of food and cash to landowner households and support for community projects. These public displays of support and involvement in community life confirm landowner patronage and help to erase the social differences that are made explicit in other contexts. At the same time, this can also compromise the ability of Lihirians to decide who resides upon their land or to remove unwanted tenants. Migrants have recognised the economic changes to customary activities and exploit the increased opportunities to establish and strengthen relationships with Lihirians. On the flip side, when rent or other contributions are not adequately distributed within the lineage, disgruntled members who fail to benefit from migrant settlers may well deny or contest their social inclusion, reflecting the vulnerable nature of migrant lives.

Vulnerable Migrants and Marginal Landowners

At various times over the past ten years, the local political elite have incited community-level fervour for large-scale evictions, highlighting the contradictions in the relationship between Lihirians and migrants. This has generated a widespread desire for immediate action, and unfortunately reinforced the view among many people that the major social problems encountered in Lihir are best addressed through a singular focus on mass evictions. While much of this attention is directed at the growing number of people living in the interior areas close to the Londolovit town site, migrants residing in other villages are sometimes equally targeted. This is regardless of whether they are working for landowner companies and legitimately renting houses, or are young families eking out a living, or are some of the mass of mobile males 'drifting without kin', money or a connection to place (Reed 2003: 70).

This deliberate homogenisation reflects the larger sense in which Lihirians feel beset by migrant-related social changes. The close correlation between unmitigated in-migration, squatter settlements, growing law and order problems, and interethnic tensions between Lihirians and migrants, and between different migrant groups, has given rise to a definite sense of civic insecurity. This reached new heights in mid-2015 when several male migrants were arrested for torturing three women from Southern Highlands Province who were accused of sorcery by their fellow migrants and their Lihirian patron. Police station records show that in recent years there has been a definite increase in the number of reported crimes and charges. The data indicate that the migrant community has contributed to a quantitative shift in the level of social disturbance as the number of charges for theft, assault, sex-related crimes, drugs and general disturbances to public order have increased annually. Lihirian protestations would also suggest that the migrant community is solely to blame for a qualitative shift in social disturbance. However, such assertions are not supported by the records, which demonstrate that Lihirians are often involved or implicated in similar activities. On another level, even though Lihirian women may experience an acute fear of outsiders, especially in some affected-area villages where there is a high ratio of migrant males to females, these fears may be somewhat misplaced, since police and hospital records suggest that the level of domestic violence inflicted upon Lihirian women by their Lihirian husbands and relatives presents a more immediate threat to their everyday well-being.

The Lihirian tendency to attribute negative influences to the migrant population masks a certain level of social disintegration within Lihirian society. It also hides the ways in which the general vilification of migrants is intersected and undermined by individual interests and the social relations of compensation. This point is well illustrated in one documented instance involving a complicated dispute over a portion of land within the lease area in the Landolam Valley. This intergenerational and interclan dispute is the principal reason why one Lihirian family group has taken up residence on the lease, and have surrounded themselves with a considerable number of migrants who help them financially and offer moral support (Burton and Onguglo 2011). Their presence on the lease land was a double act of public defiance directed at the mining company and the landowner association, who they claimed had disregarded their land rights and denied them access to benefits and compensation. The senior widow of the lineage blamed her leg ailment on the landowner association—claiming she was poisoned by someone involved in the dispute—which resulted in her leg being amputated in 2010. She was then confined to a wheelchair that had to be carried across the boggy settlement grounds. A member of the association allegedly promised support for an artificial limb, but this never materialised, and in 2011 she died (Figure 11.7). Even if there is no discernible moral to this miserable outcome, this account demonstrates that not all landowner patrons are equally empowered, that some are even more vulnerable than their migrant tenants and are just as likely to be excluded by their fellow clan members on their own ground. And in this way the blanket characterisation of migrants helps to cover over some very uncomfortable social realities.

Eviction exercises planned by the local-level government have manifestly failed because of the tendency of landowners to consider 'their' migrants as useful guests (or economic resources), while all others are deemed an illegitimate nuisance. While a handful of landowners have evicted squatters from their land and burnt their houses, the historical lack of capacity within the local government to manage and mobilise coordinated eviction exercises, or deliver basic governance, has precluded meaningful action, reducing its efforts to vitriolic statements. Most Lihirians expect that the mining company will take direct action, which partly reflects the extent to which people feel overwhelmed and frustrated by current social changes, and the belief among some people that the company is both the source and only solution to these problems. At a minimum, there is a need for strong management of contract labour and the adequate provision of accommodation for contractor employees. In most cases, this is not the sort of action that Lihirians envision. This might even generate resistance from local businessmen who have come to rely upon ready access to a pool of migrant labour that can be easily mobilised for short-term contracts. Rather, there is an expectation that the company will fund a private security firm to round up recalcitrant migrants and march them off the island. The contradiction between benefiting from migrant labour skills and simultaneously wanting to evict all outsiders is apparent.

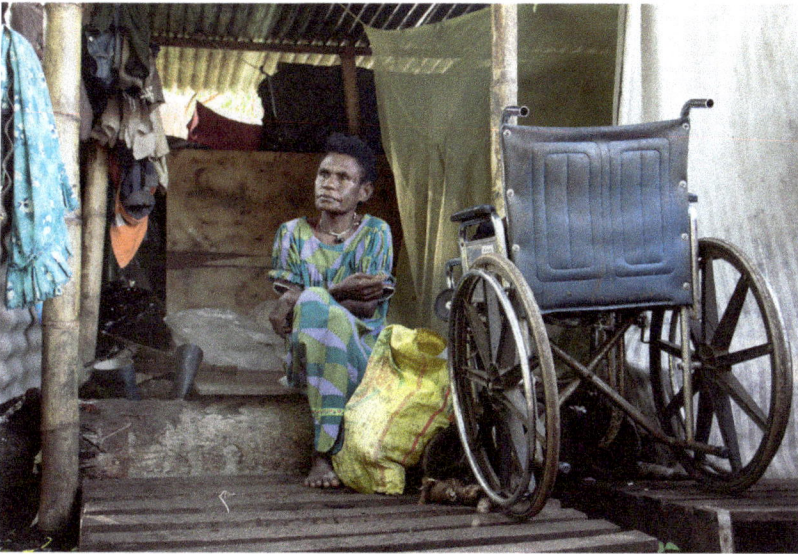

Figure 11.7 Late Anna Ikuluom of Tinetalgo clan in lower Landolam, 2011.

Source: Photo by Joyce Onguglo.

In these ways, Lihirian landowners, local business owners and the mining company provide a major pathway for in-migration. So, if it is true that some business owners have been hoodwinked into employing or entering into commercial arrangements with migrants, or have been overwhelmed by the effects of chain migration, then it is also the case that many have been blinded to these ramifications through their pursuit of business development, and that the mining company's contractor management processes have historically been inadequate for the task. The eagerness with which landowners have employed or entered into joint ventures with business-savvy migrants has only encouraged further in-migration and settlement. The inability or refusal of some landowners to accept responsibility for the welfare or actions of migrant employees and their families has directly contributed to the growing lack of control over the migrant population and the difficulties associated with retrospectively acknowledging and addressing these issues.

The other side of this scenario is the emergence of a vulnerable workforce. Classic labour theories tend to emphasise mobility as a key asset of the labourer. But in the PNG mining context it is questionable whether these theories are adequate. In Lihir, there are many migrant labourers and their families who experience entrenched conditions of unacceptable poverty and face considerable constraints upon their mobility. A lack of organised labour unions reduces their bargaining power and reinforces their reliance upon landowners for livelihoods and residence. The short-term nature of their employment contracts, the lack of suitable accommodation, the long working hours away from home, the inflated cost of living in a project enclave and their precarious social position place them under enormous pressure. These people live in a bind: they have less opportunity for investing in quality family life, better living conditions, or more sustainable livelihoods; there are few who can afford to travel back to their home village with their families; and they do not necessarily want to return to under-serviced rural areas. These migrants are perhaps better understood as a kind of 'landless rural peasantry', or a local example of the so-called 'precariat' (Standing 2011). Such vulnerability echoes Tawney's classic image of the 'man standing permanently up to the neck in water, so that even a ripple is sufficient to drown him' (Tawney 1932: 77). The total reliance upon landowners for land and livelihoods makes it difficult for migrants in this position to pursue alternative options. The deliberate conflation of migrant labourers and their families with more troublesome young mobile males ensures that the climate of fear is experienced both ways, ultimately exacerbating dependency upon landowner patrons.

Conclusion: Some Practical Considerations

The seemingly intractable challenges and impacts surrounding rural-to-resource migration, and the difficulties that developers, landowners and governments encounter when addressing this issue, point to the need for greater attention to in-migration management strategies in the planning stages of resource projects. As detailed elsewhere (Bainton et al. 2017), the basic operational capabilities, or enablers, for understanding and managing in-migration are often missing in the industry's approach to social performance in practice. The planners and managers of many large-scale resource projects in PNG have paid insufficient attention to the different forms of in-migration that can occur throughout the life of a project, or the social and geographical factors that may influence these trends. The standards of planning and compliance that were common in the 1980s and 1990s—the period when several of the major projects currently operating in PNG were being developed—are no longer suitable (IFC 2012). International standards now require prospective projects to undertake detailed studies on potential project-induced in-migration and the related risks and impacts for the operation and host communities (IFC 2009). These standards also require the development of mitigation strategies to alleviate disturbances and risks associated with in-migration.

Freedom of movement is guaranteed in PNG's National Constitution, and developers cannot stop people migrating to project areas in search of new opportunities. This does not exempt developers from addressing in-migration, especially when in-migration becomes entwined with local contracting processes, or when migrants settle upon lease areas, or when they are severely disadvantaged. A lack of attention to the poor living and working conditions of subcontractors may result in these workers becoming 'hidden' within the overall workforce. These and other 'omitted social conditions' (Roberts 1995) of large-scale resource development represent another way in which the costs of production may be externalised on to local society. Ignoring in-migration until it results in operational, financial or reputational costs invariably creates situations where developers and local actors can only hope to address the *symptoms* of in-migration. The significant amounts of financial, legal and human resources that are directed towards reactive approaches to the impacts of in-migration detract from efforts to deliver upon agreed social and economic commitments to host communities. In the long run, this may well jeopardise a developer's ability to operate, as local frustrations mount over the slow delivery of benefits and agreed-upon projects, or the seemingly limited responses to migrant-related issues. Declining levels of community trust create further challenges for establishing more proactive approaches to in-migration that positively engage host communities and other relevant actors.

Incursions on to lease land and the presence of illegal squatter settlements or relocated villages on lease areas create immediate operational risks. In 2010 the issue of project-induced in-migration at the Porgera gold mine became an international scandal when images of burning houses and homeless people appeared on the front pages of international newspapers, together with indignant statements on human rights by non-governmental organisations (Amnesty International 2010; HRW 2011). At the very least, increased scrutiny from the media and non-governmental organisations requires developers to pay careful attention to the activities of employees, security personnel or police forces engaged in any form of migrant management, particularly where this may involve involuntary movement or resettlement. The tendency among some external commentators to look for simple moral narratives often obscures the complexity of these contexts, the factors that may influence company decisions, or the ways in which various state and local actors are also implicated. Recent access to smart phones and internet connectivity across PNG has also given rise to new forms of real-time exposure through the use of social media to broadcast or contest the real and perceived impacts of resource development. As a result, the planning and execution of any sort of relocation or broader 'population management' on or around lease land should be approached in a transparent rights-based manner.

Community members who feel overwhelmed by in-migration, regardless of their role in facilitating or hosting migrants, will often argue that developers are morally bound to provide assistance, since from their perspective it is the operation that has attracted in-migrants. In their efforts to avoid playing the role of a surrogate state, it is apparent that developers cannot unilaterally address the problems surrounding in-migration, which in turn points to more complex issues surrounding the limits or boundaries of corporate social responsibility (Banks 2006), and the difficulties involved in developing strong working partnerships with private companies, local communities and the public sector. Certainly the strategic investment in community development projects outside the core project area, in neighbouring towns or provincial centres for instance, could help to direct migrants away from affected areas. But the difficulties that many developers face in fulfilling their primary community obligations may well undermine these kinds of initiatives as lease-area landowners demand that all resources be invested locally.

The Lihir context highlights the need for integrated approaches that extend beyond the immediate vicinity of lease areas and involve local, provincial and national actors. In PNG, these goals are hard to realise since developers often operate with limited input from state functionaries. This can be partly attributed to the isolation and remoteness of project locations, but is perhaps more closely related to the capacity and resourcing of state regulators. These conditions are further compounded by a lack of civil society in many mining locations. While provincial governments maintain strong interests in resource projects, their attention is usually directed towards the capture of royalties as opposed to managing site-specific issues. These tasks are often left to landowners, developers and local-level governments. But as we have seen in Southern Highlands Province (Haley and May 2007), the amplified pressures of resource development may contribute to the collapse of local governance or the inability of local governments to provide the most basic public services, let alone address complex socioeconomic phenomena like in-migration.

In the end, developers and local authorities do well to remember that migrants provide much of the necessary entrepreneurial energy for the local economy, bring capital, innovation (both economic and cultural), specific skills and technical knowledge, and make up for labour shortfalls. From a provincial or national perspective, they are crucial in the wider distribution or remittance of benefits—financial, social, political and educational (Rigg 2007: 169). If this point is often lost in local discourses on in-migration, this reflects the complex connections between in-migration, livelihood strategies, identity, landownership and economic development. By drawing attention to these conjunctions, I have sought to illuminate some of the conditions of migrant lives in resource enclaves and the qualitative and quantitative features of rural-to-resource in-migration. Host communities of resource projects frequently face major social upheavals as large numbers of migrants come and reside on customary land yet exist outside traditional mechanisms for social control. These issues are often magnified on the PNG mainland, where project corridors and enclaves encompass larger and more diverse groups. But as the Lihir case demonstrates, small islands are by no means immune; the scale may be reduced, but in some ways the impacts can be more intense as they are compressed and contained within a smaller location.

Acknowledgments

An early version of this chapter was presented at the Institute of Australian Geographers annual conference in 2009 and has benefited from helpful comments from John Connell and George Curry. A later version was also presented at the 2013 annual conference of the German Anthropology Association. Thanks to Bettina Beer and her colleagues for organising my involvement and for their feedback. Thanks to past and present staff and consultants of the Lihir gold mine social impact monitoring program, especially Elly Sawa, Walter Pondrelei, Ephraim Lenturut, Bill Sagir, Mary Arau, Bernadette Samsam, late Emma Zanahien, Jacinta Tami, Wesley Kenneth, Francis Kupe, John Vail, John Burton and Joyce Onguglo, and the team of Lihir village information officers. All have provided input into the demographic monitoring programs in Lihir over the years and their work has helped to inform this chapter. Thanks also to John Owen, Glenn Banks and Graeme Hancock for their comments on a draft version of this chapter.

References

Allen, M.G., 2012. 'Land, Identity and Conflict on Guadalcanal, Solomon Islands.' *Australian Geographer* 43: 163–180. doi.org/10.108 0/00049182.2012.682294

Amnesty International, 2010. *Undermining Rights: Forced Evictions and Police Brutality around the Porgera Gold Mine, Papua New Guinea.* London: Amnesty International.

Bainton, N.A., 2008. 'The Genesis and the Escalation of Desire and Antipathy in the Lihir Islands, Papua New Guinea.' *Journal of Pacific History* 43: 289–312. doi.org/10.1080/00223340802499609

——, 2009. 'Keeping the Network out of View: Mining, Distinctions and Exclusion in Melanesia.' *Oceania* 79: 18–33. doi.org/10.1002/ j.1834-4461.2009.tb00048.x

——, 2010. *The Lihir Destiny: Cultural Responses to Mining in Melanesia.* Canberra: ANU E Press (Asia-Pacific Environment Monograph 5).

——, 2015. 'The Lihir Language in Modern Historical Context.' In S. Ziegler (ed.), *Karl Neuhaus's Grammar of the Lihir Language of New Ireland, Papua New Guinea*. Boroko: Institute of Papua New Guinea Studies.

Bainton, N.A. and M. Macintyre, 2013. '"My Land, My Work": Business Development and Large-Scale Mining in Papua New Guinea.' In F. McCormack and K. Barclay (eds), *Engaging with Capitalism: Cases from Oceania*. Bingley (UK): Emerald Group Publishing (Research in Economic Anthropology 33). doi.org/10.1108/s0190-1281(2013)0000033008

——, 2016. 'Mortuary Ritual and Mining Riches in Island Melanesia.' In D. Lipset and E.K. Silverman (eds), *Mortuary Dialogues: Death Ritual and the Reproduction of Moral Community in Pacific Modernities*. Oxford: Berghahn Books.

Bainton, N., V. Vivoda, D. Kemp, J. Owen and J. Keenan, 2017. 'Project-Induced In-Migration and Large-Scale Mining: A Scoping Study.' St Lucia: University of Queensland, Centre for Social Responsibility in Mining.

Banks, G., 2005. 'Globalization, Poverty, and Hyperdevelopment in Papua New Guinea's Mining Sector.' *Focaal* 46: 128–143. doi.org/10.3167/092012906780786799

——, 2006. 'Mining, Social Change and Corporate Social Responsibility: Drawing Lines in the Papua New Guinea Mud.' In S. Firth (ed.), *Globalisation and Governance in the Pacific Islands*. Canberra: ANU E Press.

——, 2009. 'Activities of TNCs in Extractive Industries in Asia and the Pacific: Implications for Development.' *Transnational Corporations* 18: 43–60.

——, 2013. 'Little by Little, Inch by Inch: Project Expansion Assessments in the Papua New Guinea Mining Industry.' *Resources Policy* 38: 688–695. doi.org/10.1016/j.resourpol.2013.03.003

Bentley, K., 2011. *2010 Lihir Social Demographic Health Survey*. Canberra: Centre for Environmental Health Pty Ltd for Newcrest Mining Ltd.

BGC (Barrick Gold Corporation), 2010. *Porgera: 2010 Responsibility Report*. Toronto: BGC.

——, 2012. 'Porgera Joint Venture.' Toronto: BGC.

Burton, J., 1994. 'The Lihir VPS Database: A Tool for Human Resources Planning, Social Development Planning and Social Monitoring.' Canberra: Pacific Social Mapping (Lihir Census Project Report 1).

——, 1997a. 'The Lihir VPS Database: Collection of Data for Monitoring.' Canberra: Pacific Social Mapping (Lihir Census Project Report 4).

——, 1997b. 'C'est Qui, le Patron? Kinship and the Rentier Leader in the Upper Watut.' Canberra: The Australian National University, Resource Management in Asia-Pacific Program (Working Paper 1).

——, 2007. 'The Anthropology of Personal Identity: Intellectual Property Rights Issues in Papua New Guinea, West Papua and Australia.' *Australian Journal of Anthropology* 18: 40–55. doi.org/10.1111/j.1835-9310.2007.tb00076.x

Burton, J. and J. Onguglo, 2011. 'Social Impact Assessment of the LGL Lease Land Migrant Management Strategy.' Canberra: ANU Enterprise for Newcrest Mining Ltd.

Bury, J., 2007. 'Mining Migrants: Transnational Mining and Migration Patterns in the Peruvian Andes.' *Professional Geographer* 59: 378–389. doi.org/10.1111/j.1467-9272.2007.00620.x

Callister, G., 2008. 'Illegal Miner Study: Report on Findings.' Unpublished report by EMB Development Services Ltd and Rural Link Consultancies for Porgera Joint Venture.

Clunies Ross, A., 1984. *Migrants from Fifty Villages*. Boroko: Institute of Applied Social and Economic Research (Monograph 21).

CNS (Coffey Natural Systems Pty Ltd), 2009. Environmental Impact Statement: Million Ounce Plant Upgrade Project—Main Report, Volume 2). Melbourne: CNS for Newcrest Mining Ltd.

Connell, J., 2012. 'Population Resettlement in the Pacific: Lessons from a Hazardous History?' Australian Geographer 43: 127–142. doi.org/10.1080/00049182.2012.682292

Curry, G.N. and G. Koczberski, 2009. 'Finding Common Ground: Relational Concepts of Land Tenure and Economy in the Oil Palm Frontier of Papua New Guinea.' *Geographical Journal* 175: 98–111. doi.org/10.1111/j.1475-4959.2008.00319.x

Filer, C., 1992. 'The Lihir Hamlet Hausboi Survey: Interim Report.' Waigani: Unisearch PNG for Kennecott Explorations (Australia) and PNG Department of Mining and Petroleum.

——, 1999. 'Introduction.' In C. Filer (ed), *Dilemmas of Development: The Social and Economic Impact of the Porgera Gold Mine 1989–1994*. Boroko: National Research Institute.

——, 2012. 'The Development Forum in Papua New Guinea: Evaluating Outcomes for Local Communities.' In M. Langton and J. Longbottom (eds), *Community Futures, Legal Architecture: Foundations for Indigenous Peoples in the Global Mining Boom*. Abingdon: Routledge.

Finer, M., C.N. Jenkins, S.L. Pimm, B. Keane and C. Ross, 2008. 'Oil and Gas Projects in the Western Amazon: Threats to Wilderness, Biodiversity, and Indigenous Peoples.' *PLoS ONE* 3: e2932. doi.org/10.1371/journal.pone.0002932

Goldman, L., 2009. 'Appendix 26: LNG PNG Social Impact Assessment 2008.' In Esso Highlands Limited, *PNG LNG Project Environmental Impact Statement*. Abbotsford (Victoria): Coffey Natural Systems Pty Ltd.

Golub, A., 2007. 'Ironies of Organization: Landowners, Land Registration, and Papua New Guinea's Mining and Petroleum Industry.' *Human Organization* 66: 38–48. doi.org/10.17730/humo.66.1.157563342241q348

——, 2014. *Leviathans at the Gold Mine: Creating Indigenous and Corporate Actors in Papua New Guinea*. Durham (NC): Duke University Press.

Haley, N. and R.J. May (eds), 2007. *Conflict and Resource Development in the Southern Highlands of Papua New Guinea*. Canberra: ANU E Press (Studies in State and Society in the Pacific 3).

HRW (Human Rights Watch), 2011. 'Gold's Costly Dividend: Human Rights Impacts of Papua New Guinea's Porgera Gold Mine.' Viewed 24 April 2017 at: www.hrw.org/en/reports/2011/02/01/gold-s-costly-dividend

IFC (International Finance Corporation), 2009. *Projects and People: A Handbook for Addressing Project-Induced In-Migration*. Washington (DC): IFC.

———, 2012. *IFC Performance Standards on Environmental and Social Sustainability*. Washington (DC): IFC.

Jackson, R., 2015. *The Development and Current State of Landowner Businesses Associated with Resource Projects in Papua New Guinea*. Port Moresby: PNG Chamber of Mines and Petroleum.

Kemp, D. and J.R. Owen, 2015. 'A Third Party Review of the Barrick/Porgera Joint Venture Off-Lease Resettlement Pilot: Operating Context and Opinion on Suitability.' St Lucia: University of Queensland, Centre for Social Responsibility in Mining.

Kirsch, S., 2014. *Mining Capitalism: The Relationship between Corporations and Their Critics*. Oakland: University of California Press.

Koczberski, G., G. Curry and J. Anjen, 2012. 'Changing Land Tenure and Informal Land Markets in the Oil Palm Frontier Regions of Papua New Guinea: The Challenge for Land Reform.' *Australian Geographer* 42: 181–196. doi.org/10.1080/00049182.2012.682295

Koczberski, G., G. Curry and B. Imbun, 2009. 'Property Rights for Social Inclusion: Migrant Strategies for Securing Land and Livelihoods in Papua New Guinea.' *Asia Pacific Viewpoint* 50: 29–42. doi.org/10.1111/j.1467-8373.2009.01379.x

Koczberski, G., G. Numbasa, E. Germis and G.N. Curry, 2017. 'Informal Land Markets in Papua New Guinea.' In S. McDonnell, M.G. Allen and C. Filer (eds), *Kastom, Property and Ideology: Land Transformations in Melanesia*. Canberra: ANU Press. doi.org/10.22459/KPI.03.2017.05

Lavachery, P., S. MacEachern, T. Bouimon, B. Gouem Gouem, P. Kinyock, J. Mbairoh and O. Nkonkonda, 2005. 'Komé to Ebomé: Archaeological Research for the Chad Export Project, 1999–2003.' Journal of African Archaeology 3:175–193.

Markey, S., 2010. 'Fly-In, Fly-Out Resource Development: A New Regionalist Perspective on the Next Rural Economy.' In G. Halseth, S. Markey and D. Bruce (eds), *The Next Rural Economies: Constructing Rural Place in Global Economies.* Wallingford (UK): CABI International.

Martin, K., 2007. 'Your Own Buai You Must Buy: The Ideology of Possessive Individualism in Papua New Guinea.' *Anthropological Forum* 17: 285–298. doi.org/10.1080/00664670701637743

McGavin, P.A., L.T. Jones and B.Y. Imbun, 2001. 'In Country Fly-In/Fly-Out and National HR Development: Evidence from PNG.' In B.Y. Imbun and P.A. McGavin (eds), *Mining in Papua New Guinea: Analysis and Policy Implications.* Waigani: University of Papua New Guinea Press.

Meggitt, M. J., 1957. 'The Ipili of the Porgera Valley, Western Highlands District, Territory of New Guinea.' *Oceania* 28: 31–55. doi.org/10.1002/j.1834-4461.1957.tb00717.x

Mitjà, O., R. Hays, A. Ipai, M. Penias, R. Paru, D. Fagaho, E. de Lazzari and Q. Bassat, 2012. 'Single-Dose Azithromycin versus Benzathine Benzylpenicillin for Treatment of Yaws in Children in Papua New Guinea: An Open-Label, Non-Inferiority, Randomised Trial.' *Lancet* 379: 342–347. doi.org/10.1016/S0140-6736(11)61624-3

Mitjà, O., W. Houinei, M. Penias, A. Kapa, R. Paru, R. Hays, … Q. Bassat, 2015. 'Mass Treatment with Single-Dose Azithromycin for Yaws.' *New England Journal of Medicine* 372: 703–710. doi.org/10.1056/NEJMoa1408586

Mitjà, O., R. Paru, R. Hays, L. Griffin, N. Laban, M. Samson and Q. Bassat, 2011. 'The Impact of a Filariasis Control Program on Lihir Island, Papua New Guinea.' *PLoS Neglected Tropical Diseases* 5: e1286. doi.org/10.1371/journal.pntd.0001286

Mitjà, O., R. Paru, B. Selve, I. Betuela, P. Siba, E. de Lazzari and Q. Bassat, 2013. 'Malaria Epidemiology in Lihir Island, Papua New Guinea.' *Malaria Journal* 12: 98. doi.org/10.1186/1475-2875-12-98

Nash J. and E. Ogan, 1990. 'The Red and the Black: Bougainville Perceptions of Other Papua New Guineans.' *Pacific Studies* 13: 1–17.

NML (Newcrest Mining Ltd), 2013. *Newcrest Mining Limited Sustainability Report 2012*. Melbourne: NML.

Numbasa, G. and G. Koczberski, 2012. 'Migration, Informal Urban Settlements and Non-Market Land Transactions: A Case Study of Wewak, East Sepik Province, Papua New Guinea.' *Australian Geographer* 43: 143–161. doi.org/10.1080/00049182.2012.682293

Owen, J. and D. Kemp, 2014. 'Mining-Induced Displacement and Resettlement: A Critical Appraisal.' *Journal of Cleaner Production* 87: 478–488. doi.org/10.1016/j.jclepro.2014.09.087

Reed, A., 2003. *Papua New Guinea's Last Place: Experiences of Constraint in a Postcolonial Prison*. New York: Berghahn Books.

Rigg, J., 2007. 'Moving Lives: Migration and Livelihoods in the Lao PDR.' *Population, Space and Place* 13: 163–178. doi.org/10.1002/psp.438

Roberts, J.T., 1995. 'Subcontracting and the Omitted Social Dimensions of Large Development Projects: Household Survival at the Carajas Mines in the Brazilian Amazon.' *Economic Development and Cultural Change* 43: 735–758. doi.org/10.1086/452184

Robinson, R., 2014. 'Restoring Justice Initiative: Contributing to a Just, Safe and Secure Society in the Porgera District. Progress Report 2008–2014.' Unpublished report for the Restoring Justice Initiative.

Ryan, P., 1991. *Black Bonanza: A Landslide of Gold*. South Yarra (VA): Hyland Press.

Sidaway, J.D., 2007. 'Enclave Space: A New Metageography of Development?' *Area* 39: 331–339. doi.org/10.1111/j.1475-4762.2007.00757.x

Smith, R.C., 2005. 'Can David and Goliath Have a Happy Marriage? The Machiguenga People and the Camasea Gas Project in the Peruvian Amazon.' In J.P. Brosius, A.L. Tsing and C. Zerner (eds), *Communities and Conservation: Histories and Politics of Community-Based Natural Resource Management*. Oxford: Alta-Mira Press.

Standing, G., 2011. *The Precariat: The New Dangerous Class*. London: Bloomsbury.

Storey, K., 2001. 'Fly-In/Fly-Out and Fly-Over: Mining and Regional Development in Western Australia.' *Australian Geographer* 32: 133–148. doi.org/10.1080/00049180120066616

Strathern, M., 1975. *No Money on Our Skins: Hagen Migrants in Port Moresby*. Port Moresby: The Australian National University, New Guinea Research Unit (Bulletin 61).

Tawney, R.H., 1932. *Land and Labour in China*. London: Allen and Unwin.

Thibault, M. and S. Blaney, 2003. 'The Oil Industry as an Underlying Factor in the Bushmeat Crisis in Central Africa.' *Conservation Biology* 17: 1807–1813. doi.org/10.1111/j.1523-1739.2003.00159.x

Vail, J., 1995. 'All That Glitters: The Mt Kare Gold Rush and Its Aftermath.' In A. Biersack (ed.), *Papuan Borderlands: Huli, Duna, and Ipili Perspectives on the Papua New Guinea Highlands*. Ann Arbor: University of Michigan Press.

———, 2011. 'Lihir Demographic Summary 31/12/2010.' Unpublished report for Lihir Gold Limited.

12. Bougainville: Origins of the Conflict, and Debating the Future of Large-Scale Mining

ANTHONY J. REGAN

Introduction

The 50-year relationship between large-scale mining (LSM) and local-level politics in Bougainville has been complex and fraught. Bougainville is the only place in the world where host community violence has resulted in the long-term closure of a large-scale mine. Bougainville Copper Ltd (BCL), a subsidiary of Conzinc RioTinto Australia (CRA), operated the huge Panguna copper and gold mine from 1972 to 1989 under a 1967 agreement with the Australian colonial administration of the then Territory of Papua and New Guinea (TPNG). The first large-scale mine in what is now Papua New Guinea (PNG), it closed in 1989, early in a violent conflict that lasted from 1988 to 1997 (Regan 1998, 2011; Braithwaite et al. 2010), and it was still closed in 2016.

Widespread perceptions exist, especially outside Bougainville, that mine closure resulted mainly from generalised Bougainvillean rejection of mining. While several main strands of causal factors have been proposed—including ethnonationalism, culture and class (Regan 1998) and local cosmology (Kenema 2010)—most published accounts see the conflict originating in violent action by young mine-impacted landowners seeking permanent mine closure (e.g. Dorney 1990; Filer 1990; Connell 1991, 1992; Boege 1999; Denoon 2000; Gillespie 2009; Lasslett 2014).

Some observers (e.g. Jubilee 2014; Lasslett 2014) assert that the mine lease landowners continue to be generally opposed to any resumption of mining operations at Panguna, or that Bougainvilleans in general are opposed to large-scale mining. However, since being established in 2005 under provisions of the PNG Constitution implementing the Bougainville Peace Agreement (BPA) of August 2001, the Autonomous Bougainville Government (ABG) has considered reopening the Panguna mine as the most realistic path to achieving the fiscal self-reliance needed for either the autonomy or possible independence contemplated by the BPA, and claims widespread (though not unanimous) community support for this approach.

This chapter examines whether the conflict was in fact a consequence of mine-impacted landowner commitment to permanent mine closure, and whether broad-based opposition to mining—by either mine-impacted landowners or Bougainvilleans more generally—was central to local politics in 2016. The chapter is in three main parts. The first outlines aspects of the context in which the conflict originated that are critical to better understanding of both its origins and some new evidence (advanced here) about several distinct Bougainvillean stakeholder groups other than young mine lease landowners, all deeply involved in the conflict origins. They occupy the 'community corner' in a rectangular model of relationships involving four main sets (or 'corners') of stakeholders, the other three being the state, mining companies, and the 'fourth estate' (Ballard and Banks 2003; also Chapter 1, this volume). The second part focuses on this new evidence, outlining the origins, concerns, roles and goals of these other 'community' stakeholders, and their relationships to one another, as well as to the state and the company. The third part discusses continuities and changes in these relationships since the conflict began, and the extent to which generalised opposition to LSM now exists.

Context of Conflict Origins

Bougainville and PNG

Bougainville's population in 2016 is approximately 300,000 (less than 4 per cent of PNG's total population). Its 9,438 square kilometres is roughly 2 per cent of PNG's total land area. Pre-colonial Bougainvilleans were organised mainly around tiny stateless societies involving great diversity in language,[1] culture (Ogan 2005), and identities (Regan 2005a). Despite major social and economic changes since colonial 'rule' began in the late nineteenth century, the most significant social groups today continue to be nuclear and extended families, the localised clan-based landowning lineages to which those families belong (typically containing 50–150 members), and flexible groupings of such lineages.

While under nominal German colonial control from 1884 to 1915, the first administrative centre was only established in 1905, and Australia took control from 1914 to 1975. Under colonialism, interactions with people from elsewhere in PNG contributed to a pan-Bougainvillean identity, with the dark skin colour of most Bougainvilleans as the primary marker (Nash and Ogan 1990). Identity politicisation occurred after World War II, when:

> because of the natural affluence of their village life and the coverage of the [Bougainville] district by Christian missions (mainly Catholic and non-Australian), the administration neglected to play a conspicuous role in development almost until copper was discovered. Bougainville was known as the 'Cinderella' district not because it was poor but because it was ostensibly neglected (Griffin et al. 1979: 150).

Identity politicisation was intensified by resentment of colonial racism (Ogan 1965, 1971a, 1972) and by development of the mine, which was seen as something imposed to benefit the rest of PNG with little regard to detrimental impacts on Bougainville itself.

A major manifestation of change since 1905 has been the expanding range of groups or organisations to which Bougainvilleans belong or relate (churches, women's groups, local governments, economic enterprises, political parties, etc.). Nevertheless, the autonomy long enjoyed by local

---

1   There were 25 languages and a comparable number of sub-languages or dialects (Tryon 2005).

lineages and other pre-colonial social groupings remains the default position for Bougainvillean understandings of how to relate to these new social phenomena. This expectation of autonomy helps to explain the extent to which the diverse groups involved in the origins of the conflict expected autonomy from one another, as do groups involved in contemporary debates on the future of mining.

For most rural Bougainvilleans, PNG remains remote (Tanis 2005: 468). This was even more so in 1963, when PNG-wide politics first developed around the election of TPNG's first representative legislature, which included just one Bougainvillean representative. Concerns about national representation of Bougainville probably had little effect on voters in the 1963 and 1968 elections (Ogan 1965, 1971a; Anis et al. 1976). However, rapid changes associated with development of the mine led to much wider understanding of such matters in the 1972 elections, contributing to the election of a young Catholic priest from Buin, John Momis, a critic of the mine and the administration, who continues to be a key political figure in Bougainville.

Growing agitation for a special political and financial status saw an interim Bougainville Provincial Government established in 1974 (Ghai and Regan 1992: 55–9), and disputes over its mine revenue share precipitated Bougainville's attempted secession from PNG on 1 September 1975, just before PNG's Independence Day. The crisis was resolved in mid-1976, when the PNG Government agreed to constitutional provision for provincial government and guaranteed that the new North Solomons Provincial Government (NSPG)[2] would receive all of the royalties from the mine aside from the 5 per cent already payable to some of the Panguna mine lease landowners (Bedford and Mamak 1977). Many Bougainvilleans concluded that only intense confrontation with PNG brought results, and that the little understood process of secession (Ogan 1990: 36) and status of independence would remedy many problems.

Following the 1976 agreement to end attempted secession, Bougainvilleans had high expectations of the NSPG. In 1977, John Momis became the PNG minister responsible for the new provincial government system established under the agreement. Strong support for autonomy of the NSPG was now expected from the centre, and these expectations were reinforced by establishment of the Momis-led Melanesian Alliance (MA)

---

2   See Regan (2005b) on reasons for use of the different names, 'North Solomons' and 'Bougainville'.

party in 1980. The MA soon dominated both NSPG politics (especially from 1984) and Bougainville's four seats in the PNG Parliament. But neither Momis nor his party had a significant impact on PNG policy towards Bougainville. By the mid-1980s, the NSPG's lack of expected powers over areas of growing concern, such as mining, land and internal migration, was a source of widespread disappointment (Ghai and Regan 2000, 2006). For many, the failure to pursue secession appeared to have been a mistake.

Mine Revenue Distribution and Landowner Compensation

The social, economic and environmental impacts of mine exploration and feasibility work (1963–69), construction (1969–72) and operation (1972–89) were especially shocking for mine lease landowners but also shocked most other Bougainvilleans on many levels. The 10,000-strong construction workforce—mostly recruited from elsewhere in PNG—was more than 10 per cent of Bougainville's total population at the time. From 1972, the permanent workforce was around 3,500, about 80 per cent of them Papua New Guineans, less than a third of whom were Bougainvilleans (Quodling 1991: 37).

The limited mine employment opportunities were resented, but the far more limited mine revenue share received, not only by mine lease landowners but also by Bougainville as a whole, was of even greater concern for multiple groups involved in the origins of the conflict. Between 1972 and 1989 the four main recipients of mine-derived revenue were:

- mine lease landowners, receiving 5 per cent of royalties, occupation fees and various forms of compensation;
- from 1977 the NSPG, receiving 95 per cent of royalties, some local taxes, and other limited payments;
- the PNG Government, receiving company tax, other tax receipts, and dividends in its capacity as the largest minority (19.1 per cent) shareholder in BCL; and
- private investors (including CRA as the majority BCL shareholder) receiving dividends (see Table 12.1).

Table 12.1 Distribution of Panguna cash revenues, 1972–89.

Stakeholder	Kina (million)	%
PNG Government	1,078	61.46
Private investors	577	32.90
North Solomons Provincial Government	75	4.28
Local landowners	24	1.37
Total	1,754	100.00

Source: Author's table, based on data from Quodling (1991: 34).

The landowner revenue share was a tiny percentage of total revenues, was distributed in ways that were poorly understood by most people, and involved significant sources of inequality.

Development of the mine entailed a number of leases, the main ones being state leases to BCL of land held under customary arrangements by small matrilineages from six different language groups—Torau, Nasioi, Eivo, Nagovisi, Baitsi and Piva—the majority being Nasioi speakers. BCL's three main leases,[3] together covering 13,047 hectares, comprised a 'special mining lease' (SML) for the open-cut pit, processing plant and associated facilities; a 'port-mine access road lease' (PMARL) over land from the east coast to the mine; and a 'tailings lease' over the Kawerong-Jaba river system, covering land from the mine to the west coast. After BCL's land needs were finalised in mid-1969, colonial patrol officers (kiaps) spent much of the next three years walking every part of the lease areas with community members, demarcating the boundaries of the 829 land parcels or 'blocks',[4] of which there were 509 in the SML, 62 in the PMARL and 258 in the tailings lease.

Members of occupying lineages were henceforth generally referred to as mine lease 'landowners', though most had hazy ideas of what the term meant (Tanis 2005). The administration and BCL were just as hazy about precise identification of landowning lineage membership, as nothing akin to what is now known as social mapping was ever done to identify the members of the lineages involved. In part this was because of lack of precedents (this being PNG's first large-scale mine), but also because of pressure from CRA to finalise land demarcation as quickly as possible

---

3   There were also numerous much smaller leases (McNee 2015).
4   Origins of the term 'block' are discussed in McNee (2015: 200–8); see also AGA (1989: 31) and Filer (1990: 11–12).

from mid-1969, resulting in the kiaps involved having no time to compile the group genealogies that they were trained to use whenever acquiring land for the state.

From 1966 until about 1970, many mine lease area lineage members opposed both exploration and mining (Bedford and Mamak 1977; Denoon 2000; Vernon 2005; Brown 2014). Overt opposition ended after the deployment of significant colonial administration resources (kiaps and police), when 'protests and expectations became apparently more futile' (Connell 1990: 28), and after agreement was reached on four main types of compensation payable to affected landowners, in addition to the 5 per cent share of royalties already payable to SML landowners (Bedford and Mamak 1977). Compensation involved:

- payments for damage to land, buildings and crops, most being made during the exploration and construction phases;
- occupation fees paid at a uniform rate per hectare to the 'customary heads' of lineages owning the 829 blocks in the expectation that they would distribute the funds they received to other lineage members in accordance with custom;
- new houses for people whose land in the SML and tailings lease was actually destroyed, with some relocation costs; and
- compensation for fish populations in mine-impacted creeks and rivers, which was paid to customary right holders (known as 'fish-owners').

Acceptance of such payments did not entail landowner acquiescence. When compensation rates were initially determined, the recipients had no concept of the scale of the future impacts. As the enormity of the impacts became evident, community unhappiness grew.

In public hearings between 1969 and 1974, the PNG Land Titles Commission (LTC)[5] determined the 'customary heads'[6] of the landowning lineages and the boundaries of each of the 829 blocks, relying heavily on the prior work of the kiaps. Bedford and Mamak (1977: 71) found that 'approximately 440 registered owners ['customary heads'] … represent at least 2,000 men, women and children'.[7] The number of 'customary heads'

---

5   The files from these hearings are held by the LTC office in Port Moresby. I am grateful to the Chief Land Titles Commissioner, Benedict Batata, for locating and providing access to these records.
6   While 'customary head' was the term used almost universally by the LTC, 'the terms title holder and owners do appear' in the relevant LTC files (McNee 2015: 206).
7   The actual number of people is not known, but was probably over 4,000.

was so much smaller than the number of blocks because a person could be a member and a 'customary head' of lineages occupying more than one block. Amounts paid to 'customary heads' varied immensely, depending on the number and size of the blocks involved (see Table 12.2). The largest blocks were in coastal areas of the tailings lease, the most sparsely populated area within the three main leases, where the single largest block was 960 hectares, 27 were larger than 100 hectares, and almost two thirds of 258 blocks were over 10 hectares. By contrast, in the SML and the PMARL, no block was larger than 100 hectares, and almost three quarters were less than 10 hectares. More than half the total occupation fees for all three leases went to just 114 'customary heads' from 14 villages in the tailings lease. One quarter went to the people of just one relocated village (Bedford and Mamak 1977: 69–76).

Table 12.2 Numbers of blocks in the three main BCL leases.

Area	SML	PMARL	Tailings Lease
< 1 hectare	96	12	16
1–10 hectares	297	40	77
10–100 hectares	116	10	138
100–200 hectares	0	0	20
200–300 hectares	0	0	3
> 300 hectares	0	0	4
Total Blocks	509	62	258

Sources: PNG Land Titles Commission files and BCL compensation and occupation fee payment records.

Under the last compensation agreement between BCL and landowner representatives, signed in March 1986, K604,440.57 in occupation fees and bush compensation was payable annually to 'customary heads' from 1986 to 1989. From this total, K195,738.01 would go to the SML, K20,741.15 to the PMARL, and K487,961.41 to the tailings lease. If this money were distributed equally between the 829 blocks, it would mean an annual payment of K729 per block. An additional amount of K259,687 was to be paid each year to the 'fish-owners', and if this money were distributed equally between their 1,200 'group leaders', then each one would receive an annual payment of K216.40. However, for reasons already discussed, there were great disparities in the amounts received by the 'customary heads' or 'group leaders', and hence in the amounts available to be redistributed to lineage or group members.

Problems arising from the totality of the compensation arrangements undoubtedly contributed to the origins of the conflict. There were at least five issues that contributed to landowners having limited understanding of the arrangements:

- Most lineage members had little exposure to ideas of land ownership that would have made sense of the designation of a particular lineage as the owner of a block, or a particular person as its 'customary head' (Filer 2005; Tanis 2005; Kenema 2010).

- Expectations that 'customary heads' would distribute occupation fees and compensation payments to lineage members in accordance with 'custom' ignored the fact that neither the title nor the role of these individuals was customary, and there were no customary norms about distributing cash compensation paid for land that was used for previously unknown purposes such as industrial mining (Filer 1990).[8]

- The lack of 'social mapping' to identify the social and leadership structures or membership of 'landowning' lineages meant that there was no certainty about who might be entitled to a share of the payment for each block, leaving 'customary heads' with much discretion in distribution, some being renowned for favouritism to particular relatives while largely ignoring others with similar standing in the lineage.

- Most lineage members had limited knowledge of either English or Tok Pisin, which limited their understanding of the LTC process, which made it possible for people to be designated as 'customary heads' when this designation might otherwise be contested.

- When compensation rates were determined, few lineage members had any idea of the scale of mine impacts. What initially seemed reasonable was often questioned by the time the mine was operating.

These sources of limited understanding amplified the impacts of an additional set of five problems related to actual or perceived inequality in compensation arrangements:

- The payment of uniform rates per hectare took no account of the scale of impacts in different locations. Some communities suffered far more than others due to land destruction (by the pit or the tailings) and

---

8   This may help to explain why the term 'customary head' was soon almost universally replaced by the term 'title-holder' in virtually all contexts except LTC documentation.

village relocation.[9] Small block size and higher population densities in the SML and upper parts of the tailings lease meant that small amounts were distributed to many more people than in lower parts of the tailings lease.

- Secondary land rights were ignored, although they could be significant, and lack of compensation for their loss caused resentment, especially where land was destroyed.

- Some 'customary heads' treated payments as being made to them as individuals, with no need for distribution to other lineage members or other persons who believed they had rights to the block in question. This was partly a consequence of 'customary heads' becoming generally known as 'title-holders', which led some to assume they had individual ownership rights. However, individuals rather than customary groups could sometimes have semi-exclusive rights to particular blocks under custom (Ogan 1971b; Mitchell 1976: 22), and most 'customary heads' registered for the largest blocks in the tailings lease fell into this category.

- No account was taken of rapid population growth and the numbers of young people reaching adulthood, resulting in gradual but inexorable diminution in the size of average available payments, with little reaching most young adults. Combined with other problems, the extent of the discretion of 'customary heads' caused seriously unequal intergenerational distribution.

- The limited compensation rights of communities outside but adjacent to BCL leases, whose members often experienced significant mining impacts, generated a particular sense of injustice.

Impacts and resentment were greatest amongst SML communities, due to the significant proportions of people relocated, the small average block size and high population density, proximity of the mine pit and its noise and dust, and the presence of many outsiders on their land. Many SML

9   Relocation housing was neither of high standard nor maintained by BCL, and lacked provision for newly married couples, causing significant overcrowding and cultural stresses through the breach of post-marriage residence rules. Relocated SML villages lost all or most of their gardening land. Residents of villages near the mouth of the Jaba River were relocated to land over which they had few rights, resulting in ongoing tensions with the original owners. Such problems continue today, exacerbated by the absence of compensation payments since 1990, and by rapid population growth in both settler and host communities (personal communications, Albert Magoi, Ben Paula and Bernadine Kirra, July 2011).

community members had the mine or its facilities in plain sight, and access to all SML villages and hamlets was by means of a road through the lease.

By the late 1970s, landowner concerns grew, and BCL's slow responses caused anger. In 1979, landowners looted the Panguna supermarket, and their first representative body, the Panguna Landowners Association (PLA), was established. In 1980, landowners marched on BCL's Panguna offices and erected roadblocks, causing temporary mine closure. Despite pressure from PNG's police minister to deploy police mobile squads to clear the roadblocks, Ila Geno, then commander of the New Guinea Islands police region (including Bougainville and four other provinces) refused to do so. Instead he travelled from his base in East New Britain Province to work with the NSPG to negotiate with landowner leaders, including Francis Ona, to remove the roadblock (personal communication, Ila Geno, July 2014). These events contributed to pressure for a new compensation agreement between BCL and the PLA. Signed later in 1980, it provided for higher compensation rates, indexing of occupation fee rates, regular reviews, and a new compensation category called 'social inconvenience compensation' (SIC). The SIC was to be paid into a new landowner-controlled entity, the Road Mine Tailings Lease Trust Fund (RMTLTF), which was required to utilise the income for landowner benefit (Okole 1990; Oliver 1991: 153, 203–4; Connell 1992: 38–43; Griffin and Togolo 1997: 376). Under the subsequent 1986 Agreement, the annual SIC payment was K393,838.69.

Hopes that the 1980 Agreement would improve the landowner situation were not met. The new landowner bodies, the PLA and RMTLTF, became sources of new problems. Unresolved concerns about unequal impacts and compensation were exacerbated by control of both bodies by a few 'customary heads', all older generation males. Emerging younger generation leaders complained that the PLA failed to pursue landowner concerns with sufficient vigour.

In the 1980s, younger generation leaders from SML communities, including Perpetua Serero and Francis Ona, had particular reasons for intense resentment due to issues of the kind previously outlined. Amongst other things, Ona and Serero were from a village close to the mine pit, in an area with relatively high population density, with land rights derived from the male line rather than the usually more significant female line

(Ogan 1990: 37). More important, what rights they had were to several blocks where an uncle, Matthew Kove, was registered as the 'customary head', and ignored their claims to shares in land compensation.[10]

The RMTLTF board initially made low-interest loans to landowners, but limited funds soon saw complaints of inequitable distribution. In 1983, Severinus Ampaoi, a landowner leader and senior BCL employee, became chairman of the board. Substantial bad debts were written off and a stricter financial regime introduced, with an end to the distribution of immediate benefits to landowners. An accumulation strategy was adopted to provide a basis for long-term benefit distribution.[11] Shares were purchased in Bougainville Development Corporation (BDC), the successful business arm of the NSPG, in circumstances of great controversy. Funds were also invested in urban real estate and two plantations. Close cooperation with the BDC board contributed to perceptions that the RMTLTF board was part of a remote, wealthy Bougainvillean elite. There were mounting accusations of mismanagement, of directors 'eating' the money, of a lack of spending on landowner social amenities, of irregular board meetings, of a lack of information and of the board being dominated by 'outsiders' (Connell 1992: 42–5; Thompson and MacWilliam 1992: 36; see also Okole 1990; Turner 1990; Oliver 1991: 203–4). The central problem involved perceptions that a compensation category intended to benefit landowners in general now mainly benefited older men who were already wealthy and were too close to BCL.

The Bougainvillean Elite

A small Bougainvillean elite, mainly involved in copra production, first emerged in the 1950s (MacWilliam 2005), growing from the 1960s as smallholder cocoa production expanded. From the 1970s, burgeoning cocoa production contributed to land pressures and growing economic inequality (Mitchell 1976, 1982; Connell 1978, 1988; Tanis 2005: 456–9; MacWilliam 2013). Unequal compensation from the mine created another wealthy group (Tanis 2005: 455), as did establishment

---

10   Dorney (1990: 123) and Jubilee Australia (2014: 20) suggest that designation of males as 'customary heads' was contrary to custom. In fact, well over half of those so designated were females, and as males are members of the lineages recognised as occupying land, and may well also have individual rights to land under custom, there is no reason in principle why males should not have this designation.

11   The mine was then expected to close around the year 2000.

of mine-associated businesses. Growing inequality fuelled resentment in societies until recently highly egalitarian (Ogan 1990: 36; Regan 1998: 271–2, 274–6).

A major source of disquiet was the BDC board's response to a 1985 request from the newly elected, MA-dominated NSPG for most board members to resign. The board members included the recently ousted premier, Leo Hannett, and others close to him. The board response was to issue new shares, diluting NSPG's 57 per cent equity to 29 per cent. Moreover, without any offer being made to the majority shareholder, the new shares were allotted to six corporate entities and an individual. The new shareholders included the family companies of board members, the RMTLTF, BCL's superannuation fund, and the 'Severinus Ampaoi Group'. While BDC shares had previously been worth K9, the new shares were priced at K1.30, which entailed a major loss for the NSPG but a windfall gain for the new shareholders (Griffin and Togolo 1997: 375). The NSPG challenged this action in widely publicised National and Supreme Court proceedings. Although eruption of the conflict in late 1988 meant the case was never finally determined, a 1988 Supreme Court ruling[12] held that the NSPG 'had, prima facie, good grounds for challenging the share issue' (see Griffin and Togolo 1997: 374).

A widespread perception developed of joint action between BDC and the RMTLTF to enrich board members at the expense of other Bougainvilleans (Thompson and MacWilliam 1992; Wesley-Smith 1992a). The involvement of BCL in the BDC share issue (through its superannuation fund) and allocation of BDC shares to the RMTLTF and the Ampaoi family completed the picture. Many people, particularly MA supporters, saw the share issue as the culmination of BDC's general performance, allegedly hijacking lucrative business opportunities and blocking local Bougainvillean small and medium-sized business participation in spin-offs from the mine (Griffin and Togolo 1997: 374). As a result, BDC was 'seen in the village as a private business benefitting a few individuals', becoming 'an object of exactly the same kind of anger and resentment' as the PLA and RMTLTF (Tanis 2005: 466). 'In the popular mind, grievances against BDC, the directors of the landowners' trust fund [RMTLTF] and BCL (which had been patron of and provided advice to BDC) became intertwined' (Griffin and Togolo 1997: 375).

---

12 *North Solomons Provincial Government v Bougainville Development Corporation Ltd* [1988] PGSC 26, 21 December.

As we shall see, the MA's trenchant mid-1980s critique of Bougainvillean 'big business' was to be echoed strongly in 1988 statements and demands by Francis Ona and the 'new' PLA.

Population Growth and Youth Unemployment

Bougainville's population grew rapidly after World War II, from about 41,000 in 1950 (Nelson 2005: 196) to about 73,000 in 1966 (Laracy 2005: 125), about 109,000 in 1980 (Oliver 1991: 160–1), and about 150,000 in 1986. This means an annual growth rate of between 3.4 and 3.5 per cent (NSPG 1986a; Oliver 1991: 161). Much of Bougainville's land is unsuitable for intensive agriculture (Oliver 1991: 10–12). In the 1980s, more than 80 per cent of Bougainvilleans were concentrated in the limited areas of accessible arable land, heavily reliant on subsistence farming, with some cash crop cultivation. Both population growth and cocoa expansion from the 1960s depleted forest and other natural resources in proximity to the most densely populated areas and intensified local land shortages (Connell 1978, 1988; Mitchell 1982; Tyler 1988; Lummani 2005; MacWilliam 2013: 144–9, 184–6, 193). Economic inequality grew in many relatively densely populated parts of central and southwest Bougainville, including the Nasioi, Siwai, Nagovisi and Buin language areas (Mitchell 1982: 8; Connell 1988: 84–5; Tanis 2005: 456–7).[13]

In the mid-1980s, more than 47,000 Bougainvilleans were aged from twelve to twenty-five years and classified as 'youth'; 68 per cent of them were living in rural villages, 9 per cent in other rural settlements, and 23 per cent in urban areas (NSPG 1986a). Twenty-five per cent of these young people had no formal education, 18 per cent had some primary schooling up to grade 5, and 39 per cent only to grade 6. Only 18 per cent had completed any high school grade. Further training or education opportunities were limited, as was wage employment. Only 19 per cent of young people were in formal employment, another third being involved in agriculture. Bougainvilleans enjoyed little preference in mine-related business and employment opportunities. Young males with little or no

---

13 Pre-colonial migration into parts of the Siwai area, which had been caused by localised conflict, resulted in the presence of people who still had limited rights to land in the 1980s (personal communications, Michael Kouro 1999, Marie Tyler 2011, and Jonathan Ngati 2003). Some of them rented land for subsistence gardens and cash crops (Tyler 1988: 18), while others emigrated to seek opportunities elsewhere.

formal education faced particular problems, many having little access to land for cash cropping (Regan 1998: 274–6). Those with some education, but not enough to gain employment, often had unrealistic expectations and experienced difficulty fitting into village life (NSPG 1986a: 28–9).

Resentment of Outsiders

An unknown but significant number of residents in pre-conflict Bougainville (perhaps 20,000 out of a total population of about 150,000) originated from elsewhere in PNG. Several thousand worked as labourers on Bougainville's mainly foreign-owned copra and cocoa plantations (Lummani 2005: 253), often staying on in Bougainville when their contracts expired. Similarly, some of the 10,000-strong mine construction workforce stayed on after construction ended (Mamak et al. 1974: 11–12). Others came later, attracted by Bougainville's economic opportunities.

In the closely linked urban areas largely servicing the mine—Arawa, Panguna, Kieta and Toniva—and in 'squatter settlements' in both nearby rural areas and parts of the mine lease areas (Tanis 2005: 467), the numbers of 'outsiders' grew rapidly from the 1970s. An April 1988 household survey (NSPG 1988a) showed a population of 17,314 (many from elsewhere in PNG) in 'formal' housing areas of Arawa and nearby urban areas, and 4,259 (most from elsewhere in PNG) in nearby 'informal settlements'. Connell (1990: 48–9) reports a population of about 3,500 for Panguna, and the proportion of Bougainvilleans in the towns as being 'barely more than a third of the total'. A 1986 report discussed the many problems with informal settlements that 'caused … conflict with the local people which in turn puts greater pressure on the Government to take action' (NSPG 1986b).

Many Bougainvilleans blamed 'outsider' squatters for increased crime and land disputes, and for taking over the economy, including formal sector jobs, local fruit and vegetable markets, and small-scale mine-related and other businesses (NSPG 1986b, 1988a; Tanis 2005: 466). In the 1988 NSPG election, a non-Bougainvillean was elected to the provincial assembly by votes from the growing concentration of non-Bougainvilleans. In his home area of Nagovisi, James Tanis (2005: 466) reports worries that this showed outsiders 'were taking over the government, and would control power, protect the squatter settlements and threaten customary land ownership'.

In March 1988, in response to such concerns, the NSPG announced a joint operation with the police force, known as Operation Mekim Save ('Make Them Take Notice'). However, this did not lead to the expected expulsion of unemployed non-Bougainvilleans and destruction of squatter settlements. Instead, an NSPG report talked of PNG constitutional rights protecting outsiders (NSPG 1988a). Tanis describes how frustrated villagers felt that the right to freedom of movement under the PNG Constitution:

> was now being exercised by outsiders at the expense of the rights of Bougainvilleans to the freedom to enjoy their customary land ... Whereas the category of 'threatened' landowners was once restricted to those living in mine lease areas, increasingly there was a sense in which all Bougainvillean landowners felt threatened by, and had grievances against, outsiders (Tanis 2005: 466–7).

Church Allegiances

The three Christian churches with earliest access to Bougainville gained significant influence there. The Catholics arrived first in 1901, followed by the Methodists in 1917 and Seventh Day Adventists in 1924 (Oliver 1991: 58–64). By the 1960s, the Catholics had 'converted up to 80 per cent of the people ... with the Methodists—later the Uniting—Church 15 per cent and the Seventh Day Adventists five per cent' (Griffin 1995: 10). All three main churches combined spiritual 'formation' with provision of health and education services. An almost complete lack of government services in most areas of Bougainville contributed to commonly expressed grievances about colonial neglect.[14] Catholic Church activity was especially pervasive: an amazing 30 stations operated across Bougainville in 1966 (roughly one per 2,000 Catholics), most with associated schools and clinics (Laracy 2005: 125; see also Oliver 1991: 58–68).

Catholic education was partly directed to training local priests in an elite seminary system. Though most of the trainees never became priests, they constituted a linked and articulate elite, familiar with Catholic social teachings. While the Methodist and Seventh Day Adventist churches were generally supportive of the colonial administration, the Catholics were more independent. From the early 1960s, they sponsored economic

---

14   There were no government schools at all until the 1950s.

development activities, initially largely in reaction to the grievances of the Buka-based Hahalis Welfare Society, which accused the church of neglecting the material well-being of its adherents (Rimoldi and Rimoldi 1992; Laracy 2005: 131). Some priests publicly criticised the administration's treatment of 'natives' in relation to forestry and other projects (Fingleton 1970; Laracy 2005: 131).

In a series of press releases, church publications and a strong letter to the colonial administrator, all in 1966, the head of the Catholic Church in Bougainville, Bishop Lemay, criticised the administration's actions in the early stages of mine exploration, describing colonial mining law as 'unsuitable and as contravening native land customs and traditions' (Lemay 1966, cited in Laracy 1999: 583). This stand caused considerable concern in the colonial administration (Downs 1980: 350–2; Laracy 2005).[15]

In 1975, Lemay's successor, Gregory Singkai, the first Bougainvillean bishop, supported secession, saying 'we realised that most of the people involved in the independence movement … are Catholic and we decided it is better for us to be with them as Catholics and have influence than be on the outside' (Woolford 1975: 208; see also Griffin et al. 1979: 216–17). By this time, however, the Methodist and Seventh Day Adventist churches also supported secession. This consolidation of church support for Bougainvillean political initiatives helped to legitimise broader demands arising from the politicisation of Bougainvillean identity.

The Impetus for Conflict

While there was no planning for the major violent conflict that occurred from 1988, the elements of the situation just surveyed combined to contribute to a widespread sense of grievance amongst many Bougainvilleans. They resented a remote Port Moresby–based government for the imposition of the mine, for the limited share of mine revenue

---

15 Amongst the vocal clergy at the time was an American priest, Father Bob Wiley, then based at Tunuru, near Arawa. He was the parish priest at Panguna when the conflict began in 1988 and, in 1991, he claimed to have warned BCL 'big men' in the 1960s that it would be the then schoolchildren who would fight back later, when they grew up. He quoted one such schoolboy, Philip Takaung, who later told him that that the church personnel 'were the only ones who told us the truth … no one could believe that the mountain was going to be removed' (Wiley 1991, cited in Laracy 2005: 132). In 2016, Philip Takaung was the vice-president of the Me'ekamui Government of Unity, based at Panguna.

received by Bougainvilleans, for the influx of people from elsewhere in PNG, and for the problems that phenomenon was thought to be causing. The mine contributed to growing economic inequality, which was a particular source of tension in a situation of rapid population growth, where a high proportion of young adult males had little or no education and few opportunities for economic advancement, and where outsiders were seen as taking opportunities that should have been provided to Bougainvilleans.

Mine lease landowners—in particular younger adults—had even stronger feelings of grievance: about inadequate and unfairly distributed compensation, unfair access to mine employment and business opportunities, and environmental and related impacts. The rejection by younger adult leaders of key figures amongst their older compatriots resonated with concerns amongst many other Bougainvilleans about growing inequality and an emerging Bougainvillean elite. Grievances were articulated by leaders emerging from a small Catholic Church–educated elite. Church sympathy for landowner grievances gave strong legitimacy to their cause. While the NSPG and Bougainvillean MPs in the PNG parliament were strongly sympathetic to Bougainvillean concerns, they were powerless to deliver change. This combination of circumstances provided fertile ground for the emergence of an alternative leadership voicing the grievances and demands of various elements of the Bougainvillean population.

Conflict Origins: Groups, Concerns and Agendas[16]

The sequence of events leading to mine closure in 1989 began with the theft of BCL explosives, damage to mine offices, and destruction of pylons carrying electricity to the mine from the coast. These were widely reported

16   The account in this part of the chapter draws on multiple sources, including many documents from the 1980s held in my files; personal communications over the past 20 years from key actors; and my observation of, or participation in, discussions amongst Bougainvillean leaders during preparations for, and conduct of, negotiations about the content and implementation of the BPA. Personal communications of particular importance came from Theodore Miriung (1994–96), Stephen Monei (1994–96), Joseph Kabui (1998–2008), Cornelius Besia (1999–2016), Jonathan Ngati (1999–2005), James Tanis (1999–2016), Damien Dameng (2002–05), Dennis Kuiai (2003–16), Albert Magoi (2011–12), William Mungta (2011), Ben Paula (2011–16), Lawrence Uakai (2011–14), Eddie Mohin (2013–16) and John Amuna (2015).

as the actions of 'young generation' mine lease landowners—a view echoed in most subsequent accounts of the conflict origins. The 'young generation' mainly comprised adults under forty years of age, who were too young to have participated in the 1980 compensation negotiations.

Key young generation leaders included the subsequent Bougainville Revolutionary Army (BRA) leader, Francis Ona, and his cousin-sister, Perpetua Serero, respectively secretary and chair of the 'new PLA' (NPLA), which emerged from wider attempts to wrest PLA control from older generation leaders in August 1987. The NPLA's objectives are typically described as being to 'close the mine and arrest social disintegration' (Denoon 2000: 196). However, accounts emphasising the concerns and objectives of young generation leaders are almost invariably unbalanced. They ignore the extent to which these leaders represented the concerns of other members of their communities, including older generation leaders who had long been pressuring BCL about compensation issues. They also ignore the extent to which key motivating grievances arose from internal community conflicts only partially related to compensation-related issues, and significant linkages between young landowner leaders and the NSPG, the Bougainvillean MPs, and leaders of other Bougainvillean groups with a variety of grievances and objectives, only some of which were related to the mine.

Mine Lease Landowners: 'Young' and 'Older' Generations

As we have seen, Ona and Serero had a particular grievance originating in family disputes from long before mining started, whose significance is widely recognised in SML communities but otherwise little known. In 2014, a relative of Ona published an account of the issues involved (Roka 2014: 23–6), describing how Ona and his family (siblings and maternal first cousins) resented the unfair distribution of compensation by their uncle, Matthew Kove, a key landowner leader from Guava village. The source of the conflict was the control of land belonging to the lineage of Ona's grandfather's wife, Hali. Control of this land had passed through the matrilineal line to Hali's daughter from a relationship prior to her marriage to the grandfather. That daughter was Kove's mother. The result was that most of the descendants of Hali's marriage to the grandfather were largely excluded from decisions about their lineage land. They believed that this had enabled Kove to be designated as 'customary head' for five blocks, including one of the largest, in the SML.[17] Roka describes how Kove and his immediate family became wealthy through the distribution of compensation at his discretion, ignoring Ona and his relatives. The latter pursued:

> their education with what little they earned from the sale of vegetables. When they asked Matthew Kove for financial help, he was known to burn banknotes before their eyes (Roka 2014: 24–5).

Kove had influential roles in the PLA and RMTLTF, contributing to Ona's negative views of other leaders involved in control of those bodies. Roka describes how:

> Serero, Ona and some other younger people began political sabotage to topple Matthew Kove and his cronies, who they claimed were corrupt and not landowner-oriented. Nearly all executives in the PLA had become rich men with high standards of living whilst the landowners felt they were in a backwater subjected to environmental pollution from the mine and harassment and exploitation by the rising population of Papua New Guineans brought to Bougainville as mine-workers. The Ona-Serero group's call for change around 1986–87 did not produce any results and its members rebelled … in 1987 (Roka 2014: 25).

At the same time, Kove was a respected landowner leader and a signatory to the 1980 and 1986 compensation agreements between BCL and the PLA. His views on mining impacts and compensation—similar to those of the young generation leaders—were expressed to BCL in his capacity as a PLA executive member throughout the 1980s (Auna 1989: 10). In May 1987, together with other PLA leaders, including Ona (Griffin and Togolo 1997: 377), Kove attended a meeting with BCL's managing director at which he supported the 'Bougainville Initiative' adopted by John Momis. The speeches made at this meeting 'were very heated and contained threats to the effect that BCL could not expect to continue to conduct enormously profitable mining operations without a major redistribution of wealth amongst local … landowners' (Auna 1989: 17).

Even in late 1988, well after the emergence of the NPLA, Kove appeared to support preparations for action against BCL when he attended a customary ceremony conducted by senior Nasioi men from the SML area to 'spiritually' prepare the small group that Ona was organising to begin such action (personal communication, Albert Magoi, July 2011). Feelings

---

17 I am grateful to Greg McNee for identifying the blocks in question.

against Kove amongst his young relatives were so strong, however, that his was probably the first death arising from the conflict. He was killed by Ona's close relatives, in what was described as a 'traditional' manner, in January 1989 (Oliver 1991: 210).

The roles of Ona and Serero in the origins of the conflict have been variously characterised, but the most remarkable—and deeply misleading—depiction is that of 'young activists at the vanguard of ... [an] anti-capitalist movement' challenging the 'old' PLA, and doing so 'to operationalise their [radical] social agenda' (Lasslett 2014: 49). There is in fact little credible evidence of their being so motivated, and much indicating that, in their cases, the impacts of common landowner grievances were dramatically intensified by personal grievances.

Ona certainly played a key leadership role in mobilising pressure against BCL, the PNG Government and the NSPG, being well equipped to do so because of the force of his personality and his own certainty about the rightness of whatever he did. Many close to him say that he was also remarkably stubborn and often closed to alternative views. When faced with criticism, he tended to blame others. He also drew inspiration from Catholicism. From at least late November 1988, when he took to the bush with a small support group to avoid capture by police mobile squads deployed in response to the damage then being done to BCL facilities, Ona focused heavily on prayer. His associates were required to join him, whether they were Catholic or not. He had a particular devotion to the Blessed Virgin Mary (Hermkens 2007), and told associates that he received guidance from both Mary and God.

The Ona–Serero group rebellion must also be seen in the context of an earlier and wider mobilisation of young landowners, which is evident in Joseph Kabui's early career. A young generation landowner from the upper part of the tailings lease, Kabui returned to Bougainville from the 'senior' Catholic seminary in Port Moresby in the late 1970s, and became an organiser of the Bougainville Mine Workers Union (BMWU). His relative, Michael Pariu, who played key roles in establishing and then leading both the PLA and the RMTLTF, was elected to the NSPG assembly in 1980 to represent the Ioro constituency, which included both SML and tailings lease areas. Growing dissatisfaction with the PLA and RMTLTF leadership on the part of young generation landowners led Kabui to campaign on their behalf and defeat Pariu in the 1984 election (personal communication, Joseph Kabui, October 1998).

Kabui then supported Ona, Serero and others who (unsuccessfully) sought to extract more benefits for the younger generation from the PLA and the RMTLTF (Auna 1989: 18). Both Kabui and Momis supported them in the August 1987 PLA special general meeting of about 100 landowners that ousted the older leaders and elected a new executive, with Serero as chair and Ona as secretary. Members of the old executive rejected the outcome, insisting that they remained in control. To the frustration of the young leaders, BCL continued to recognise the old executive. Communication between the NPLA and BCL was initially limited to letters seeking company recognition, but in March 1988 it shifted to increasingly strident demands against BCL.

Despite the predominance of younger landowners, the NPLA did work closely with some key older generation leaders from both mine lease and adjacent areas. One of the most significant, Damien Dameng, was considerably older than most of the older PLA leaders. Though he came from an area just outside the mine lease boundaries, he had a significant following, held strong views on mining, and participated actively in most of the NPLA's written communications with BCL and actions against the company.

The concerns and demands of the NPLA were expressed from March to November 1988 in a variety of ways—through speeches made and a petition handed to BCL at a demonstration on 11 March 1988, through a series of letters to BCL, the NSPG and the PNG Government, and through demands made in at least nine meetings between those various entities. Copies of the petition and letters, and of minutes from several of the meetings, still exist, but difficulties arise in their selective use. First, NPLA demands varied over time, reflecting the concerns and goals of associated groups, not just those of young landowners. Second, 'outrageous demands are part and parcel of the Melanesian Way of doing politics' (Filer 1992: 441). Finally, many of Ona's letters were prepared by others whose particular concerns and personal styles could influence the content—a factor contributing to some of Ona's more radical assertions (personal communications, Lawrence Uakai and James Tanis, 2013).

As for their concerns, a useful summary appears in the minutes of a 26 May 1988 meeting between NSPG and PNG Government officials and NPLA representatives (NSPG 1988b; AGA 1989; Connell 1991). Six main categories were identified. Significantly, the first was 'environmental', involving loss of the previously 'happy life' of the people due to the destruction of land and environment. The second was 'relocation', involving inadequacy and lack of maintenance of the houses, which were by now becoming health risks, lack of housing for newly married couples, lack of garden land for relocated villages, inadequate septic tanks and water supplies, and lack of electric power. The third was 'business opportunities', involving BCL's use of mainly foreign contractors, and its involvement in non-mining businesses that unfairly competed with local businesses. The fourth was 'compensation', relating to outstanding payments for a particular village. The fifth was 'services', involving poor and unsafe roads for village access, and poor standards of health and education facilities in mine lease areas. And the sixth was 'inconvenience', involving the presence of squatter settlements in mine lease areas.

The demands made in this document included:

- dramatic increases in both compensation to mine lease landowners and mine revenue shares of landowners and NSPG;
- reimbursement to the NSPG of 50 per cent of mine revenue generated since mining began in 1972;
- priority in mine-related employment and contracts for landowners;
- much improved roads and services for landowners from both BCL and NSPG;
- much improved village relocation arrangements; and
- replacement of RMTLTF by NPLA as recipient of BCL SIC payments.

Both the concerns and the demands were broadly consistent with issues that had been raised by older generation PLA leaders in their 1980s engagement with BCL, including the demands made when Momis presented his Bougainville Initiative to BCL May 1987.

An earlier petition dated 11 March 1988, signed by Ona, Serero and Damien Damen (Dameng), included a demand for mine closure within 28 days (PLA 1988). It was in a letter to BCL dated 5 April 1988, signed by Ona and Serero, that a demand for payment of K10 billion (then equivalent to far more than US$10 billion) was first made. In that letter, Ona and Serero stated that mining could continue if this demand were met, and if BCL:

- paid 50 per cent of its profits to landowners and NSPG;

- consulted the NPLA before any mine expansion; and
- became a company owned by landowners and other Bougainvilleans within five years.

Informants close to Ona at the time advise that demands for mine closure were intended to pressure the PNG Government and BCL to negotiate new terms for mining. Permanent mine closure could not have been a serious demand when claims for additional payments, employment and contract preferences could only be met if mining continued. Nor can the K10 billion demand be seen as an indirect demand for mine closure. Then NSPG premier, Joseph Kabui, described this and related demands as 'ways my people are trying to express their fear and the feeling of total loss of their environment and lifestyle' (Kabui 1988, cited in Oliver 1998: 462–3). Moreover, by the time Ona started to demand secession from PNG, as in some of his statements in November 1988, he clearly envisaged the mine becoming independent Bougainville's main revenue source (Oliver 1991: 208)—a position he advanced in statements to BRA leaders as late as 1993 (personal communication, Eddie Mohin, July 2014).

Furthermore, the demands made by young landowners did not involve the repudiation of deals made previously by the older generation, as argued by Filer (1990). Rather, they involved a rejection of the unfair revenue distribution under the Bougainville Copper Agreement (BCA) and the 1980 and 1986 compensation agreements, and of the extent to which a small fraction of the older generation seemed to benefit from those arrangements. Although the old PLA leaders agreed that the arrangements were unfair, the NPLA leaders blamed the old leaders for both enriching themselves under these arrangements and failing to be sufficiently active in pursuit of change.

Even when Ona and his small support group went into the bush in late 1988, he remained open to negotiation.[18] Early in December 1988, Momis and Kabui met Ona in his 'bush camp' on several occasions, seeking to persuade him to meet a new national government ministerial committee dealing with the emerging crisis. Ona agreed to a meeting in Guava on 8 December. At this stage, the mine had closed due to the destruction of a power supply pylon on 5 December. The tense meeting was attended by

---

18   This account draws on personal communications from Joseph Kabui (1998–2005) and John Momis (2014–15), copies of minutes and reports of the meeting held in my files, and a national newspaper article (Anon. 1988).

four national government ministers, including Momis and Deputy Prime Minister Akoka Doi, representatives of the NSPG, including Kabui, and numerous landowners. Ona spoke of landowner concerns, insisting on renegotiation of the BCA and the withdrawal of police mobile squads. When the deputy prime minister agreed to both demands, Ona agreed to the mine reopening. Ona and other key figures proceeded to a hotel on the small island of Arovo, near Kieta, to sign the new agreement. On leaving that event, however, and contrary to the directions of the deputy prime minister, police arrested key landowner representatives. Ona felt betrayed by the national government, violent conflict intensified and negotiations ceased for some months.

In summary, there is no clear evidence that Ona was committed to either permanent mine closure or secession just before, or in the early stages of, the violence. Rather, the evidence suggests that he was mainly seeking to renegotiate the terms and conditions of mining.[19] More generally, there is no evidence that Ona planned the violent conflict that actually occurred.

Young Mineworkers

The key roles of young Bougainvillean mineworkers (generally under thirty years of age) in the origins of the conflict have drawn little attention from commentators, but have long been acknowledged by other Bougainvillean actors. BRA leader Sam Kauona has described how Ona 'organised a core group comprising concerned Bougainvilleans … working in the mine with him' (NZine 2000a). James Tanis, briefly employed by BCL during the university holidays in late 2008, reports that young mineworkers from all over Bougainville made 'bitter complaints that BCL was favouring the non-Bougainvilleans and giving limited opportunities to Bougainvilleans' (Tanis 2005: 467). Many of those involved later reported concerns that BCL unfairly ignored unemployed age-mates in their home areas.

The BMWU represented all of BCL's PNG citizen employees (Hess and Maidment 2014; Hess et al. 2016). As the majority were from elsewhere in PNG, young Bougainvillean mineworkers felt that the union could not

---

19   Writing early in 1990, anthropologist Gene Ogan reports the receipt of a letter from Bishop Singkai saying that Ona's main aim, even after the conflict developed, was 'to keep the Mine closed, thus punishing the Papua New Guinea Government that has taken so much from the landowners and the Province without putting anything back in return' (Ogan 1990: 39).

represent their particular concerns, and they had little in common with most of the older Bougainvillean employees who accepted the company's policies too readily.

About 50 young men from all over Bougainville met irregularly for 18 months before the conflict began to discuss their grievances. A core group of seven or eight, led by Ona and later known as the BRA 'founders', met more regularly, gradually taking on planning and organisational roles. Members of the larger group made contributions to a fund intended to support action against BCL. Early in 1988, one was sent to Honiara with several thousand kina to buy weapons for use in a proposed rampage through the mine. Both he and the funds disappeared. Subsequently, a group of workers and landowners waited some days at Marau, on Bougainville's southwest coast. A submarine supposedly organised by a relative of Ona's, then studying overseas, was expected to deliver weapons and ammunition, but they never arrived. In both instances, the aim was to force negotiations with BCL and the government.

When these efforts failed, a core group of young landowners joined the 'founders' in organising the theft of BCL explosives. Power pylons were targeted because they were not protected by BCL security guards, and serious damage to even one of them would cause temporary mine closure. A series of joint operations destroyed a succession of pylons over several months from early December 1988. To plan and support those operations, and to avoid suspicion of their involvement, most of the 'founders' continued in employment with BCL until the mine closed in May 1989.

The young mineworkers sought temporary mine closure to force negotiations on new mining conditions that would be far more beneficial to Bougainvillean interests, including those of BCL employees. They saw no need to make public their particular concerns as they assumed that, once negotiations began, interests such as theirs could be joined with those of the young landowners at the negotiating table. Initially, they sought neither secession nor permanent mine closure. It was police mobile squad violence (from December 1988), followed by that of the PNG Defence Force (from April 1989), which provided the basis for leaders of other Bougainvillean groups to mobilise support for other goals, notably secession.

The Arawa Mungkas Association

Although Ona was on the executive of this important Arawa-based organisation, its major focus was on the impacts of the influx into Bougainville of people from elsewhere in PNG and the limited mine-related opportunities for Bougainvilleans in general. The name, Mungkas Association, originated with Bougainvillean students at the University of PNG who met to discuss Bougainville's political future in the late 1960s (Griffin et al. 1979: 210–12). Various manifestations subsequently emerged, in Port Moresby and other urban centres, as well as in Bougainville, to address specific problems. The Arawa Mungkas Association, set up in 1987, comprised representatives of communities from many parts of Bougainville then resident in mine-related urban areas. During 1988, its members held public meetings and demonstrations to express grievances about squatter settlements.

The minutes of an executive meeting in held in Arawa on 24 March 1988, with Ona and Serero in attendance, provide insights into their concerns and demands. Squatters from elsewhere in PNG were said to be committing crimes, taking over customary land, running small businesses and taking jobs and business opportunities that should have been reserved for Bougainvilleans. The NSPG administrator, Peter Tsiamalili, reported on the Operation Mekim Save taskforce that had just been set up to respond to such problems. He counselled moderation and collaboration with the NSPG.

Some members wanted 'to take violent action now' (Mungkas 1988: 2), while others spoke of the need to save Bougainville from crime, or called for an end to the PNG plantation labour scheme, under which labourers were recruited from elsewhere in PNG, because 'it is the vehicle of death' (ibid.: 6). As an indication of the broad but unfocused sympathy amongst Bougainvilleans for secession from PNG, the minutes record unanimous agreement on two points: 'in view [of] all the suffering, misery and lose facing [sic] we have been made to bear', to 'put it in strongest terms to the Prov. [Provincial] and Nat. [National] Governments that North Solomons DEMANDS SUCCESSION [sic]'; but 'if they object to this demand', then the two governments must 'act immediately to remove all unemployed squatters from the whole of North Solomons', and the PNG Constitution must be amended to require employers to return workers to their place of origin at the end of their contracts. Furthermore:

This meeting strongly felt that our inclusion with PNG is a direct route to Bville neutralisation. This is starting. This plan is not because of BCL, but for our own safety and wellbeing. (ibid.: 3)

Additional calls were made for BCL and the NSPG to remove outsiders; for 80 per cent of BCL employees to be Bougainvilleans; for BCL to terminate non-Bougainvillean contractors undertaking jobs that could be handled by Bougainvilleans; and for 'all tradestores [to] be returned to landowners'.[20] The minutes also record the presentation of a plan by 'landowners' (presumably the NPLA) that included 'mine shutdown', secession, and 'BCL demanded to pay K10 billion for the damage'. However, it is most unlikely that permanent closure was envisaged here, or that the Mungkas executive would have supported such a proposal, for the minutes also advocated pressure on BCL to preference Bougainvillean employees and contractors.

'Pressure Groups' in Southwest Bougainville

The Mungkas leaders had close links with the Bana Pressure Group,[21] the Siwai Pressure Group and similar, though less well organised, groups from the Buin area. Bana and Siwai were relatively densely populated areas, with intensive cultivation of cash crops and associated land shortages, close to the mine, and hence experiencing some of its impacts. Little has ever been written about these groups and their roles in the conflict (but see Tanis 2005: 453–4, 463). Francis Ona received strong support from them, both before and after the conflict started.

These groups were mostly led by older men. Some held land adjacent to the tailings lease, receiving little or no compensation despite the impacts, leading to particular resentment of BCL. They were also concerned about the broader problem of environmental degradation, the lack of employment and business opportunities with the mine, and the number of non-Bougainvilleans now settling on their land or running small businesses nearby (Tanis 2005). Pressure group leaders thought that independence was the most realistic way of getting a fairer deal from BCL. They rejected the earlier generation of 'nationalist' Bougainville leadership

---

20 The mid-1988 NSPG report on law and order later assessed such proposals as 'impractical' (NSPG 1988a), only adding to Mungkas leadership frustration.

21 Bana is a term applied to the combined areas occupied by speakers of the Banoni, Baitsi and Nagovisi languages, being derived from the first two letters of the names of these languages.

that had abandoned the secessionist cause in 1976. Their position was also a reaction to PNG's failure to agree to 1981 NSPG proposals for changing the BCA (Wesley-Smith 1992a).

As Bana Pressure Group chairman from mid-1988, James Singko expressed its position in ever more radical terms. From late 1988, he assisted Ona with funds raised from the sale of bananas and other crops from his plantation. He accompanied Ona into the bush in late 1988, bringing young men from his group, and soon becoming Ona's deputy in the BRA. He is also regarded as being the strongest proponent of secession, and central to that cause being agreed as a core goal of the rebellion at a major meeting at Sipuru, in central Bougainville, in late February 1989. In later tensions with Ona over the direction of the rebellion, Singko claimed the right to negotiate with the PNG Government on the issue since he was the sponsor of that particular aspect of their agenda (personal communication, James Tanis, July 2015).

Marginalised Youth and *Raskol* Gangs

Initial damage to BCL offices and power pylons in November 1988 led to indiscriminate violence by PNG police mobile squads against mine lease landowners. Young mineworker leaders then recruited disaffected agemates—many involved in criminal (*raskol*) gangs—to join what was soon being called the BRA,[22] in the expectation that gang members would be more ready than others to meet the unexpected police violence with a violent response. This recruitment campaign helps to explain both the rapid expansion of Ona's support from early 1989, and some of the BRA's subsequent indiscipline.

Because the marginalised youth recruited in this way came from all over Bougainville, it is difficult to be specific about their goals. Information from some who undertook the recruiting indicates that few had political goals. Many were undisciplined and continued with their criminal activities (Oliver 1991: 214). Moreover, many were later involved in the chaotic situation that followed an agreement made on 2 March

---

22   In 1989, they were more frequently known as militants or 'Rambos'. Rambo was a character played by actor Sylvester Stallone in a popular 1982 movie, *First Blood*, which was very popular in Bougainville before and during the conflict.

1990, under which PNG security forces withdrew from Bougainville by 16 March. On 8 March, Bishop Singkai expressed concern that the proposed withdrawal meant:

> a power vacuum is being created which invites misguided youth to terrorise innocent persons. We believe that the BRA will influence the so-called 'Rambo Gangs' not to steal or destroy private property (Singkai 1990).

His hopes were not realised. Ona was concerned about the damage to BCL facilities and equipment because he envisaged an intact mine reopening as the main source of revenue for an independent Bougainville (personal communications, Joseph Kabui 2003, James Tanis 2005, Eddie Mohin 2015). In a radio broadcast on 18 March, he talked of receiving many complaints: 'Those you fought to protect now regard you as worse than the security forces … a lot of you I don't consider as BRA'. He directed the return of vehicles and goods stolen 'against BRA law' and threatened that he could 'surrender now and everything will fall apart' (Oliver 1991: 237).

In a 2000 interview, former BRA leader Sam Kauona described the situation as follows:

> Another problem arose from Bougainvilleans making claims to a lot of materials, including hundreds of vehicles abandoned by the mining company … The soldiers in the BRA refused to listen to Sam. 'It was madness' said Sam. 'They didn't know how to drive the vehicles but they went from one to another—there were so many. There was virtually no control. They took all the money that was left in the drawers. Because of all the material goods Bougainvilleans started fighting against each other … The problem became so complicated that it became impossible to address this internal instability' (NZine 2000b).

Indigenous Political-Religious Movements

The leaders of several highly autonomous local movements strongly supported the NPLA. Once regarded as 'cargo cults', they have more recently been described in Bougainville as 'indigenous political-religious movements' (BCC 2004: 32–3). The most significant was Me'ekamui Pontoku Onoring, led from the early 1960s by Damien Dameng, a man from Irang village, just a few kilometres southwest of Panguna (Oliver 1991: 180–1; Regan 2002; Tanis 2005: 461).

Dameng sought the autonomy necessary for Bougainvillean communities to pursue their own 'development' paths based on their own choice to accept or reject elements of custom, Christianity, and other 'modern' institutions. He opposed mining as an uncontrollable force damaging culture and autonomy, and supported permanent mine closure. He had links with, and some influence over, similar movements elsewhere (BCC 2004: 32–3), most of which had supported secession in 1975, and did so again from early 1989.

Dameng associated closely with the NPLA. In March 1988, he signed the initial petition presented to BCL, along with Ona and Serero, being described in the document as 'Chairman of the Demonstration' (PLA 1988). He attended most of the meetings between the NPLA and representatives of BCL and the national and provincial governments between March and October 1988. He was the only person recorded as speaking for landowners at the 26 May meeting between NSPG and PNG Government officials (NSPG 1988b). And some of his supporters were amongst the young men who accompanied Ona into the bush late in 1988.

When police mobile squads began their active pursuit of Ona and his band, the latter moved southeast to the rugged and largely trackless Kongara Mountains. Most of the inhabitants were followers of the Methodist (now Uniting) church, but many also followed Dameng. Once in the area, Ona secured support from local clan leaders to recruit more young men to their group, and similar approaches were soon made to clan leaders and chiefs in other areas.[23] This form of recruitment is the basis for recent assertions by some former combatant leaders of customary rights (to compensation) to be taken into account if the mine ever reopens.

Church Blessings

From an early stage there was a strong sense that the Catholic Church, in particular, supported the NPLA and, by extension, the BRA. From before the conflict, religious beliefs provided a source of inspiration to many of those involved. Church social justice teachings influenced many with Catholic seminary training, while other Catholics, including Ona, drew inspiration from a range of more basic Catholic teachings and practices.

---

23   In February 2016, families from the Mangkaa area in Kongara asked the Bougainville authorities for a goodwill payment for having provided food and shelter to Ona from August 1989.

As violence spread from late 1988, Bishop Singkai and the Justice and Peace Committee of the Bougainville Diocese spoke publicly against PNG security force violence (Oliver 1991: 216, 219, 232–4; see also Griffin 1995). Singkai maintained close communication with Ona in his bush hide-out, amongst other things sending rosary beads to Ona and other members of his core group (personal communication, James Tanis, February 2016). Father Bob Wiley, a critic of the mine in 1966, was parish priest at Panguna in 1988, and was known as a BRA supporter (see Wiley 1992).[24] Perceptions of church support were reinforced from mid-1990 when Bishop John Zale, head of the Uniting Church in Bougainville, joined Bishop Singkai as a member of the Bougainville Interim Government established by the BRA.[25]

Some Pentecostal Protestants also had their own biblical explanations for Bougainville's special place in God's plans. Such views were prevalent in the 1980s and continue to be widely discussed amongst Pentecostal Bougainvilleans in 2016.[26] A variety of religious beliefs therefore contributed to the ideological emphasis on the primacy of Bougainvillean rights that emerged after 1989.

Mainstream Politics

While both the NSPG and the Bougainvillean representatives in PNG's national parliament were generally critical of BCL in the 1980s, they tended to support revision of the BCA rather than mine closure. John Momis attacked the unfairness of the BCA in a five-page letter known as 'The Bougainville Initiative' while he was campaigning in the 1987 national elections. When he presented it to BCL's managing director in May 1987, he was accompanied by Kove, Ona, and others who were later members of the NPLA (Auna 1989; Griffin and Togolo 1997: 377).

---

24 Early in 1989, he blessed BRA members before they went into operation and remonstrated strongly with one of Ona's local supporters who questioned whether this was appropriate priestly behaviour (personal communications, John Amuna and Wendelinus Bitanuma, 2015).

25 Both were members of the Bougainvillean delegation that negotiated a ceasefire with PNG Government representatives on the New Zealand naval ship Endeavour from 29 July 1990 (Oliver 1991: 248–53).

26 They were central to an 'awareness' package, regularly delivered from about 2010 by former BRA leader Sam Kauona, that concerned the 'stolen rights' to the gold of Bougainville. The theft was said to have begun with gold taken by the ships of the biblical King Solomon for the construction of the temple in Jerusalem. A 2010 paper by a former bible college lecturer close to the Me'ekamui leadership sets out similar views (Taruna 2010).

This document (soon rejected by BCL) has been dismissed as an effort to prevent a reduction in the MA party's vote in the impending election (Griffin 1995: 16; Griffin and Togolo 1997: 377). BCL's former managing director later labelled it the 'major catalyst' of the conflict (Quodling 1991: 51–2), but Momis was acting in a way that more likely 'reflected rather than provoked the impatience of his electorate' (Filer 1992: 441). After all, he was close to the leaders of the groups discussed here, and his demands were similar to the 1981 NSPG proposals to the national government, made in advance of the failed review of the BCA, to increase royalties from 1.25 to 5 per cent (Wesley-Smith 1992b: 100), as well as to later positions taken by both the PLA and the NPLA.

Any confidence Ona and his supporters might have had in the NSPG and the MA members of parliament to advance NPLA concerns was undermined by internal MA conflict in 1988. Alexis Sarei, elected premier in 1984, unexpectedly resigned in 1987 (Oliver 1991: 194–5). The NSPG constitution required a province-wide 'presidential' by-election, but the MA-dominated NSPG assembly amended the constitution to enable election of a caretaker premier from among the sitting members. Joseph Kabui was the MA nominee who was elected. His NPLA associates saw this as a major advance for their cause, but the party's plan was that Kabui, who was seen as young and inexperienced, should merely 'keep the seat warm' for a more senior MA figure to be elected as premier in the 1988 provincial elections. To the anger of Momis, Kabui decided to run as a pro-MA candidate without MA endorsement. In a three-way contest he defeated both the former premier, Leo Hannett, and the MA candidate, Anthony Anugu. Momis and his party allies could not forgive Kabui, who in turn resented MA election campaign attacks on himself.

An unfortunate consequence of this was the much-reduced capacity of Bougainville's mainstream political leadership to cooperate as crisis became conflict. The standing of MA, and of Momis as the previously dominant figure in Bougainvillean politics, were both damaged: 'it was clear that people were no longer following party politics' (Tanis 2005: 469). The failure of the NSPG's 1988 Operation Mekim Save was also a grave disappointment to many people, and contributed to loss of faith in the NSPG's ability to protect Bougainvillean rights and interests as opposed to the constitutional rights of all PNG citizens (Tanis 2005: 466–7). That was undoubtedly a factor in the push for radical action amongst the coalition of groups identified here in the period from late October 1988 to February 1989.

Kabui had a long history of links to Ona and his cause, and maintained these links after Ona went into the bush, but references to Kabui in Ona's frequent public statements in 1989 were increasingly dismissive. Other members of Kabui's government, while sympathetic to Ona, were soon deeply concerned about the escalating violence. They sought moderate solutions at the same time as Ona and the BRA became more radical in their demands. By the time of the March 1990 withdrawal of PNG security forces, the BRA felt that much of the NSPG had opposed them, and had little interest in sharing power with it.

Linkages and Networks

Extensive linkages existed between the leaders of most community groups involved in the origins of the conflict origins, and also between them, the state and the company. But, by late 1988, most of these groups were to varying degrees opposing both the national government and the NSPG, despite their links to individual MA members. Intergroup linkages meant that the NPLA, the Mungkas Association, the young mineworkers, the Bana and Siwai pressure groups, and Me'ekamui Pontoku Onoring operated as a loose coalition.

Bishop Singkai and several of his priests, notably Bob Wiley and Mark Roberts, also had close links with the NPLA and pressure group leaders. A significant proportion of the leaders of all groups came from the Catholic educated elite and so shared less formal links. Church support legitimised increasingly radical actions as being in the interests of Bougainville as a whole.

The core leadership of each group was small, and there were overlaps between them. Most of the key leaders were active MA members, which provided another set of linkages, yet each key one influenced and mobilised significant and to some extent distinct communities. Their influence was concentrated in central and southwest Bougainville, but for some it extended much further. Those stretching to almost all parts of Bougainville included the overarching church (especially Catholic) and MA networks, the Mungkas network (through Arawa-based communities from all over Bougainville), the 'indigenous political-religious movements', the disaffected young males, and the young men recruited through clan leaders. Hence, during 1988 and into 1989, the leadership reflected the concerns of, spoke to, and mobilised support from, a significant proportion of the whole population. Once violence on the part of (mostly non-Bougainvillean) police mobile squad members began in December 1988, it was a small step to build wider and more committed support networks, largely by mobilising people around ethnic divisions. Community concerns and demands rapidly moved beyond mine-related impacts, temporary mine closure, and the renegotiation of mining agreements. Support for secession was consolidated, and mining issues became secondary.

The groups increasingly worked together from late 1987 onwards. They played different roles, however, in the origins of the conflict. The NPLA, young landowners (some NPLA activists and others not) and young mineworkers initially took the lead. A few key members of Mungkas, the pressure groups and Me'ekamui Pontoku Onoring worked closely with the NPLA in the lead-up to the first violence against BCL in November 1988, and then actively supported Ona as the conflict developed. Others only became actively engaged once police mobile squad violence started. Some marginalised youth and *raskol* gang members gravitated to Ona soon after the violence began, while others were actively recruited by young mineworkers from early 1989. As violence emerged, Mungkas and pressure group leaders mobilised support for Ona, some later becoming key BRA or Bougainville Interim Government leaders.

The group leaders had a range of attitudes to mining. Dameng wanted mining ended—a view attracting some, though not all, of his NPLA associates. Most of the young mineworkers, pressure group and Mungkas members sought much fairer conditions for mining. While almost all soon favoured secession, the pressure groups were initially most strongly committed to combining secession with continued mining, thus providing the revenue needed for independence.

From at least late 1987, Ona was a key link between the different groups. His own attitudes to future mining were ambivalent—sometimes probably for permanent mine closure, and sometimes more supportive of mining provided it was under radical new conditions. Undoubtedly, though, continued mining was key to the agenda of many in the wider coalition, and often for Ona's agenda as well. Needing to keep that coalition together, Ona could neither ignore the positions of other groups nor dominate their leaders, many of whom had the strength of personality to match him. Indeed, there were always some strong tensions in relationships between Ona and both Dameng and Singko.[27]

Of course, this coalition did not reflect the views of all Bougainvilleans. Other groups and leaders held different views on Panguna, and on Bougainville's future relations with PNG. However, part of the loose coalition's strength was that virtually all Bougainvilleans had sympathy for at least some aspects of their key concerns and demands.

It is most unlikely that any of the leaders of the various groups involved envisaged anything like the violent conflict that actually developed in 1989. Each wanted negotiations on a variety of related issues. Strong demands were initially directed towards negotiations over mining. Damage to BCL property occurred when frustration levels became high, and was intended to demonstrate the degree of frustration. The dangers of escalating violence involved in these actions were recognised by Ona and Singko when they took to the bush. But even when they had done so, in the early stages they were still prepared to negotiate. Undoubtedly, it was the deployment of the police riot squads, followed by the PNG Defence Force, and their often random and extreme violence, that transformed the issues and tactics, and led to active recruitment of young men willing to fight violence with violence.

An Emergent Ideology[28]

The shared discussion of problems amongst the groups discussed here contributed to the beginnings of a shared ideology about the primacy of the rights of Bougainvilleans over those of outsiders. These were rights to mineral revenues and mine-related employment, contracts and businesses. Central was the idea that Bougainville's land, so central to clan identities, needed special protection from both mining and from people from elsewhere in PNG. Bougainvilleans deserved respect from outsiders, just as a Bougainvillean clan regarded as original to a particular area should be accorded respect by other clans, even those with some land rights in the area.

---

27 Some key figures found Ona frustrating to deal with, and one described it as like dealing with a *maleo* (an eel) (personal communication, Theodore Miriung, 1995).
28 The following analysis summarises information from personal communications by Theodore Miriung (1995–96), James Tanis (1998–2016), and Damien Dameng (2002).

Emerging in the mid to late 1980s, these ideas were elaborated between 1990 and 1991. James Singko, by now Ona's deputy, took a lead role. Articulated in the Nasioi and Nagovisi languages, the ideas were seen as derived from both Bougainvillean culture and Christianity. Church support for Ona and the BRA added to their legitimacy. The emphasis was on the sacred nature of Bougainville and the traditions of its true people, and their concern about damage to both their land and traditions wrought by outsiders. The corollary was the duty of true Bougainvilleans to assert control over, or perhaps oust, outsiders causing harm, and to establish an original Bougainvillean government to protect the land, its people and their traditions.

The key principles were encapsulated in Nasioi words *me'ekamui, sipungeta* and *osikaing. Me'ekamui* means something akin to 'sacred land'. Just as the land of any particular Bougainvillean group is sacred to it, so the whole of Bougainville is sacred to all Bougainvilleans. *Sipungeta* literally means 'hearth'. In this context it meant something that was both 'original' and 'inventive', as well as being strengthened by fire, and therefore suggested the idea of control from a strong base. A government for the sacred land needed similar qualities, involving reliance on traditional strengths to respond to new challenges.

Osikaing means 'original inhabitants'. The original inhabitants of the sacred land had special duties towards that land, as well as corresponding rights. To be original, they must control what is theirs, especially their land. Outsiders entering it should respect *osikaing* ways. *Osikaing* contrasts with *taboranku*, meaning 'strangers' or 'aliens'. A stranger does not have the same rights as the original inhabitants. BCL, the national government and people from elsewhere in PNG were all strangers who had turned Bougainvillean *osikaing* into *taboranku* in their own land. Ousting outsiders and reasserting *osikaing* control would substitute social harmony for discord.

This ideology, emerging from debates preceding the conflict, and forged by the early heat of violence, continues to resonate in post-conflict debates about the future of LSM.

The Rectangular Model

Discussion so far has focused mainly on the local community 'corner' in the rectangular model of stakeholder relationships (see Chapter 1, this volume), highlighting the complexity of these relationships. Though the 'fourth estate' was almost completely absent in 1988, brief comments can be made about the state and company 'corners' and the complex links between them and the community 'corner'.

The state 'corner' involved both the PNG Government and the NSPG. While the latter had no formal powers over mining, it was a significant factor in the local politics of LSM, with close links to the community sector. Although the national government was the regulator, it was largely seen as a remote presence, focused on exploiting Bougainville's mineral wealth to benefit the rest of PNG, and so mainly interested in the collection of mining revenues through dividends and taxation. As a result, state and company interests coincided. But key national government actors had deep links with the Bougainvillean community sector, especially John Momis and the MA's other Bougainvillean MPs.

As the only LSM entity in the province, BCL dominated the company 'corner', largely because of an official 'moratorium' on mining exploration or development in other parts of Bougainville that had been in place under PNG law from 1971. There were, however, many other mining-related companies providing services to BCL. The mining giant also had close links with local corporate entities that had close community links, as well as significant Bougainvillean entities like BDC and the RMTLTF. Some Bougainvilleans, like Severinus Ampaoi, were senior BCL employees, yet some employees of both BCL and its suppliers had close links to elements in the community 'corner'. In particular, several such employees provided active support to mine lease landowner groups during the critical period from 1987 to 1989.

Current Views on Resumption of Large-Scale Mining

The current (2016) relationship between local-level politics and large-scale mining in Bougainville has both similarities and differences to that which existed between 1987 and 1989, especially with respect to the extent of community opposition to LSM. The current situation has of course been deeply influenced by dramatic changes arising from the conflict itself. Although the Panguna mine has been closed since 1989, the possibility of reopening it ensures that LSM is a factor in political relationships between PNG and Bougainville, and also within Bougainville.

The differences between the two situations include a major shift in the state's 'corner', largely resulting from the 2001 BPA, which made legislative authority over mining available to the ABG. There is also much greater complexity in each of the company, state and community 'corners' and in their mutual relationship, along with emergence of a new, multifaceted 'fourth estate' corner. Major differences arising from the complex post-conflict situation include: the national government's much-reduced role in Bougainville as compared to 1988; the ABG's limited capacity and reach in a situation where several parallel 'governments' claim 'sovereignty' over all or part of the autonomous region; armed groups remaining a factor to contend with; and numerous fringe foreign mining interests seeking to play significant roles in alliance with different local factions. A major similarity is that multiple Bougainvillean groups with considerable autonomy from one another, aware that important decisions on the future of mining may be made in the foreseeable future, all seek to ensure that their voices are heard.

There is some continuity, too, in aspects of the issues being dealt with, and even in the identity of some key individuals involved in debate. ABG President John Momis, always a strident critic of BCL, who supported Francis Ona in 1987 and 1988, now supports the reopening of Panguna or possible development of a maximum of two other large-scale mines— that limit being set by the ABG's *Bougainville Mining Act 2015*. This does not involve a change in the president's long-term position, since he demands entirely new and fair mining conditions.

New actors in the political process include some of the leaders of nine new mine lease landowner associations established since 2011 to replace the PLA and NPLA, and others associated with the Me'ekamui Government of Unity. More familiar actors include figures formerly associated with both the PLA (Michael Pariu and Lawrence Daveona) and the NPLA (Wendelinus Bitanuma, John Amuna, Philip Miriori and Philip Takaung). For the most part, their positions are similar to that of Momis and the ABG. Nevertheless, the very different social, economic and political context of Bougainville in 2016 has resulted in dramatic changes in debates on the future of mining.

The State 'Corner'

The state 'corner' now involves both the PNG Government and the ABG. Although the ABG became the sole regulator from 2014 under its own mining legislation, national government roles continue. Amongst other things, the PNG Government is the second largest shareholder in BCL, and its legislation continues to cover some mining-related subjects, including occupational health and safety, land and environment. Tension between the two governments centres on the national government's efforts to remain involved in Bougainville mining matters by taking over Rio Tinto's equity in BCL.

Soon after it was established in mid-2005, the ABG began considering the possible resumption of LSM, which represented a significant shift from the positions of most Bougainvilleans early in the peace process, when resumption seemed unlikely (Regan 2014: 78). Concern about the revenue needed for either autonomy or independence led the first ABG president, Joseph Kabui, to examine this possibility from mid-2005, as did both his successors—James Tanis (2009–10) and John Momis (from 2010). The two main options under consideration have been the reopening of Panguna or the lifting of the 1971 'moratorium' on mineral exploration (preserved under the ABG's mining laws) with a view to possible developments elsewhere in Bougainville.

Kabui initially considered the first option, favouring potential developers other than BCL —initially a small Australian company called Ord River Resources Ltd. But intense pressure from former BRA commander Sam Kauona shifted Kabui and his cabinet to a second option, favouring a small Canadian company called Invincible Resources Ltd (IRL), which was established by Kauona's close associate Lindsay Semple, a dual Australian-Canadian citizen, solely for the purpose of mining investment in Bougainville. Semple raised funds from a number of high-risk investors and used the money for a number of 'development' projects directed towards gaining ABG support for IRL to establish several new mines, smaller than Panguna, each in partnership with local landowners. The irresistible lure of significant revenue flows, ending the ABG's reliance on PNG Government grants, seemed to be underwritten by IRL's 2006 advance of a K20 million loan to the ABG to support 'capacity building'. In 2008, the ABG signed an agreement with IRL whereby the latter would hold a 70 per cent stake in an entity called Bogenvil Resources

Development Corporation (BRDC), supposedly established to implement a multisectoral Bougainville development plan using revenues generated by a virtual monopoly of mining activity beyond Panguna.

Controversies about IRL and BRDC centred on details of the K20 million loan repayment agreement, the failure of the 'development' projects, the opacity of the 30 per cent minority shareholding in BRDC,[29] the near monopoly vested in IRL and the extent of indirect IRL control of the ABG. After Kabui died unexpectedly in June 2008, IRL and its ABG backers strove to ensure continued implementation of the development plan, but these efforts ended when Tanis became president in January 2009.

From early 2010, the ABG again examined the possibility of reopening Panguna. High commodity prices, proven reserves, and some existing mine-related infrastructure sustained optimism that reopening could occur within perhaps six years of negotiations commencing. By contrast, experience elsewhere in PNG showed that new mines take more than 15 years to get from exploration to operation. So, while open to other possibilities, the ABG under Tanis and Momis decided that Panguna offered the best prospect of significant revenues around the time of the referendum on independence that is due to be held in 2019. Paradoxically, reopening Panguna also offered the best prospect for meeting the expected high costs of remediating mine-related environmental damage and negative social impacts, such as those experienced by the relocated villages.

After Momis took office in June 2010, initial uncertainty about BCL as the preferred operator of a reopened Panguna mine was resolved in part because mine lease landowner community leaders argued in favour of BCL, preferring the 'devil they knew' to a 'new devil', mainly because BCL accepted some responsibility for the legacy issues. Uncertainty was also reduced as the result of a three-day ABG workshop on mining policy options, held in March 2011, that clarified the choice between the option of BCL reopening Panguna, an alternative developer doing so, and consideration of alternative prospects.

The ABG has consistently emphasised that no LSM will be permitted without landowner consent. Over three years from mid-2011, the ABG supported the establishment of nine independent associations to represent

---

29   Fifteen per cent was to be held by the ABG, 9 per cent by an 'ex-combatant's' company and 6 per cent by a company representing un-named Bougainvillean 'pioneer politicians'. The shareholders in these two companies have never been revealed.

landowner communities impacted by the Panguna mine communities in considering possible resumption. Over 3,000 landowners voted for association executives at nine separate general meetings.[30] The ABG has consulted the associations but has not sought to influence their positions on mining issues—indeed, there have been occasional tensions between the ABG and some association leaders. Five years on, in 2016, no decision had been made to even initiate negotiations with BCL or any other mining company about reopening Panguna.

Engagement with BCL has mainly been about steps preparatory to possible negotiation, inclusive of arrangements needed for a reconciliation process proposed by more than 50 mine lease landowner leaders who attended the first formal post-conflict meeting with the ABG and a senior BCL representative in July 2012. Extensive discussions have occurred within and between associations, and between them and the ABG, about possible conditions for reopening the mine, which reflects ideas about the primacy of Bougainvillean rights.

The ABG also undertook extensive public consultations about the future of mining, which included consultations about its mining policy in 2010–11, about the future of Panguna and other possible LSM activities in 2013–14,[31] and about its draft mining legislation in 2014–15. In the absence of opinion polling, it is difficult to be certain as to whether there has been a preponderance of views one way or another. Nevertheless, those involved in organising the consultation reported broad community support for resumption of mining. That is consistent with regular feedback to the ABG from the nine Panguna landowner association executives, as well as from the elected ABG members from their own constituencies, including the mine lease areas.

From about 2014, the ABG began seriously examining options other than reopening Panguna. The reasons for this included:

- falling commodity prices, which made it less likely that the reopening could be financed (at an estimated cost of US$6–8 billion);
- growing realisation of the extent to which negative sovereign risk assessment could make financing even more difficult;

---

30 This contrasts with roughly 100 attending the 1987 meeting that elected the NPLA executive.
31 This involved five regional public forums, and separate forums for women leaders and ex-combatants.

- the high expectations of mine lease landowners, former combatants, and others about compensation, not just for mining impacts but also a range of conflict impacts; and
- the increasingly disruptive activities of small foreign companies linked to different Bougainville factions.

In taking over full regulatory responsibility for mining, the ABG established its own mining department in 2007, negotiated transfer of national government powers (2006–08) and developed its own mining policy (2009–14). It developed its own mining law in two main stages. The first was the Bougainville Mining (Transitional Arrangements) Act, proposed in 2012 and enacted in 2014 (Regan 2014: 87–91). This first law was based on PNG's existing Mining Act, but with significant changes to meet Bougainville's needs. These included:

- vesting both the ownership of minerals and powerful veto rights over exploration in the owners of customary land;
- divesting BCL of its mining tenements, leaving only an exploration licence over the former SML; and
- imposing a strict prohibition on the operation of more than two very large mines (like Panguna) at one time.

The second stage involved passage of the Bougainville Mining Act in March 2015. It was the product of the policy development process initiated in 2009, incorporated many of the initiatives included in the 2014 Act, and extended landowner rights to include the power of veto over mining development tenements, as well as exploration licences.

While reluctantly accepting the validity of this ABG legislation, the national government was attempting (from 2013) to purchase Rio Tinto's 53.8 per cent equity in BCL, thus raising its own stake in the company to 72.8 per cent. Opposing this move, the ABG argued that:

- PNG majority ownership of the mine at the heart of the conflict was unacceptable in Bougainville;
- the PNG Government was seeking to control Panguna despite the transfer of mining powers to the ABG; and
- PNG would gain significant control of the Bougainville economy, which would be a sensitive issue in the lead-up to the referendum.

The ABG demanded that, if Rio Tinto divested, then its equity should be 'gifted' to the ABG and the landowners, and it should provide significant additional funding to meet corporate responsibilities for mine legacy issues, including extensive environmental, social and human rights impacts, especially for the thousands of relocated people who were now living in appalling conditions (Momis 2015). These issues remain unresolved at the time of writing.

The Community 'Corner'

Just as in 1988–89, this part of the matrix of stakeholder relationships extends well beyond the immediate vicinity of Panguna. Since the 1960s, the significant social and economic impacts of the mine have ensured that most Bougainvilleans have felt that they have some stake in Panguna's mineral resources. These feelings grew far stronger after the conflict, in which all communities shed blood, many at Ona's explicit request. Under customary principles applicable in most Bougainvillean cultures, if blood is shed when supporting landowners in land-related conflict, the clan whose blood is shed gains some rights over the land in question. Such principles are referred to widely in public discussions about Panguna.

The three main groups of former combatants are a significant new addition to the 'community corner'.[32] Theirs have become increasingly prominent voices in public debates since Momis was elected as ABG president in 2010, partly because of suspicions of his PNG Government links. Two of the most significant voices are those of the former BRA leaders Sam Kauona and Ishmael Toroama. Kauona presents as a strident critic of BCL, opposed to the reopening of Panguna and the ABG's mining policies, despite his own close association since 2004 with foreign corporate mining interests, most notably IRL. For Toroama, a key issue in Panguna-related debate is recognition of the rights of former combatants on the basis of blood shed during the conflict.

Despite the major logistical and financial difficulties involved, the ABG has made extensive efforts to consult widely about the future of LSM, not just with mine lease landowner communities, but amongst Bougainvilleans more generally. Elected executives of the nine landowner

32　The three groups are the original BRA, the Bougainville Resistance Forces that opposed them, and the Me'ekamui Defence Force, most of whose members defected from the BRA to remain with Ona when he refused to support the peace process in 1998.

associations, representing what could now be as many as 15,000 people in the mine lease areas, have consulted widely with their own communities as well as with the ABG. Consistent with the outcomes of the ABG's wider process of community consultation, they report broad community support for mining resumption, but only under fair conditions, including redress for past injustices.

The Company 'Corner'

In June 2016, Rio Tinto announced a plan to transfer its 53.8 per cent stake in BCL to an 'independent trustee' that would be jointly owned by the PNG Government and the ABG. Under the terms of this arrangement, Rio aimed to transfer 36.4 per cent of its equity in BCL to the ABG, and 17.4 per cent to the PNG Government. This meant that both would hold an equal proportion (36.4 per cent) of the shares in BCL, since the PNG Government already held 19 per cent. This announcement aroused a storm of protest from nearly all of the Bougainvillean actors mentioned in this chapter, most of whom demanded that the PNG Government should transfer the 17.4 per cent to the ABG so that it would become the majority shareholder (Anon. 2016a). This move was opposed by PNG Prime Minister Peter O'Neill, who announced instead that the 17.4 per cent would be transferred to a trust for the benefit of all Bougainvilleans, including the Panguna landowners (Anon. 2016b, 2016c). This left the PNG Government with its original 19 per cent stake, the ABG with 36.4 per cent, the trust with 17.4 per cent and several thousand small shareholders with a little over 27 per cent.

An account of the details and ramifications of these developments, and the new political process engendered by this turn of events, is beyond the scope of this chapter, but it has not so far led to any significant change in the prospects of reopening the Panguna mine. What it might have done is to limit the scope for continued interference by other foreign companies in the prospective development of LSM projects in Bougainville.

In 1988 there was a small local corporate sector linked to the mine, notably the BDC and the RMTLTF, but both of these entities ceased to operate soon after the conflict began, and it is still unclear what happened to their funds and some of their other assets. By 2016, however, several other sets of corporate interests had emerged, with strong linkages to different Bougainvillean groups.

As previously mentioned, former BRA leader Sam Kauona, supported by a small group of other former combatants, has had close links to dual Australian-Canadian citizen Lindsay Semple since 2004–05. Semple and Kauona have been financially supported by a series of at least three main external investors: the Canadian company IRL from 2004 to 2010; another Canadian company, Morumbi Resources, from 2010 to 2013 (Regan 2014: 86–9); and from 2015, New York investment firm Kuhns Brothers. They lost most of their influence over the ABG when Joseph Kabui died in 2008, and their cause then seemed to be lost in 2010 when the investors in IRL lost faith in Semple. However, they soon re-entered the field with Morumbi, seeking exclusive rights to exploration and development in prospective areas through agreements made between Morumbi subsidiaries and small groups of landowners, which they unsuccessfully sought to persuade the ABG to recognise. After Canadian police visited Bougainville in late 2013 to investigate the activities of both IRL and Morumbi, the latter cut its ties to Semple. Since Kuhns Brothers became involved, their activities have included the establishment of a small gold-buying and refining operation, the creation of an investment fund called Bougainville Fund Management LLC,[33] and various proposals for funding development projects outside the mining sector in what appears to be an approach similar to that attempted by IRL.

Since Ona's death in July 2005, one of the two main sets of self-proclaimed successors to his leadership of the 'Me'ekamui' groups claiming to be the legitimate Bougainville government (as alternatives to the ABG) has been the Panguna-based Me'ekamui Government of Unity (MGU). The MGU has also had extensive involvement with a succession of small foreign companies. The first was Australian registered company Tall J Foundation Pty Ltd, involving a group of American citizens who attempted mechanised extraction of gold from the Panguna riverine tailings in 2008. It was followed by Cefeida SA, a company established in 2009 by former Tall J investors, Stewart Sytner and Thomas Megas, to attract investment funds by claiming 'that they possessed exclusive rights to develop and mine in [Bougainville] … and that such rights "were extremely valuable and rare for outsiders"'.[34] Next was a Canadian company, Transpacific Ventures Ltd, also linked to Sytner. It produced an information memorandum for prospective investors in July 2013, claiming an agreement with the MGU

---

33 See www.secinfo.com/d1yT3c.we.htm (viewed 24 April 2016).
34 *Amiron Development Corporation v Sytner* [2013] E.D. New York 12-CV-3036, 29 March.

'on an exclusive basis for 20 years renewable, to advise the customary landowners (the Me'ekamui) in developing their natural resources sector' under a new mining law 'expected to transfer all land and mineral rights on the island of Bougainville and its territorial waters to the Sovereign Me'ekamui Tribal Government (the Me'ekamui)' (Transpacific Ventures 2013: 2). In 2014, Transpacific was succeeded by Australian company, United Resources Management Pacific, which was succeeded in 2016 by another Australian company, Central Exploration, both of which were introduced to the MGU by an Australian, Ian Renzie Duncan, who was previously involved in Transpacific Ventures. The companies involved, from about 2012 until at least the end of 2015, paid monthly 'salaries' to senior MGU members.

These are just a few of the small corporate entities that have sought access to Bougainville's mineral resources since the ABG began considering the possibility of renewed LSM. Many have sought to advance their interests with no reference to the ABG. Rather, they have used links with the leaders of local factions (some armed) or have engaged in direct dealings with small landowner groups from prospective areas. The need to control such entities provided a major impetus for the ABG to develop its initial mining legislation in 2012.

The 'Fourth Estate'

From 1989, the Bougainville conflict attracted considerable international attention, particularly in Australia. Several NGOs and other groups supported the BRA, or Bougainvilleans suffering the impacts of the PNG Government's air and sea blockade of Bougainville imposed from mid-1990. These supporters included existing organisations and new ones established for the purpose, such as the Independent Bougainville Information Service, Australian Humanitarian Aid for Bougainville and Bougainville Freedom Movement (BFM).

Since the conflict ended, a few additional groups and some individuals have become involved, particularly with the emergence of debate on the future of LSM. They have included 'mediators' offering services to the ABG, as well as some journalists, lawyers, advisers or consultants mainly seeking to advise or influence various Bougainvillean groups—especially Panguna lease landowner associations.

Since the ABG began developing its mining legislation in mid-2012, a concerted campaign against its mining policy has been mounted by a network of fourth estate actors whose central point appears to be Australian 'activist' academic, Kristian Lasslett. He is involved with the International State Crime Initiative, whose website includes material purporting to document the responsibility of Rio Tinto and BCL for the conflict and its impacts, and demanding that they be brought to account.[35] This view is advanced in Lasslett's own publications (e.g. Lasslett 2014) and in other publications in which he acknowledges a central role (e.g. Jubilee 2014).

The main PNG node in this network is the Bismarck Ramu Group, which operates two blogs that feature stories about mining on Bougainville— 'PNGexposed',[36] which is intended to expose political corruption, and 'Papua New Guinea Mine Watch', which was established to campaign against environmental damage by the Ramu nickel mine in PNG's Madang Province, but has since expanded its remit to other mining projects.[37] Both blogs regularly publish strong attacks on the ABG, some under Lasslett's name, but many anonymous, though very much in Lasslett's style. According to Lasslett, the Bismarck Ramu Group is 'one of Papua New Guinea's unsung treasures, and their support over the years has been truly amazing' (Lasslett 2014: x).

The Australian branch of the network includes Jubilee Australia, an NGO that has published two reports highly critical of the ABG's mining policy and legislation (Jubilee 2014, 2015). The research for its 2014 report was overseen by Lasslett, who has had close links with Jubilee since his student days. The two Bougainvilleans who undertook the research were close associates of Lasslett, and are known to share his views on BCL and Rio Tinto.

The Australian branch also includes a few MPs from the Greens and Labour parties. Greens Party senator Lee Rhiannon has spoken at the launches of the Jubilee reports, and has regularly asked questions of the Australian Department of Foreign Affairs and Trade in Senate Estimates Hearings about alleged Australian Government influence on ABG mining policy. Antony Loewenstein, an Australian activist journalist and author,

---

35   See: www.statecrime.org/testimonyproject/bougainville/.
36   For stories about Bougainville, see: pngexposed.wordpress.com/tag/bougainville/.
37   For stories about Bougainville, see: ramumine.wordpress.com/tag/bougainville/.

has written articles critical of ABG mining policy, and covers such issues in his book on 'vulture capitalism' (Loewenstein 2013). Loewenstein acknowledges the support and assistance of Lasslett in his writings on Bougainville.

Finally, the Australian node includes the BFM, the last of the 'support groups' from the 1990s that still has a public voice. It regularly attacks the ABG on the PNG Mine Watch blog, as well as in other outlets such as 'New Dawn on Bougainville', in pieces by written by Vikki John. Lasslett acknowledges the 'encouragement and guidance from activists and advocates involved in the Bougainville anti-war and independence movements' that he received while doing his PhD research on 'state-corporate decisions and motivations that underpinned the crimes on Bougainville' (from 1988–89), including support from key BFM figures, Vikki John amongst them (Lasslett 2014: ix).

This 'fourth estate' network does have some links in Bougainvillean communities, but mainly with a few people who share the views of other network members. Curiously, Lasslett's connections include a key personal staff member of Jimmy Miringtoro, one of Bougainville's MPs in PNG's national parliament, who, despite being a strident critic of the ABG's mining policies, has his own interests in the selection of companies that might redevelop the Panguna mine.

All the members of this network advance strident claims about what they believe to be the general opposition to mining on the part of either the people of Bougainville in general, or of mine lease communities in particular. The evidence offered in support of these claims is weak or flawed. For example, the basis for Jubilee's 2014 claims that 'the mine affected communities' are 'opposed to any discussion of the mine's reopening' (Jubilee 2014: 46) was a set of 'semi-structured' interviews with just 65 individuals selected by a 'snowballing' technique, plus a single focus group discussion with 17 males in a village just outside the mine lease area who 'refused to be interviewed separately, instead stating that they preferred to reach a consensus … and then present one common position for each question' (ibid.: 49). In a situation where there are perhaps 15,000 people in former mine lease area communities, where it is well established that a large (but unknown) proportion support the resumption of mining, subject to strict conditions, the most likely explanation for the unanimous opposition reported from this supposed 'sample' is the so-called 'snowballing' technique adopted and supervised by persons who are themselves deeply opposed to resumption.

None of the members of this supposed 'civil society' network has ever communicated with the ABG or the landowner associations about the issues on which they so regularly make public comment. They do not countenance the possibility that, in considering resumption, the ABG and the landowner associations find themselves with very limited options as they seek realistic means of not only providing a secure basis for Bougainville's autonomy or possible independence, but also creating new livelihoods for the former mine lease communities.

The sustained attacks by members of the network on President Momis, the ABG, advisers to the ABG[38] and the Australian Government (for funding such advice) are based on the assumption, unsupported by credible evidence, of a strong consensus against the reopening of Panguna or any other form of LSM. They make such assumptions in the process of projecting their own theories, ideologies or needs onto Bougainville, convinced that they are finding or providing the supporting evidence. None has the motivation to understand the facts and the complexity of Bougainville, either in 2016 or in the period from 1987 to 1989.

Their activities have had little impact in Bougainville or on Bougainvilleans, other than providing support to groups seeking control over the region's mineral resources in partnership with foreign investors other than Rio Tinto. Their analysis has, however, influenced the positions of some donor countries, and those of some international NGOs involved in Bougainville. In these ways, they contribute to tensions, divisions and conflicts of which they have little understanding.

Conclusion

Returning to the central issues identified in the introduction to this chapter, I address four main questions:

- Are the origins of the conflict to be found primarily in the concerns of young mine lease landowners committed to permanent mine closure?

---

38   As an adviser to the ABG, funded by Australia at the ABG's request, I have been the subject of sustained criticism and attack from these quarters.

- What was the relationship between Francis Ona, or the NPLA, and the other Bougainvillean groups involved in the origins of the conflict?
- Was there generalised opposition to mining amongst these other groups in the period 1987–89?
- Is there generalised Bougainvillean opposition to the resumption of LSM in 2016?

Having addressed the questions, I shall consider the whether the rectangular model of stakeholder relationships advocated in the introduction to this book effectively captures the complexity of stakeholder relationships in the Bougainville case.

The evidence presented here establishes that the conflict did not originate mainly in the concerns of young mine lease landowners to put an end to LSM. The dominant public voices of Ona and the NPLA have confused the picture, especially where there has been selective reading of their public statements. The local-level politics of LSM was never confined to the mine leases or adjacent areas directly affected by mining. The peculiar features of its island geography, history and political economy mean that Bougainville constitutes a quite distinct political arena within PNG. It is an arena within which LSM is regarded as having had various impacts on all Bougainvilleans, not just mine lease landowners, generating a range of different concerns and demands. There is no evidence of either generalised resistance to LSM per se, by either Bougainvilleans in general or local landowner communities in particular, or of general support for permanent mine closure from either group. Indeed, it is not possible to find a consistent position on the future of mining even in the various documents stating the concerns and demands of the NPLA. There is clear evidence, however, of general (and resounding) rejection by both landowners and the other groups discussed here of the unfairness of the BCA as the basis for mining.

Francis Ona was not a class hero in the vanguard of an anti-capitalist movement. Rather, he was a key actor whose own interests, concerns and networks placed him at a critically important intersection of wider interests, concerns and influences that included Christian and customary values. He arrived at that intersection, and was inspired to take on a leadership role, not only because of the unfairness of his own situation as a young adult landowner, gravely exacerbated by a bitter family dispute, but also because of his own personality. What evidence we have of his motivations and roles points to the importance of individual agency in the relationship between LSM and local-level politics, while the evidence of the roles of other key figures in the various groups considered here indicates the difficulties of mapping the influence of multiple individual motivations.

The linkages between the various groups involved in the origins of the conflict were developing from at least as early as 1987 and, as a result, by early 1988 Ona increasingly saw himself, and was seen by the leaders of other groups, as the key spokesperson for a coalition of interests. While some of his statements could be interpreted as expressions of a desire for permanent mine closure, many were clearly more directed to pressuring BCL to meet specific demands for a fairer mining dispensation which could only be met if mining continued. Those demands reflected the views of the loose coalition with which Ona and the NPLA were working.

The existence of that coalition in 1988 helps to explain some key aspects of the early stages of the conflict that have not previously been well understood. One is the rapid mobilisation of mainly young men in all parts of Bougainville from early 1989. Another is the fact that violence was not directed only against BCL and the PNG security forces, but also against other Bougainvilleans—some wealthy, or educated, or in middle-ranking to senior government or company jobs. The generalised resentment of marginalised young males was the key here. Further, the chaotic situation that emerged after the PNG security forces withdrew in March 1990 was largely due to the BRA's recruitment of criminal gang members and disaffected youth.

With a few notable exceptions, such as Damien Dameng's group, there is no evidence that the other Bougainvillean groups involved in the origins of the conflict were seeking permanent mine closure. In this respect, there are similarities between the situation that existed in 1987–89 and that which exists in 2016. Undoubtedly, in 2016 some people oppose the resumption of LSM under any circumstances. But many other opinions on mining also exist. In the absence of either representative opinion polling or rigorous research, it is difficult for even the ABG, despite its extensive public consultation, to be certain about the preponderance of views on the future of LSM, either on the part of mine lease landowners or Bougainvilleans more generally. It is clearly a mistake to take 'noise' created by fourth estate actors, especially those in the 'Lasslett network', as in any way accurately representing a majority position.

Just as in 1988, there is a multiplicity of positions on this issue. That should be no surprise, given the history of mining, the expectations of mine lease landowners, the ABG's revenue needs, the referendum timetable, and the multiple agendas of foreign actors who have developed links with Bougainvillean groups in the hope of economic advantage from either the reopening of Panguna or the exploration of other prospects.

Again, as in 1988, some Bougainvilleans oppose any form of LSM, including some in the former BCL lease areas strongly opposed to Panguna reopening. However, there is certainly some evidence that major voices in the debate are more concerned with the conditions under which a limited resumption of mining might be permitted—a situation that is also similar to 1988. Many groups and leaders, sometimes claimed by outside commentators to be 'anti-mining' (Anon. 2015), are in fact strong supporters of mining but have their own reasons for attacking the ABG on the issues involved (Regan 2014: 85–93).

To some extent, what appears to be happening in the 'community corner' involves multiple interest groups staking out positions in advance of any decision on the future of mining. Without the existence of political parties conducting policy debate, Bougainvilleans generally expect decisions on major issues to be made by some form of consensus. To influence such decisions, voices in favour of particular positions jostle to be heard. The situation points to difficulties in the operation of systems of elected representative government in a cultural context such as that found in Bougainville.

The ideology of primacy of Bougainvillean rights remains a lasting legacy of the Bougainville conflict, heavily influencing much of the *Bougainville Mining Act 2015*, including provisions on the ownership of minerals by owners of customary land, the distribution of preferences for employment and business opportunities, and the resettlement of communities impacted by mining. These provisions are remarkably similar to the NPLA demands of 1988.

The same principles are also brought to the fore in ABG discussions with the nine landowner associations about the possible future of Panguna. The kinds of landowner benefit packages negotiated for mining projects elsewhere in PNG are unlikely to be sufficient to gain landowner community support for future LSM in Bougainville. Whether any responsible investor will be able to envisage a profitable enterprise in such circumstances has yet to be tested.

As to the reasons for the hitherto very incomplete understanding of the origins of the conflict, analysis of the composition and roles of the network of Bougainvillean groups that were involved underlines the complexity of the 'community corner' in 1988. It also illuminates some of the main reasons for the difficulties involved in gaining accurate understandings of the groups involved and their relationships one with another. These relationships were opaque even to most Bougainvilleans at the time. In large part this was because just one group—the NPLA—was making most or the public statements. Most of the others had no reason to draw attention to themselves. Indeed, some had good reason to avoid scrutiny.

The mainly oral traditions of Bougainvillean communities are also a factor here, because they help to explain why so few of those involved produced written records of what they were doing. After 1988, the situation changed very quickly, as the key issues no longer involved the future of mining. Even so, few—if any—of those previously involved now had any reason to record what they had done or why. Furthermore, there was a long period from the early 1990s when many had good reasons not to discuss their involvement—at least until immunity from prosecution for 'crisis-related' offences was provided under the 2001 BPA.

As to whether the rectangular model of stakeholder relationships effectively captures the complexity of the relationships on Bougainville, evidence from 1988–89 emphasises the danger in seeing one or more communities or voices as constituting a single community 'stakeholder'. The many groups then involved had multiple, and often quite distinct, concerns and agendas, and complex relationships one with another, as well as with actors in the state and company 'corners' at that time. The image of a rectangle, with distinct sets of stakeholders in their separate 'corners', may well be useful as a classificatory device, but, in itself, it does not help us to understand the complexities of relationships within or between the actors in each category. Those relationships are perhaps more akin to complex webs made up of interpersonal and intergroup linkages.

References

AGA (Applied Geology Associates Ltd), 1989. *Environmental, Socio-Economic, Public Health Review of Bougainville Copper Mine, Panguna.* Wellington: AGA.

Anis, T., E. Makis, T. Miriung and E. Ogan, 1976. 'Towards a New Politics? The Elections in Bougainville.' In D. Stone (ed.), *Prelude to Self-Government: Electoral Politics in Papua New Guinea.* Canberra: The Australian National University, Research School of Pacific Studies.

Anon., 1988. 'Rebels Agree to End the Sabotage.' *Post-Courier*, 10 December.

——, 2015. 'Momis Loses the Vote in Panguna Heartland.' PNG Mine Watch Blogpost, 4 June. Viewed 12 October 2016 at: ramumine. wordpress.com/2015/06/04/momis-loses-the-vote-in-panguna-heartland/

——, 2016a. 'Outrage over BCL Shares.' *Post-Courier*, 1 July.

——, 2016b. 'B'ville Majority Owner of Copper Ltd.' *The National*, 18 August.

——, 2016c. 'PM Adament [sic] on Shares.' *Post-Courier*, 19 August.

Auna, J., 1989. 'Affidavit of Joseph Lawrence Auna.' Filed in the Supreme Court of Victoria in CL220 of 1989, *Bougainville Copper Limited v Metals and Minerals Insurance Pte Ltd and Ors.*

Ballard, C. and G. Banks, 2003. 'Resource Wars: The Anthropology of Mining.' *Annual Review of Anthropology* 32: 287–313. doi.org/10.1146/annurev.anthro.32.061002.093116

BCC (Bougainville Constitutional Commission), 2004. 'Report of the Bougainville Constitutional Commission: Report on the Third and Final Draft of the Bougainville Constitution.' Arawa and Buka: BCC.

Bedford, R. and A. Mamak, 1977. *Compensating for Development: The Bougainville Case.* Christchurch: University of Canterbury, Department of Geography (Bougainville Special Publication 2).

Boege, V., 1999. 'Mining, Environmental Degradation and War: The Bougainville Case.' In M. Suliman (ed.), *Ecology, Politics and Violent Conflict*. London: Zed Books.

Braithwaite, J., H. Charlesworth, P. Reddy and L. Dunn, 2010. *Reconciliation and Architectures of Commitment: Sequencing Peace in Bougainville*. Canberra: ANU E Press.

Brown, W., 2014. 'Telefomin and Panguna: A Kiap's View.' In C. Spark, S. Spark and C. Twomey (eds), *Australians in Papua New Guinea 1960–1975*. St Lucia: University of Queensland Press.

Connell, J., 1978. *Taim Bilong Mani: The Evolution of Agriculture in a Solomon Island Society*. Canberra: The Australian National University, Development Studies Centre (Monograph 12).

——, 1988. 'Temporary Townsfolk? Siwai Migrants in Urban Papua New Guinea.' *Pacific Studies* 11(3): 77–100.

——, 1990. 'Panguna Mine Impact (1).' In P. Polomka (ed.), *Bougainville: Perspectives on a Crisis*. Canberra: The Australian National University, Strategic and Defence Studies Centre (Canberra Papers on Strategy and Defence 66).

——, 1991. 'Competition and Conflict: The Bougainville Copper Mine, Papua New Guinea.' In J. Connell and R. Howitt (eds), *Mining and Indigenous Peoples in Australasia*. Melbourne: Oxford University Press.

——, 1992. '"Logic is a Capitalist Cover-up": Compensation and Crisis in Bougainville, Papua New Guinea'. In S. Henningham and R.J. May (eds), *Resources Development and Politics in the Pacific Islands*. Bathurst (NSW): Crawford House Press.

Denoon, D., 2000. *Getting under the Skin: The Bougainville Copper Agreement and the Creation of the Panguna Mine*. Melbourne: Melbourne University Press.

Dorney, S., 1990. *Papua New Guinea: People, Politics and History since 1975*. Sydney: Random House.

Downs, I., 1980. *The Australian Trusteeship: Papua New Guinea 1945–75*. Canberra: Australian Government Publishing Service.

Filer, C., 1990. 'The Bougainville Rebellion, the Mining Industry and the Process of Social Disintegration in Papua New Guinea.' In R.J. May and M. Spriggs (eds), *The Bougainville Crisis*. Bathurst (NSW): Crawford House Press. doi.org/10.1080/03149099009508487

——, 1992. Review of P. Quodling's *Bougainville: The Mine and the People*. *Contemporary Pacific* 4: 440–442.

——, 2005. 'The Role of Land-owning Communities in Papua New Guinea's Mineral Policy Framework.' In E. Bastida, T. Wälde and J. Warden-Fernández (eds), *International and Comparative Mineral Law and Policy: Trends and Prospects*. The Hague: Kluwer Law International.

Fingleton, W., 1970. 'Bougainville: A Chronicle of Just Grievances.' *New Guinea Quarterly* 5(2): 13–20.

Ghai, Y. and A. Regan, 1992. *The Law, Politics and Administration of Decentralisation in Papua New Guinea*. Port Moresby: National Research Institute (Monograph 30).

——, 2000. 'Bougainville and the Dialectics of Ethnicity, Autonomy and Separation.' In Y. Ghai (ed.), *Autonomy and Ethnicity: Negotiating Competing Claims in Multi-Ethnic States*. Cambridge: Cambridge University Press.

——, 2006. 'Unitary State, Devolution, Autonomy, Secession: Dialectics of State Building and Nation Building in Bougainville, Papua New Guinea.' *The Round Table* 95: 386, 589–608. doi.org/10.1080/00358530600931178

Gillespie, W.R., 2009. *Running With Rebels: Behind the Lies in Bougainville's Hidden War*. Port Kembla (NSW): Ginibi Productions.

Griffin, J., 1995. *Bougainville: A Challenge for the Churches*. North Sydney: Catholic Commission for Justice and Peace.

Griffin, J., H. Nelson and S. Firth, 1979. *Papua New Guinea: A Political History*. Richmond (VA): Heinemann Educational Australia.

Griffin, J. and M. Togolo, 1997. 'North Solomons Province, 1974–2000.' In R. May and A. Regan (eds), *Political Decentralisation in a New State: The Experience of Provincial Government in Papua New Guinea*. Bathurst (NSW): Crawford House Press.

Hermkens, A., 2007. 'Religion in War and Peace: Unravelling Mary's Intervention in the Bougainville Crisis.' *Culture and Religion* 8: 271–289. doi.org/10.1080/14755610701652111

Hess, M. and E. Maidment, 2014. 'Industrial Conflict in Paradise: Making the Bougainville Copper Construction Agreement 1970.' *Economic and Labour Relations Review* 25: 271–289. doi.org/10.1177/1035304614533625

Hess, M., E. Maidment, and G. Keimelo, 2016. 'Establishing the Bougainville Mineworkers' Union, 1969–1976.' *Journal of Pacific History* 51: 21–42. doi.org/10.1080/00223344.2015.1095274

Jubilee, 2014. *Voices of Bougainville: Nikana Kangsi, Nikana Dong Damana (Our Land, Our Future).* Sydney: Jubilee Australia.

——, 2015. *The Devil in the Detail. Analysis of the Bougainville Mining Act 2015.* Sydney: Jubilee Australia.

Kenema, S., 2010. 'An Analysis of Post-Conflict Explanations of Indigenous Dissent Relating to the Bougainville Copper Mining Conflict, Papua New Guinea.' *Intersections* 1.2 and 2.1, April.

Laracy, H., 1999. 'Maine, Massachusetts, and the Marists: American Catholic Missionaries in the South Pacific.' *Catholic Historical Record* 85: 566–590. doi.org/10.1353/cat.1999.0139

——, 2005. '"Imperium in Imperio"? The Catholic Church in Bougainville.' In A. Regan and H. Griffin (eds), *Bougainville Before the Conflict.* Canberra: Pandanus Books.

Lasslett, K., 2014. *State Crime on the Margins of Empire: Rio Tinto, the War on Bougainville and Resistance to Mining.* London: Pluto Press.

Loewenstein, A., 2013. *Profits of Doom: How Vulture Capitalism is Swallowing the World.* Melbourne: Melbourne University Press.

Lummani, J., 2005. 'Post-1960s Cocoa and Copra Production in Bougainville.' In A. Regan and H. Griffin (eds), *Bougainville Before the Conflict.* Canberra: Pandanus Books.

MacWilliam, S., 2005. 'Post-War Reconstruction in Bougainville: Plantations, Smallholders and Indigenous Capital.' In A. Regan and H. Griffin (eds), *Bougainville Before the Conflict*. Canberra: Pandanus Books.

——, 2013. *Securing Village Life: Development in Late Colonial Papua New Guinea*. Canberra: ANU E Press.

Mamak, A., R.D. Bedford, L. Hannett and M. Havini, 1974. *Bougainvillean Nationalism: Aspects of Unity and Discord.* Christchurch: University of Canterbury, Department of Geography (Bougainville Special Publication 1).

McNee, G., 2015. 'Social Mapping and Landowner Identification Desktop Study.' In *Final Report Prepared for Joint Panguna Negotiations Coordination Committee.* Perth: Social Sustainability Services Pty Ltd.

Mitchell, D.D., 1976. *Land and Agriculture in Nagovisi, Papua New Guinea.* Boroko: Institute of Applied Social and Economic Research (Monograph 3).

——, 1982. 'Frozen Assets in Nagovisi.' *Oceania* 53: 57–65.

Momis, J., 2015. 'ABG Position on Panguna Mine Future, and PNG Proposals to Purchase Rio Tinto Shares in BCL.' Statement by President John L. Momis to the Bougainville House of Representatives, 22 December.

Mungkas, 1988. 'Mungkas Ass. Meet 2/88 Prov. Commerc. Conf. Room 24/3/88, 1800.' Handwritten meeting minutes.

Nash, J. and E. Ogan, 1990. 'The Red and the Black: Bougainvillean Perceptions of other Papua New Guineans.' *Pacific Studies* 13(2): 1–17.

Nelson, H., 2005. 'Bougainville in World War II.' In A. Regan and H. Griffin (eds), *Bougainville Before the Conflict*. Canberra: Pandanus Books.

NSPG (North Solomons Provincial Government), 1986a. 'North Solomons Provincial Policy for Youth in Their Communities.' Arawa: NSPG Division of Community Affairs (unpublished report).

——, 1986b. 'Department of North Solomons: Information Paper No.3 July 1986.' In A. Sarei (ed.), *Department of North Solomons, Information Papers: Provincial Task Force; Provincial Disaster Relief Programme; Squatter Settlements.* Arawa: Department of North Solomons (unpublished report).

——, 1988a. 'North Solomons Provincial Government 1988 Law and Order Report.' Arawa: NSPG Planning Office (unpublished report).

——, 1988b. 'Special Meeting between the National Government and North Solomons Provincial Government Held at the Provincial Assembly Hall on the 26th May 1988.' Unpublished meeting minutes.

NZine, 2000a. 'Conflict in Bougainville—Part 2: Interview with Sam Kauona Sirivi.' Blogpost by 'Dorothy', 23 June. Viewed 16 March 2017 at: www.nzine.co.nz/features/bville2.html

——, 2000b. 'Conflict in Bougainville—Part 3: Successes of the Bougainville Revolutionary Army.' Blogpost by 'Dorothy', 23 June. Viewed 16 March 2017 at: www.nzine.co.nz/features/bville3.html

Ogan, E., 1965. 'An Election in Bougainville.' *Ethnology* 14: 397–407. doi.org/10.2307/3772789

——, 1971a. 'Charisma and Race.' In A. Epstein, R. Parker and M. Reay, (eds), *The Politics of Dependence: Papua New Guinea 1968.* Canberra: Australian National University Press.

——, 1971b. 'Nasioi Land Tenure: An Extended Case Study.' *Oceania* 42: 81–93. doi.org/10.1002/j.1834-4461.1971.tb00306.x

——, 1972. *Business and Cargo: Socio-Economic Change among the Nasioi of Bougainville.* Port Moresby and Canberra: The Australian National University, New Guinea Research Unit (Bulletin 44).

——, 1990. 'Perspectives on a Crisis (5).' In P. Polomka (ed.), *Bougainville: Perspectives on a Crisis.* Canberra: The Australian National University, Strategic and Defence Studies Centre (Canberra Papers on Strategy and Defence 66).

——, 2005. 'An Introduction to Bougainville Cultures.' In A. Regan and H. Griffin (eds), *Bougainville Before the Conflict.* Canberra: Pandanus Books.

Okole, H., 1990. 'The Politics of the Panguna Landowners' Organization.' In R. May and M. Spriggs (eds), *The Bougainville Crisis*. Bathurst (NSW): Crawford House Press.

Oliver, D., 1991. *Black Islanders: A Personal Perspective of Bougainville 1937–1991*. Melbourne: Hyland House.

Oliver, M., 1998. 'July–December 1988.' In C. Moore and M. Kooyam (eds), *A Papua New Guinea Political Chronicle 1967–1991*. Bathurst (NSW): Crawford House Publishing.

PLA (Panguna Landowners Association), 1988. '11th March Mine Closer [sic] Demonstration—1988: The Main Topics to be Highlighted in the Speech by Members of Panguna Landowners Association.' Unpublished document.

Quodling, P., 1991. *Bougainville: The Mine and the People*. Sydney: Centre for Independent Studies (Pacific Paper 3).

Regan A., 1998. 'Causes and Course of the Bougainville Conflict.' *Journal of Pacific History* 33: 269–285. doi.org/10.1080/00223349808572878

——, 2002. 'Bougainville: Beyond Survival.' *Cultural Survival* 26: 20–24.

——, 2005a. 'Identities among Bougainvilleans.' In A. Regan and H. Griffin (eds), *Bougainville Before the Conflict*. Canberra: Pandanus Books.

——, 2005b. 'A Note for Readers.' In A. Regan and H. Griffin (eds), *Bougainville Before the Conflict*. Canberra: Pandanus Books.

——, 2011. *Light Intervention: Lessons from Bougainville*. Washington (DC): United States Institute of Peace Press.

——, 2014. 'Bougainville: Large-Scale Mining and the Risks of Conflict Recurrence.' *Security Challenges* 10(2): 71–96.

Rimoldi, M. and E. Rimoldi, 1992. *Hahalis and the Labour of Love: A Social Movement on Buka Island*. Oxford: Berg.

Roka, L., 2014. 'Bougainville Manifesto.' Hervey Bay: Puk Puk Publications.

Singkai, G., 1990. 'Diocese of Bougainville, March 08, 1990: Press Release.' Unpublished document.

Tanis, J., 2005. 'Nagovisi Villages as a Window on Bougainville in 1988.' In A. Regan and H. Griffin (eds), *Bougainville Before the Conflict*. Canberra: Pandanus Books.

Taruna, J., 2010. 'A Bougainville Israeli Link: A Completion of the New Covenant Signed by the President, Dr. John Momis on the [sic] October 28th 2010.' Unpublished paper.

Thompson, H. and S. MacWilliam, 1992. *The Political Economy of Papua New Guinea: Critical Essays*. Manila: Journal of Contemporary Asia Press.

Transpacific Ventures, 2013. 'Updated Information Memorandum to Subscribe for 8,000,000 Shares at A$0.10 per Share in Transpacific Ventures Limited.' Unpublished document.

Tryon, D., 2005. 'The Languages of Bougainville.' In A. Regan and H. Griffin (eds), *Bougainville Before the Conflict*. Canberra: Pandanus Books.

Turner, M., 1990. *Papua New Guinea: The Challenge of Independence*. Ringwood (VA): Penguin Books.

Tyler, M., 1988. 'Nutrition & Subsistence Food Production Project, Siwai North Solomons Province, September 1985 – June 1988.' Arawa: North Solomons Provincial Government (unpublished report).

Vernon, D., 2005. 'The Panguna Mine.' In A. Regan and H. Griffin (eds), *Bougainville Before the Conflict*. Canberra: Pandanus Books.

Wesley-Smith, T., 1992a. 'The Non-Review of the Bougainville Copper Agreement.' In M. Spriggs and D. Denoon (eds), *The Bougainville Crisis: 1991 Update*. Canberra: The Australian National University, Research School of Pacific Studies, Department of Political and Social Change (Monograph 16).

——, 1992b. 'Development and Crisis in Bougainville: A Bibliographic Essay.' *Contemporary Pacific* 4: 407–432.

Wiley, B., 1992. 'Bougainville: A Matter of Attitude.' *Contemporary Pacific* 4: 376–378.

Woolford, D., 1975. *Papua New Guinea: Initiation and Independence*. St Lucia: University of Queensland Press.

13. Between New Caledonia and Papua New Guinea

COLIN FILER AND PIERRE-YVES LE MEUR

In this concluding chapter, we return to the question posed in the introduction—the question of how to explain the similarities and differences in the relationship between large-scale mines and local-level politics in New Caledonia (NC) and Papua New Guinea (PNG), or in different parts of these two countries. The subtitle of this book (and the title of this chapter) could be read as a sign of our belief that the relationship between large-scale mines and local-level politics is one whose form and content mainly varies between countries or jurisdictions, largely as a result of differences in their political history and the current legal and policy frameworks in which the relationship is embedded. However, while there clearly are some points to be made on this score, the contributions to this volume also show that the relationship varies between projects and places as much as it does between countries, so there is no reason to assign a special kind of power to what we described (in Chapter 1) as the 'state corner' in our rectangular model of stakeholder relationships. Instead, we may conceive of the space 'between' NC and PNG as one that contains a number of different relationships between large-scale mines and local-level politics whose variation can partly be understood in terms of the 'balance of power' between the four corners in that model, but partly also in terms of other dimensions of difference.

Between Version One

Although the contributors to this volume have a good deal to say about political conflict, they have eschewed the well-worn concept of the 'resource curse' (Ross 2003) and the related question of whether political actors in mine-affected communities are motivated by 'greed or grievance' (Collier and Hoeffler 2004). Anthony Regan's review of the history of political conflict around the Panguna mine (Chapter 12) does not lead us to conclude that local political reactions to large-scale mining projects are generally more intense or more violent in PNG than they are in NC, and even if we had evidence to this effect, we could not readily link it to the extent of each country's economic dependence on the mining industry, nor to the motivations of political actors who belong to mine-affected communities. If it barely makes sense to say that large-scale mines are 'more political' in one country than in the other, then it might make more sense to argue that large-scale mining operations are not so much *causes* of political conflict as things or spaces that *attract* it, in which case the form and extent of conflict in each country needs to be considered as a sort of independent variable (Banks 2008).

We are also reluctant to conclude that variations in the relationship between large-scale mines and local-level politics can simply be explained by reference to the fact that PNG and NC represent or possess two different kinds of 'society'—say a 'post-colonial' one as opposed to a 'late colonial' or 'semi-colonial' one. In this respect, we agree with actor-network theorists who would say that this does not count as an explanation so much as a summary of what it is we are trying to explain (Latour 2005). To emphasise this point, we have used the noun 'society' in a very different way, to designate a recently and somewhat poorly assembled group of 'stakeholders' in our model of political relationships. The question then is whether the four groups of stakeholders in our model have been assembled in different ways in the two countries under discussion. On this score, it could be argued that the rectangular model is better 'fitted' to the politics of the large-scale mining industry in PNG than to its counterpart in NC because there is more power, or a group of more powerful actors, located in PNG's 'social corner', and less power, or a group of less powerful actors, located in PNG's 'state corner' (Chapter 9).

Of course, there is no escaping the significance of what Le Meur (Chapter 5) calls the 'meta-conflict' of settler colonialism in any account of what sets the two countries apart. All five of our chapters on NC deal with mine-affected communities that are primarily Kanak communities in a country (or territory) where Kanaks still account for a minority of the total population. In some respects, the political complexion of the mining industry in NC is more like that found in other countries where indigenous people are still subordinate to people of European descent than like that which obtains in the independent and 'indigenous' state of PNG. Even so, the fact that Kanaks account for roughly 40 per cent of NC's population explains why their leaders have been able to pursue the path of decolonisation and independence, rather than simply seek to advocate their rights as an indigenous minority.

Although PNG has its own colonial legacy, the Autonomous Region of Bougainville is the only part of the country where this legacy continues to cast a shadow across the politics of the mining industry, and that is because the Panguna mine was conceived and built as an Australian colonial enterprise. Even in nine decades of colonial rule, PNG never witnessed the wholesale expropriation of the native population, nor the creation of 'tribal reserves' of the kind established in NC, nor even the establishment of an effective system of indirect rule on the part of the colonial administration. And while PNG's achievement of national independence took place at the same time that Kanaks began to demand the right of self-determination, the slow process of decolonisation in NC has resulted in a form of local-level politics that is dominated by a deepening division between French and Kanak political institutions. Furthermore, this division has been complicated by a further split between those Kanak political institutions that were part of the colonial system of indirect rule and those that have since been created in response to the demand for self-determination (Chapters 4 and 6). The only parallel to be found in PNG's post-colonial form of local-level politics is the one between the laws inherited from Australia and those created to accommodate what the National Constitution nominates as 'Papua New Guinean forms of social, political and economic organisation'.

The key point about the power of the state in NC is that local (Caledonian) institutions are opposed to metropolitan (French) institutions at the same time that 'neotraditional' (Kanak) institutions are opposed to 'modern' (European) institutions (Demmer and Salomon 2017). This is less a form of 'hybridity' (Clements et al. 2007) than a sort of double duality that diminishes the power of 'the state as such' at the same time that it makes more space for the practice of local-level politics in the suspended moment of the Nouméa Agreement preceding the referendum on independence to be held in 2018 (Le Meur 2017). This should not lead us to think that there is any significant difference between the two countries in the number of roles (per capita or per project) that make up the political structure of the mining industry, but it does lead us to wonder if the practice of politics has escaped, avoided or even subverted the practice of government—or governmentality—in PNG's version of this structure to an extent that is not so evident in NC.

If this means that community actors in PNG now wield more power than their (indigenous) counterparts in NC, it also seems to be consistent with the observation that this power is exercised within a neotraditional political order that is rather more chaotic and contestable, mainly because the institution of chieftainship is less significant. However, it is hard to tell how much of this difference in the form of local-level politics is due to the impressions left by the colonial legacy, to the relative strength or weakness of contemporary state institutions, or to differences in pre-colonial 'political cultures' (Chapters 4, 8 and 9).

When it comes to the politics of the large-scale mining industry, what clearly does matter is that all large-scale mining operations in PNG have required the state's acquisition of large areas of customary land, whereas those in NC tend to exclude customary claims to the state land on which they are almost invariably located (Chapters 3 and 4). While this explains the difference in the power that community actors can exercise over the mining industry in their capacity as 'customary landowners', it also explains why community actors in both countries have been slow to represent themselves as 'indigenous peoples' whose land rights need to be reinforced by appeals to international norms. In the PNG case, political actors in all corners of our rectangular model seem to believe that the 'customary landowner' is already a more powerful character than any kind of indigenous person (Chapter 9). In the NC case, by contrast, the limits previously imposed on the extent of customary land rights seem to have discouraged Kanak political leaders from using such rights as the basis of their demand to exert greater control over the operation of the mining industry (Chapter 6), even though mining agreements have included a recognition that the original occupants of any land should take precedence over those who followed (and sometimes dispossessed) them (Chapters 3 and 4).

If the 'ideology of landownership' therefore has much greater political weight in PNG than it has in NC, there is also a notable difference in the qualities and quantities of the 'benefit streams' that flow to mine-affected communities. This is not just a difference in the extent to which communities are entitled to compensation for the use of their customary land, but also a function of the fact that NC lacks any counterpart to the institutional machinery through which PNG's national government collects an output-based royalty from each large-scale mining project and then redistributes the whole of this income to different groups of political actors (including groups of customary landowners) within the province where the project is located. Furthermore, NC lacks any obvious counterpart to the policy by which PNG's national government obliges mining companies to make special efforts to train and employ the members of mine-affected communities. As a result, Kanak political leaders have laid a greater emphasis on what PNG's policy regime designates as 'equity benefits' and 'business development opportunities', which means that the distribution of shares in mining companies and their suppliers becomes a matter of paramount concern (Chapters 2 and 6). This type of benefit is also a matter of great concern in PNG, but the wider range and volume of 'landowner benefits' or 'community benefits' in PNG seems to have created more of a stimulus for the process that Glenn Banks and colleagues (in Chapter 7) call 'immanent development', and more of an incentive for the local beneficiaries to exclude 'outsiders' from their realm of entitlement (Chapter 11).

In one sense, this would suggest that PNG has a more intense form of identity politics, if the key political question is who counts as a landowner or community member in respect of any particular mining project. However, that question is clearly related to the intensity of local-level political contests over the distribution of material costs and benefits, and it could be argued that this type of contest is less intense in NC because there is less at stake. In that case, it would seem to follow that NC does not have a *less intense* form of identity politics, but a *different form* of identity politics—one that is more concerned with what we have called (in Chapter 1) the representational issue than the distributional issue, and more concerned with the establishment of a Kanak identity than a strictly local identity (Chapters 4 and 5).

These considerations lead us to wonder whether and how it makes sense to regard PNG and NC as countries located at different points along one or more historical trajectories, and if so, how it makes sense to regard PNG as a country that is more 'advanced' in the political construction of its mining industry, despite the enduring paradox of human poverty in the midst of mineral wealth (Filer et al. 2016). If PNG is more advanced in the process of decolonisation, and also more advanced in the double-sided process of modernisation and globalisation, how does the relationship between these different trajectories produce a national contrast in the relationship between large-scale mines and local-level politics?

Although we have postulated a general process of modernisation and globalisation within the large-scale mining industry, this should not be taken to mean that we subscribe to a version of 'modernisation theory' (or 'dependency theory') that would allow us to arrange our two countries or their component parts, including their mining enclaves, along a single path that leads to a single goal, whether that goal be conceived as a condition of 'modernity', 'development' or 'neoliberal governmentality'. As we pointed out in the introduction to this volume, the process has several different aspects that are not invariably found in combination with each other, and each of these has different political effects. We have not suggested that this process is one variant of an even more general process that has taken the same form over the same period of time in all other branches of the global capitalist economy. We have not even suggested that the process is permanent and irreversible in those places where its effects can be observed, since the mining industry is notorious for its booms and slumps, and one feature of the process we have described is that the mining industry now moves more rapidly from one location to another, often leaving desolation in its wake.

When we say that PNG contains a more 'advanced' form of the process, we partly refer to the extent of the gulf between multinational mining companies and mine-affected communities (Chapter 7), to the way in which *representations* of this gulf contributed to the 'enlightenment' of the global mining industry in the 1990s (Chapter 9), and to the role of international actors like the World Bank in attempting to repatriate the lessons of this enlightenment to the country from which they had been drawn (Filer et al. 2008). In all such respects, it could be said that NC was left behind for a while, even if it is now catching up with some of the trends, as new 'social' actors and political narratives have entered the mineral policy domain. The discourse of environmental protection, supported by non-governmental and community-based organisations, is an obvious case in point (Chapters 5 and 6), but this is only one part of the broader discourse of corporate social responsibility.

We might also be tempted to argue that PNG's large-scale mining industry is also very modern because 'society' has as much weight as the state in building the metaphorical roads or bridges that connect big mining companies with groups of indigenous people or customary landowners. If we take that path, we might then say that mine-affected communities in NC are also 'state-affected communities' to an extent that is not true of PNG because the PNG Government is well known for its habit of disappearing from mine-affected areas once development agreements have been signed (Chapters 8 and 9). In that case, the political complexion of PNG's mining industry appears to possess a 'neoliberal' character that is not (yet) evident in NC, since the 'selective absence of the state' (Szablowski 2007) appears to have been the result of an ongoing transfer of power from state actors to both corporate and social actors. However, we also need to bear in mind that NC's oldest mining company (SLN) already held quasi-governmental powers before the emergence of neoliberal forms of governance or governmentality in the period since 1980 (Chapter 5).

If PNG and NC are still to be conceived as jurisdictions located at different points along a single historical trajectory, then this path may also be construed in terms of a shift from the social relations of employment to the social relations of compensation, so long as 'compensation' is understood to be something broader than compensation for the loss of customary land. The broader concept belongs to the general model of *reflexive* modernisation (Beck 1992), characterised by the double internalisation of risk as a social construct and a way of ordering material reality (Dean 2010). This form of 'risk society' is obviously connected with the practice of environmental (and social) impact and risk assessment and with norms of environmental (and social) justice (Dupuy 2002; Walker 2012). From this point of view, we may then proceed to ask whether the transformation of social relations as the subject of political action has still had somewhat different political effects in each of our two countries because of the manner in which the social relations of compensation have come to encompass the mining industry.

On this score, we would say that PNG acquired a resource-dependent economy in the process of dispensing with the legacy of colonial rule and, for this reason, the modernisation of the mining industry was encompassed by a national ideology—the ideology of landownership— that served to intensify the social relations of compensation and extend their reach to parts of the country that were not directly affected by any large-scale mining operations (Filer 1997). In NC, by contrast, a 'traditional' large-scale mining industry already dominated the colonial economy at the time when PNG achieved its independence in 1975, and the process of modernisation did not begin to have political effects until PNG had already acquired its new national ideology. What happened instead is that the exercise of political control over the mining industry became the means by which representatives of the Kanak population have sought compensation, not for the loss of specific areas of customary land, but for the broader historical experience of dispossession and subordination under the French colonial regime. As a result, the political relationship between ideas of indigenous identity and environmental justice seems to have a very different complexion (Ali and Grewal 2006; Le Meur 2010). In this respect, the 'half-way house' would be the case of Bougainville, where a modern mine was established before PNG became an independent country, but then became a symbol of neocolonial dispossession for a new generation of secessionists whose actions were triggered by demands for environmental (and social) justice (Chapter 12).

Between Version Two

Once we think of Bougainville as a 'mineral province' that lies between PNG and NC in both a political and a geographical sense, we can immediately see that a comparison between two national jurisdictions is only one type of comparison that can be made when we seek to understand variations in the relationship between large-scale mines and local-level politics. In one political respect, Bougainville and NC have more in common with each other than either has with (the rest of) PNG, since both are 'partial countries' confronting the prospect of a referendum on independence, and Bougainvillean nationalism bears some comparison with Kanak nationalism as an ideology that plays out in the politics of the mining industry. But if we go one step further and think of Bougainville as one of *several* 'mineral provinces' in PNG, then the questions previously posed about the difference between two national jurisdictions, two sets of political institutions, or two historical trajectories, can be turned into questions about the difference between a larger number of smaller political entities and the mining operations that they have hosted.

As noted in the introduction to this volume, PNG is a much bigger country than NC, both in terms of the scales used by cartographers to measure surface areas and in terms of the size of their respective populations. The whole of the resident population of NC is of a similar order of magnitude to that of the Autonomous Region of Bougainville and to each of PNG's five provinces that currently host at least one large-scale mining project. While NC has been divided into three provinces, two of which can be counted as 'mineral provinces', one of them (North Province) has a much smaller population than the other one (South Province), comparable to the population of what officially counts as a 'district' in PNG. Yet there is no simple correlation between the size of a mineral province and the size or shape or scale of the mining operations located within it. In both of NC's mineral provinces, large-scale processing plants derive their raw material from mining operations that vary a good deal in their scale and location. This type of variation is not found in PNG's five mineral provinces, where each big mining project consists of a single processing plant that derives its raw material from a single mining operation. Yet these projects also vary in scale and location and impact, and this variation in the 'forces of production' is related to variations in way that 'mine-affected communities' are constructed as political entities in their own right and the way that 'localities' are to be defined in any assessment of local-level politics.

The entire population of Bougainville has come to be constructed as a single 'mine-affected community' because of the conflict that initially led to the closure of the Panguna mine and then escalated in the wake of that event, even if the customary owners of the mine lease areas still count as a more specific collection of political actors with a greater sense of grievance and entitlement (Chapter 12). Two-thirds of the population of PNG's Western Province has been officially recognised as a community or collection of communities affected by the operation of the Ok Tedi mine because of the scale of the environmental damage it has caused (Chapter 8). In contrast, the community affected by the Lihir project in New Ireland Province is generally defined—and certainly defines itself— as a much smaller collection of people with customary rights to land in the offshore group of islands where the mine is located (Chapters 10 and 11). And the communities affected by some of the mining projects in NC may be smaller still, comprising the resident populations of single communes or municipalities (Chapters 4, 5 and 6). But then again, as in the case of Bougainville, it could be argued that the whole of the Kanak population counts as one 'mine-affected community' because of the way that control of the mining industry has come to be defined as a focal point in their struggles for autonomy and independence. Furthermore, NC has the hallmarks of a single mining enclave or mineral province because of the density of the physical and social networks linking all of the actors in the local nickel industry, despite the segmentary form of local politics.

In our previous comparison between national jurisdictions, we were mainly concerned with the difference between (national-level) political *institutions* that have influenced or constrained the relationship between large-scale mines and local-level politics. This type of comparison bestows a measure of primacy on the 'state corner' in our rectangular model of stakeholder relationships, even when the relative 'power of the state' is understood to be something that varies from one jurisdiction to another. We are now moving in a rather different direction, where each mineral province is not only conceived as a sub-national level of political organisation, but also as a geographical space within which state actors have developed a distinctive set of political relationships with actors in the corporate, social and community corners. And at this juncture, we need to be wary of an ambiguity commonly found in academic discussion of the 'politics of scale', whereby scales and levels of political (or economic) organisation are not clearly differentiated because both are treated as artefacts of political (or economic) activity (Leitner et al. 2008; Huber and Emel 2009; Allen 2017).

Once this ambiguity is resolved, two additional dimensions can be added to the institutional dimension of difference in the politics of the mining industry. On one hand, there is a vertical or *organisational* dimension in which all four corners of our rectangular model, including the community corner, are internally stratified into levels or layers, so that some positions or roles are notionally 'higher up' than others. On the other hand, there is a horizontal or *spatial* dimension in which the four corners are also internally divided by the relative geographical proximity of different positions or roles to the project that constitutes the focal point of political activity. In both dimensions, there is room for variation in the number of political roles, and hence the number of political actors, in each of the four corners, including the number that belong to the 'local' level in an organisational hierarchy or the 'central' zone in a sort of spatial hierarchy.

The point of making this distinction is to understand the powers exercised by actors who occupy one position in the organisational dimension and another position in the spatial dimension. If we look at the corporate corner, for example, we commonly find that a multinational mining company's 'country manager', operating at the national level, typically has less power than the mine manager who operates at the local level. Both are separately accountable to the company's global or regional headquarters, but the country manager is typically responsible for the management of relationships with state actors who also operate at the national level, in the national capital, whereas the mine manager is responsible for the actual operation of the mine, as well as the management of relationships with mine-affected communities.

Even in the state corner, where the powers formally exercised by politicians and public servants commonly seem to be a function of their level of election or appointment, the powers that they actually exercise at a local level often turn out to be a function of their political and social connections to the place where a mining project is located. What this means is that the distance between the project and the capital of the mineral province in which it is located, let alone the national capital, generally turns out to be a critical variable in the exercise of state power. And in the community corner itself, where the emergence or elaboration of an organisational hierarchy often counts as one of the effects of a large-scale mining project, there is no reason to assume that the actors who occupy the most senior positions will be those who occupy the innermost zone of maximal environmental impact. In other words, the practice of local-level politics may be partly motivated by a mismatch between the internal organisation of 'the community'—or its form of representation— and the relative proximity of different groups of people to the site of the mining operation, or the degree of 'impact' that they experience as a result of it, or the types and amounts of benefits (or costs) that they derive from it.

The exercise of power in the practice of local-level politics is not simply a function of the position that an actor happens to occupy in one of the four corners of our rectangular model, or in one of the levels in an organisational hierarchy, or one of the concentric zones of impact or benefit, grievance or entitlement, that surround a specific mining project. As stated in the introduction to this volume, it is primarily revealed in the capacity of actors to engage in specific forms of action: to move from one corner to another, from one level to another, from one zone to another; to control, support or attack the movements of other actors between different positions; and to challenge or defend the legitimacy of the positions themselves, as well as the manner of their occupation.

The history of mine-related political action in Bougainville (Chapter 12) presents us with an extreme case of the fluidity and complexity that can result from this exercise of power. The closure of the Panguna mine was not only accompanied by the evacuation of the corporate corner and the collapse of the organisational hierarchy of the provincial government, but also by the emergence of new (contested) hierarchies along the axis linking the community and social corners within the province. This in turn entailed the emergence of new (contested) distinctions between these two corners, and hence in the definition of who has how much of a claim to be part of the community affected either by the mine or by the conflict that engulfed it. While Anthony Regan argues that the social corner was barely occupied until foreign activists began to campaign against the reopening of the mine, it is clear from his own account that it was already occupied by the Christian churches and other organisations during the original campaign over the distributional issue. More recently, the state and corporate corners have been repopulated in new forms, but the distributional and representational issues are still deeply contentious.

The Ok Tedi mine was the second of PNG's mining projects to be designated as a project of 'national significance', but despite its size and the physical extent of its impact, this mine has acquired a form of local-level politics that is quite unlike its counterpart in Bougainville. For one thing, there has been much greater interference by national-level state actors, partly to offset or reinforce the marginal role of the Fly River Provincial Government. Although this aspect of the balance of power within the state corner has not prevented the mining company or its corporate shareholders from performing a quasi-governmental role, the recent nationalisation of the mine has only served to make the company seem even more like a branch of the national government (Chapter 8). The consolidation of a multifunctional organisational hierarchy along this axis has been accompanied by a division of the mine-affected area into zones whose boundaries have been largely uncontested, and the communities contained within these zones are notable for the absence of organisational hierarchies in which positions of leadership or seniority are the subject of alternative forms of political contest. As a result, the organisational (and institutional) complexity that surrounds the operation of the mine and the management of 'community affairs' has no counterpart in a community corner that seems to be as 'flat' as it is 'wide' (Burton 1997). And despite the activation of the social corner during the (relatively brief) period when some community members were engaged in litigation against the mining company, this is a case that shows the absence of any correlation between the scale of a mining project and the amount of local-level politics that it generates. It is more like an extreme case of a mine that functions as an 'anti-politics machine' (Ferguson 1990).

The Koniambo project is the closest approximation to a project of national significance in NC, even if the nation in question is the one imagined by the Kanak nationalist movement. Its scale relative to the size and population of North Province is comparable to that of the Panguna and Ok Tedi mines in their respective provinces, but because it has only been developed over the course of the past decade, there remains a good deal of uncertainty about the direction in which the practice of local-level politics is moving. In the planning and construction of the main processing plant, provincial state actors have played a mediating role between the corporate and community corners (Chapter 2), which is not unlike the role played by some national state actors in the original planning and construction of the Panguna and Ok Tedi projects during the 1960s and 1970s. This means that the risk of 'social disintegration' within the community corner has been addressed by the application of 'normative frameworks' generated at the provincial (state) level (Chapters 2 and 3). Nevertheless, these efforts have not been sufficient to conceal horizontal lines of political cleavage between zones of engagement (or entitlement) in the 'logic of proximity' to the processing plant, since community actors at the centre of the action appear to be less satisfied with the provincial solution to the distributional issue, and more inclined to support or oppose the project on the grounds of their 'traditional' political divisions (Chapter 3).

As previously noted, the factor that complicates the mediating role of (provincial) state actors in this context is the dual form of organisational hierarchy in which some of these actors occupy positions in the neotraditional political system as well as the formal system of electoral politics and bureaucratic administration. The split between chieftaincies and municipalities in the practice of local-level politics seems to have been less significant in the recent history of the Koniambo project than in the contest surrounding the closure and possible reopening of the Boakaine mine in the same province (Chapter 4). If this mine were to reopen, then it would be part of the network of relatively small mining operations that supply additional raw material to the Gwangyang plant in South Korea (Chapter 3), but its continued closure signals the presence of a barrier between the contested politics of the local community—the municipality and chieftaincies of Canala—and the entanglements of corporate and state actors operating at the level of North Province (Chapter 2).

If this form of political disengagement has been a function of the physical distance between the mine-affected community and the provincial capital, the same point would seem to apply to the neighbouring community of Thio in South Province (Chapter 5). In this case also, mining operations were disrupted by an episode of political conflict, yet the causes and consequences of that conflict were different because the practice of local-level politics in this community has reflected the relative subordination and marginalisation of Kanak political actors in a different provincial context. In this case, the French company that 'traditionally' dominated the whole of the mining industry in NC was unable to maintain its customary practice of dealing privately with community protest, while community leaders deliberately made the conflict public as part of their demand for state actors to take more responsibility for validating their new deal with the company. However, this process itself has resulted in new forms of community engagement with actors operating in the social corner, whose actions tend now to be justified by reference to global, rather than provincial or national, normative frameworks.

These new forms of engagement are even more prominent in the politics of the Goro project, which is the second of the truly modern large-scale projects in NC, albeit one whose 'mine-affected community' is no larger than those attached to the medium-scale mining operations in Canala and Thio (Chapter 6). Here the social corner has been occupied by international and Caledonian non-governmental organisations (NGOs) concerned with environmental protection, operating in an unstable (and sometimes ruptured) alliance with community organisations and 'customary authorities' (Horowitz 2012). These actors have combined different forms of expression, protest and action, from the mobilisation of legal competencies and scientific counter-expertise to demonstration, blockage and a limited use of violence. While the NGOs have taken on some aspects of the mediating role played by provincial (state) actors in the organisation of the Koniambo project, their involvement has also created the opportunity for community actors to adopt a wider range of strategies and tactics in their own attempts to extract a 'sustainable development agreement' from the mining company (even if the company has since failed to uphold its own side of that bargain).

On most counts, the Lihir project in PNG is bigger than both the Goro and Koniambo projects, but has never been officially designated as a project of national significance, and is isolated from the capital of its mineral province, not only in terms of physical distance but also by virtue of its location on a small island. The insularity of the mine-affected community has exercised a profound influence on the practice of local-level politics, in a manner that would no doubt be replicated if a large-scale mine were to be developed on one of the islands in NC's Loyalty Islands Province. While this community is internally divided by the strength of claims to ownership of the customary land leased to the mining company, and hence by a 'logic of proximity', it is also united by an ideology of landownership that denies membership of the community to anyone who lacks customary land rights in any part of the Lihir island group. While the status of the 'immigrant outsider' has become a political issue in the vicinity of other mining projects (Chapters 2 and 5), and was indeed one of the issues behind the eruption of the conflict that closed the Panguna mine (Chapter 12), this issue has taken on a peculiar form in the Lihirian context because of the way that such people have become the clients of 'landowner patrons' competing with each other for positions of seniority in the community's 'socioeconomic hierarchy' (Chapter 11). In effect, community leaders have reconstructed the social corner in our rectangular model to be one that is not occupied by 'civil society', but by members of a national society who have no formal rights to occupy any of the other three corners, and therefore constitute a rather different kind of nuisance.

The community–society axis has another kind of significance in this context, because of the way that the PNG Government has been persuaded (primarily by actors from the World Bank Group) to adopt the principle of 'gender mainstreaming' or 'gender equity' as a feature of its own mineral policy regime. This has created a space or corridor within the social corner through which a few Lihirian women have departed the shores of their island community in order to participate in national and global debates about the implementation of this policy. However, as Susan Hemer explains (in Chapter 10), this opportunity has been taken at the cost of their status and influence within a community in which men seek to maintain a 'customary' monopoly over the practice of local-level politics, and women can only be granted a form of authority if it is not 'political'.

Four-Wheel Drive?

While a case can be made for limiting the definition of 'local-level politics' to the political practice that takes place within mine-affected communities and along the three channels of communication between community actors and actors in the other three corners of our rectangular model, the internal constitution of the other three corners, and the three distinctive channels of communication between them, can also have local dimensions and effects (Chapter 9). If the roles or positions assembled in each of these four corners constitute the political structure of the mining industry in a formal sense, the manner in which they are created and occupied is closely related to what we have called the representational issue, which is one of the two big issues on which our political actors act (Chapter 1). However, we can now see why the representational issue is somewhat less significant than the distributional issue, since the positions that our actors occupy can mostly be assigned to the organisational dimension of difference between the sites of their political activity. The distributional issue, by contrast, is a matter of substance with which they engage from positions in all three of the dimensions of difference that we have so far identified—not just the organisational dimension, but also the spatial and institutional dimensions.

This leads us to wonder whether there is a fourth dimension of difference or variation in the relationship between large-scale mines and local-level politics that could be just as significant. One obvious candidate would be the gender dimension, since it is widely recognised that big mining projects normally have a significant impact on gender relations in mine-affected communities (Lahiri-Dutt 2011). There is clearly a good deal of scope for variation between different projects, not only in the extent to which this impact has become a political issue for different groups of actors, but also in the consequences of the policies or activities by which it is addressed. The main reason that we have not treated this as the fourth dimension of difference in this volume is that only one of our contributors has dealt with the gendered nature of local-level politics in any detail (Chapter 10), so we do not have much additional light to cast on the difference between NC and PNG, or the variation between mining projects or mineral provinces, in this particular respect.

The second obvious candidate for recognition as a fourth dimension is the temporal dimension inherent in the concept of the mining 'project cycle' that leads from exploration to closure. The addition of this dimension is certainly consistent with our argument that what really counts in any analysis of local-level politics is the way that actors shift themselves and others from one position to another, since these forms of mobility have their own particular time scales. It might also be argued that the balance of power between different groups of actors—and especially between the four corners of our rectangular model—tends to change over the different phases of the mining project cycle in ways that are independent of any particular institutional context (Chapter 2). This type of systematic shift has been ascribed to a form of 'capital logic' that applies to the relationship between large-scale mines and local-level politics in many different countries (Gerritsen and Macintyre 1991), and could therefore be seen as one aspect of the general process of modernisation and globalisation. The logic in question is one that operates around changes in the relationship between the costs borne or felt by mine-affected communities and the financial capacity of mining companies to compensate them in different phases of the project cycle (Howitt 2001). However, we would argue that the existence of three other dimensions of variation is precisely what makes it difficult for any group of actors to predict—and hence to manage— the temporal transformations of power. Indeed, this is one of the main reasons why large-scale mining projects are subject to moments of rupture that come as a surprise to most of the actors who think they understand the local political trajectory. The forced closure of the Panguna mine is an obvious case in point (Chapter 12), but other contributions to this volume provide examples of 'nasty surprises' that disrupted the smooth passage of the mining project cycle (Chapters 4, 5 and 8). Whether or not these moment of rupture count as episodes of political conflict, the complexity of structural transformation is what enables actors to have the kind of agency that defies the power of management, and often leads to unintended and unwelcome outcomes.

References

Ali, S.H. and A.S. Grewal, 2006. 'The Ecology and Economy of Indigenous Resistance: Divergent Perspectives on Mining in New Caledonia.' *Contemporary Pacific* 18: 361–392.

Allen, M.G., 2017. 'Islands, Extraction and Violence: Mining and the Politics of Scale in Island Melanesia.' *Political Geography* 57: 81–90. dx.doi.org/10.1016/j.polgeo.2016.12.004

Banks, G., 2008. 'Understanding "Resource" Conflicts in Papua New Guinea.' *Asia Pacific Viewpoint* 49: 23–34. doi.org/10.1111/j.1467-8373.2008.00358.x

Beck, U., 1992. *Risk Society: Towards a New Modernity* (transl. M. Ritter). London: Sage.

Burton, J., 1997. 'Terra Nugax and the Discovery Paradigm: How Ok Tedi Was Shaped by the Way It Was Found and How the Rise of Political Process in the North Fly Took the Company by Surprise.' In G. Banks and C. Ballard (eds), *The Ok Tedi Settlement: Issues, Outcomes and Implications*. Canberra: The Australian National University, National Centre for Development Studies (Pacific Policy Paper 27).

Clements, K., V. Boege, A. Brown, W. Foley and A. Nolan, 2007. 'State Building Reconsidered: The Role of Hybridity in the Formation of Political Order.' *Political Science* 59: 45–56. doi.org/10.1177/003231870705900106

Collier, P. and A. Hoeffler, 2004. 'Greed and Grievance in Civil War.' *Oxford Economic Papers* 56: 563–595. doi.org/10.1093/oep/gpf064

Dean, M., 2010, *Governmentality: Power and Rule in Modern Society* (2nd edition). London: Sage.

Demmer, C. and C. Salomon, 2017. 'À Propos du Sénat Coutumier: De la Promotion Mélanésienne à la Défense des Droits Autochtones.' In C. Demmer and B. Trépied (eds), *La Coutume Kanak dans l'État. Perspectives Coloniales et Postcoloniales sur la Nouvelle-Calédonie*. Paris: L'Harmattan.

Dupuy, J.-P., 2002. *Pour un Catastrophisme Éclairé: Quand l'Impossible est Certain*. Paris: Seuil.

Ferguson, J., 1990. *The Anti-Politics Machine: 'Development', Depoliticization, and Bureaucratic Power in Lesotho*. Cambridge: Cambridge University Press.

Filer, C., 1997. 'Compensation, Rent and Power in Papua New Guinea.' In S. Toft (ed.), *Compensation for Resource Development in Papua New Guinea*. Boroko (PNG): Law Reform Commission (Monograph 6). Canberra: The Australian National University, National Centre for Development Studies (Pacific Policy Paper 24).

Filer, C., M. Andrew, B.Y. Imbun, P. Jenkins and B.F. Sagir, 2016. 'Papua New Guinea: Jobs, Poverty, and Resources.' In G. Betcherman and M. Rama (eds), *Jobs for Development: Challenges and Solutions in Different Country Settings*. Oxford: Oxford University Press. doi.org/10.1093/acprof:oso/9780198754848.001.0001

Filer, C., J. Burton and G. Banks, 2008. 'The Fragmentation of Responsibilities in the Melanesian Mining Sector.' In C. O'Faircheallaigh and S. Ali (eds), *Earth Matters: Indigenous Peoples, the Extractive Industries and Corporate Social Responsibility*. London: Greenleaf Publishing. doi.org/10.9774/GLEAF.978-1-909493-79-7_11

Gerritsen, R. and M. Macintyre, 1991. 'Dilemmas of Distribution: The Misima Gold Mine, Papua New Guinea.' In J. Connell and R. Howitt (eds), *Mining and Indigenous Peoples in Australasia*. Sydney: Sydney University Press.

Horowitz, L.S., 2012. 'Translation Alignment: Actor-Network Theory, Resistance, and the Power Dynamics of Alliance in New Caledonia.' *Antipode* 44: 806–827. doi.org/10.1111/j.1467-8330.2011.00926.x

Howitt, R., 2001. *Rethinking Resource Management: Justice, Sustainability and Indigenous Peoples*. London: Routledge.

Huber, M. and J. Emel, 2009. 'Fixed Minerals, Scalar Politics: The Weight of Scale in Conflicts over the "1872 Mining Law" in the United States.' *Environment and Planning A* 41: 371–388. doi.org/10.1068/a40166

Lahiri-Dutt, K. (ed.), 2011. *Gendering the Field: Towards Sustainable Livelihoods for Mining Communities*. Canberra: ANU E Press (Asia-Pacific Environment Monograph 6).

Latour, B., 2005. *Reassembling the Social: An Introduction to Actor-Network Theory*. Oxford: Oxford University Press.

Le Meur, P.-Y., 2010. 'La Terre en Nouvelle-Calédonie: Pollution, Appartenance et Propriété Intellectuelle.' *Multitudes* 41: 91–98. doi.org/10.3917/mult.041.0091

——, 2017. 'Le Destin Commun en Nouvelle-Calédonie: Entre Projet National, Patrimoine Minier et Désarticulations Historiques.' *Mouvements* 91(3): 35–45.

Leitner, H., E. Sheppard and K.M. Sziarto, 2008. 'The Spatialities of Contentious Politics.' *Transactions of the Institute of British Geographers* 33: 157–172. doi.org/10.1111/j.1475-5661.2008.00293.x

Ross, M., 2003. 'The Natural Resource Curse: How Wealth Can Make You Poor.' In I. Bannon and P. Collier (eds), *Natural Resources and Violent Conflict: Options and Actions*. Washington (DC): World Bank.

Szablowski, D., 2007. *Transnational Law and Local Struggles: Mining Communities and the World Bank*. Portland (OR): Hart Publishing.

Walker, G., 2012. *Environmental Justice: Concepts, Evidence and Politics*. London: Routledge.

www.ingramcontent.com/pod-product-compliance
Lightning Source LLC
Chambersburg PA
CBHW040154270326
41929CB00041B/3384